Cotton, Water, Salts and Soums

Christopher Martius • Inna Rudenko
John P.A. Lamers • Paul L.G. Vlek
Editors

Cotton, Water, Salts and Soums

Economic and Ecological Restructuring in Khorezm, Uzbekistan

Editors
Christopher Martius
Center for Development Research (ZEF)
Ecology and Natural Resources
Walter-Flex-Strasse 3
53113 Bonn
Germany
gcmartius@gmail.com

John P.A. Lamers
ZEF/UNESCO Khorezm Project
Khamid Olimjan 14
220100 Urgench, Khorezm
Uzbekistan
j.lamers@uni-bonn.de

Inna Rudenko
ZEF/UNESCO Khorezm Project
Khamid Olimjan 14
220100 Urgench, Khorezm
Uzbekistan
inna@zef.uznet.net

Paul L.G. Vlek
Center for Development Research (ZEF)
Ecology and Natural Resources
Walter-Flex-Strasse 3
53113 Bonn
Germany
p.vlek@uni-bonn.de

ISBN 978-94-007-1962-0   e-ISBN 978-94-007-1963-7
DOI 10.1007/978-94-007-1963-7
Springer Dordrecht Heidelberg London New York

Library of Congress Control Number: 2011937451

© Springer Science+Business Media B.V. 2012
No part of this work may be reproduced, stored in a retrieval system, or transmitted in any form or by any means, electronic, mechanical, photocopying, microfilming, recording or otherwise, without written permission from the Publisher, with the exception of any material supplied specifically for the purpose of being entered and executed on a computer system, for exclusive use by the purchaser of the work.

Printed on acid-free paper

Springer is part of Springer Science+Business Media (www.springer.com)

Preface

This book reports on the first 7 years of research in a long-term science and capacity building project in the Khorezm region of Uzbekistan, located in the heart of Central Asia. This project of the Center of Development Research (ZEF), University of Bonn, and the University of Urgench, Uzbekistan is funded since the year 2000 by the German Federal Ministry of Research and Higher Education (BMBF) and carried out under endorsement of the Ministry of Agriculture and Water Resources of Uzbekistan. The project also benefits from a close collaboration with the United Nations Educational, Scientific and Cultural Organization (UNESCO), the German Space Agency (DLR). This book summarizes the findings in natural and social science disciplines during the period 2001–2007.

This project takes a long-term interdisciplinary research approach to address land and water use in irrigated areas of the Aral Sea basin and seeks options to restructure the region in ecological and economic terms. It combines natural and human sciences to jointly elucidate problems of economic transition in post-Soviet countries. This transition affects the use of land, water and biological resources, ecological sustainability, economic efficiency and the livelihoods of the people of the region. The environmental problems are reflected in the increasing land degradation, which is mainly caused by excess water use and subsequent soil salinization in an area dominated by cotton, wheat and rice production. These unsolved resource management problems threaten food security, agricultural incomes, rural livelihoods and thus, human security.

The research program is focused on Khorezm, a region in Uzbekistan sitting amidst the Kyzylkum and Karakum deserts. It receives all the water it needs for its main economic activity, agriculture, from the Amudarya river, the ancient Oxus. Khorezm is a region which probably looks back on several thousand years of irrigated agriculture. For centuries it provided a decent livelihood to the Khorezmians. However, it was under Soviet rule that the irrigation system was scaled-up to an extent that triggered the environmental and economic problems the region faces today. The Uzbek state has inherited these problems, which were exacerbated by the economic transition.

v

The authors in this book describe and analyze the biophysical environment and the aspects of society and institutions that shape land and water use. Options for improving land and water management are then discussed. Based on economic studies and modeling, opportunities are elucidated how to reform economic system management, for the benefit of humans and nature in the region.

The project benefited and still benefits from the willingness of the BMBF to grant the project the space to experiment multi- and interdisciplinary science, over a period that by far exceeds the conventional project lifetime of 2–3 years. This long-term commitment, today in its 11th year, allowed the implementation of a structured and comprehensive research program. It involved setting up the necessary infrastructure to support advanced scientific field research activities in natural and social sciences in Khorezm. It also allowed the training, over the course of the years, of a large group of young, enthusiastic Uzbek/Khorezmian researchers that carried out the research, many as part of the requirements to obtain the Bachelor or Master degree at the State University of Urgench, or a Ph.D. degree in the Bonn Interdisciplinary Graduate School for Development Research (BiGS-DR) of the Center for Development Research (ZEF) at the University of Bonn or at other Universities in Germany or Uzbekistan. The intensive collaboration of Uzbek scientists with their international counterparts enabled the project to develop a unique learning atmosphere.

This book complements the numerous monographs and publications in refereed scientific journal articles, contributions to national and international conferences and symposia, discussion and work papers, and short science communications (http://www.khorezm.zef.de/) prepared by project members. With this manuscript we summarize our knowledge and innovations for improving land and water use in the degraded, salt-effected croplands of the Khorezm region. Once a world center for sciences that produced mathematicians such as Al-Khorezmi, Khorezm will hopefully regain this strength to cope with the challenges of a modern and rapidly changing world.

Acknowledgements

This book would not have been completed without the time and hard efforts contributed by many. For these contributions, the editors thank in particular the following persons and organizations:

- The German Ministry for Education and Research (BMBF; project number 0339970A) for funding the project. Without their vision to support such a long-term program, the project and, consequently, the book would not have been possible;
- UNESCO, and in particular Vefa Moustafaev, Barry Lane and Jorge Espinal, for their unwavering support during project preparation, dealing with administration and accounting and much more.
- The project is fortunate to work also under the *aegis* of the Uzbekistan Ministry of Agriculture and Water Resources.
- The German Academic Exchange Service (DAAD), the *BMBF*'s International Postgraduate Studies in Water Technologies (IPSWAT) program, and the Robert-Bosch Foundation for scholarships provided to many project students;
- Thanks are due to the German Space Agency (DLR) and University of Würzburg for partnership and support;
- Furthermore, we are thankful to the farmers, householders, WUA members and other collaborators in the field, for their active participation in research;
- All partners, staff and colleagues from Germany and Uzbekistan for having made the research of numerous people and consequently this book possible. Without the cooperation partners and colleagues we could not have achieved this successfully;
- To Alma van der Veen and Katharina Moraht at the Center for Development Research (ZEF) for their great support with project reports and web publications;
- Mrs. Margaret Jend from ZEF, Bonn for her invaluable support in proofreading the English texts, and for the discussion of many linguistic matters;
- We would like to give our sincerest thanks to the dedicated staff members in Khorezm, Uzbekistan who overcame numerous logistic challenges and have helped compiling the book. And in particular, Guzal Matniyazova, Davran Abdullaev, Kudrat Nurmetov, and Sasha Lee for their never-ending patience with formatting the text and arranging figures, tables, and references;

- The authors for their contributions, delivering their best and working hard on comments;
- Co-authors and supervisors of various Ph.D. candidates and junior scientists from the project for their valuable comments and suggestions during the completion of the chapters.

Urgench, Accra, São José dos Campos
April 2012

Christopher Martius
Inna Rudenko
John P.A. Lamers
Paul L.G. Vlek

Contents

Part I   Introduction

**1   Cotton, Water, Salts and *Soums* – Research and Capacity
Building for Decision-Making in Khorezm, Uzbekistan** 3
Paul L.G. Vlek, John P.A. Lamers, Christopher Martius,
Inna Rudenko, Ahmad Manschadi, and Ruzumbay Eshchanov

Part II   The Background (Physical Setting)

**2   Agro-Meteorological Trends of Recent Climate Development
in Khorezm and Implications for Crop Production** 25
Christopher Conrad, Gunther Schorcht, Bernhard Tischbein,
Sanjar Davletov, Murod Sultonov, and John P.A. Lamers

3   Soils and Soil Ecology in Khorezm 37
Akmal Akramkhanov, Ramazan Kuziev, Rolf Sommer,
Christopher Martius, Oksana Forkutsa, and Luiz Massucati

**4   Spatial Distribution of Cotton Yield and its Relationship
to Environmental, Irrigation Infrastructure
and Water Management Factors on a Regional
Scale in Khorezm, Uzbekistan** 59
Gerd Ruecker, Christopher Conrad, Nazirbay Ibragimov,
Kirsten Kienzler, Mirzahayot Ibrakhimov, Christopher Martius,
and John P.A. Lamers

**5   Water Management in Khorezm: Current Situation
and Options for Improvement (Hydrological Perspective)** 69
Bernhard Tischbein, Usman Khalid Awan, Iskandar Abdullaev,
Ihtiyor Bobojonov, Christopher Conrad, Hujiyaz Jabborov,
Irina Forkutsa, Mirzahayot Ibrakhimov, and Gavhar Poluasheva

ix

Part III   The System Setting (Society and Institution)

6   Farm Reform in Uzbekistan... 95
Nodir Djanibekov, Ihtiyor Bobojonov, and John P.A. Lamers

**7   Mapping and Analyzing Service Provision for Supporting
Agricultural Production in Khorezm, Uzbekistan**............................. 113
Davron Niyazmetov, Inna Rudenko, and John P.A. Lamers

**8   Politics of Agricultural Water Management
in Khorezm, Uzbekistan**.. 127
Gert Jan Veldwisch, Peter Mollinga, Darya Hirsch, and Resul Yalcin

**9   Water and Sanitation-Related Health Aspects
in Khorezm, Uzbekistan**.. 141
Susanne Herbst, Dilorom Fayzieva, and Thomas Kistemann

**10   Price and Income Elasticity of Residential
Electricity Consumption in Khorezm**... 155
Bahtiyor Eshchanov, Mona Grinwis, and Sanaatbek Salaev

Part IV   Land Management Improvement Options

**11   Optimal Irrigation and N-fertilizer Management
for Sustainable Winter Wheat Production
in Khorezm, Uzbekistan**.. 171
Nazirbay Ibragimov, Yulduz Djumaniyazova, Jumanazar Ruzimov,
Ruzumbay Eshchanov, Clemens Scheer, Kirsten Kienzler,
John P.A. Lamers, and Maksud Bekchanov

**12   Groundwater Contribution to N fertilization
in Irrigated Cotton and Winter Wheat
in the Khorezm Region, Uzbekistan**.. 181
Kirsten Kienzler, Nazirbay Ibragimov, John P.A. Lamers,
Rolf Sommer, and Paul L.G. Vlek

**13   Introducing Conservation Agriculture on Irrigated
Meadow Alluvial Soils (Arenosols) in Khorezm, Uzbekistan**.............. 195
Alim Pulatov, Oybek Egamberdiev, Abdullah Karimov,
Mehriddin Tursunov, Sarah Kienzler, Ken Sayre, Latif Tursunov,
John P.A. Lamers, and Christopher Martius

**14   Crop Diversification in Support of Sustainable
Agriculture in Khorezm**.. 219
Ihtiyor Bobojonov, John P.A. Lamers, Nodir Djanibekov,
Nazirbay Ibragimov, Tamara Begdullaeva, Abdu-Kadir Ergashev,
Kirsten Kienzler, Ruzumbay Eshchanov, Azad Rakhimov,
Jumanazar Ruzimov, and Christopher Martius

Contents xi

**15 Conversion of Degraded Cropland to Tree Plantations
for Ecosystem and Livelihood Benefits** 235
Asia Khamzina, John P.A. Lamers, and Paul L.G. Vlek

**16 Abundance of Natural Riparian Forests and Tree Plantations
in the Amudarya Delta of Uzbekistan and Their Impact
on Emissions of Soil-Borne Greenhouse Gases** 249
Clemens Scheer, Alexander Tupitsa, Evgeniy Botman,
John P.A. Lamers, Martin Worbes, Reiner Wassmann,
Christopher Martius, and Paul L.G. Vlek

Part V Land and Water Management Improvement Tools

**17 Economic-Ecological Optimization Model of Land
and Resource Use at Farm-Aggregated Level** 267
Rolf Sommer, Nodir Djanibekov, Marc Müller, and Omonbek Salaev

**18 Water Patterns in the Landscape of Khorezm, Uzbekistan:
A GIS Approach to Socio-Physical Research** 285
Lisa Oberkircher, Anna Haubold, Christopher Martius,
and Tillmann K. Buttschardt

**19 Modeling Irrigation Scheduling Under Shallow
Groundwater Conditions as a Tool for an Integrated
Management of Surface and Groundwater Resources** 309
Usman Khalid Awan, Bernhard Tischbein, Pulatbay Kamalov,
Christopher Martius, and Mohsin Hafeez

**20 Estimation of Spatial and Temporal Variability
of Crop Water Productivity with Incomplete Data** 329
Maksud Bekchanov, John P.A. Lamers, Aziz Karimov, and Marc Müller

Part VI Economic System Management Reform Options

**21 A Computable General Equilibrium Analysis of Agricultural
Development Reforms: National and Regional Perspective** 347
Maksud Bekchanov, Marc Müller, and John P.A. Lamers

**22 State Order and Policy Strategies in the Cotton
and Wheat Value Chains** 371
Inna Rudenko, Kudrat Nurmetov, and John P.A. Lamers

**23 Prospects of Agricultural Water Service Fees
in the Irrigated Drylands Downstream of Amudarya** 389
Nodir Djanibekov, Ihtiyor Bobojonov, and Utkur Djanibekov

Glossary of Uzbek and Russian words 413

Index 415

Abbreviations and Acronyms

ADB	Asian Development Bank
ANOVA	Analysis of variance
ASB	Aral Sea Basin
ASP	Agricultural service providing organizations
BCR	Benefit cost ratio
CA	Conservation agriculture
CAC	Central Asian countries
CDM	Clean Development Mechanism
GEF	Global Environment Facility (UNDP/GEF)
CES	Constant elasticity of substitution
CFU	Colony forming units (a unit in microbiology)
CGE	Computable General Equilibrium model
CIS	Commonwealth of Independent States
CP	Crude protein
CR	Conveyance ratios
CT	Conventional tillage
DF	Depleted Fraction
DLR	German Aerospace Center
DM	Dry matter
DW	Dry weight
EC	Electrical conductivity
EC-IFAS	Executive Committee of the Interstate Fund for the Aral Sea
ET	Evapotranspiration
ET_{act}	Actual evapotranspiration
ETc	Crop evapotranspiration
ET_o	Reference evapotranspiration
FAO	Food and Agriculture Organization of the United Nations
FAR	Field application ratio
FBS	Food Balance Sheets
FC	Faecal coliform bacteria
FLEOM	Farm-Level Economic-Ecological Optimization

FPAR	Fraction of Photosynthetically Active Radiation
GDD	Growing Degree Days
GDP	Gross Domestic Product
GHG	Greenhouse gases
GIS	Geographical Information System
GM	Gross margin
GRP	Gross Regional Product
IOC	Input-output coefficients
IOM	Input-output matrix
IOT	Input-output table
IPCC	Intergovernmental Panel on Climate Change
LP	Linear-programming
LSD-test	The least significant difference
MAWR	Ministry of Agriculture and Water Resources of Uzbekistan
MDG	Millennium Development Goal
ME	Metabolizable energy
MEM	Mixed Estimation Method
MODIS	Moderate Resolution Imaging Spectroradiometer
NDVI	Normalized Difference Vegetation Index
Neftebaza	Unitary Company for distribution of fuel and lubricants in Uzbekistan
NPV	Net Present Value
NT	No tillage
O&M	Operation and Maintenance
OblSES	Regional Centre of Sanitation and Epidemiology
OblStat	Regional Statistical Department
OblVodkhoz	Regional Department of Agriculture and Water Resources Management
OGME	The Khorezmian Hydrogeological Melioration Expedition
PB	Permanent bed planting
PET	Potential evapotranspiration
QishloqXo'jalikKimyo	Regional Joint Stock Company for distribution of mineral fertilizers in Uzbekistan
RB	Raised beds
SAM	Social Accounting Matrix
SANIIRI	Central-Asian Research Institute of Irrigation Science named after Jurin
SNA	System of National Accounts
SOC	Soil organic carbon
SOM	Soil organic matter
STD	Standard deviation
TDS	Total dissolved solids
TSS	Total soluble salts
UNESCO	United Nations Educational, Scientific and Cultural Organization

Abbreviations and Acronyms

UPRADIK	Irrigation and Drainage System Management
USD	United States Dollars
USSR	Union of Soviet Socialist Republics
UzbekEnergo	State-owned monopole electricity supplier for all customers
Uzgiprozem	State Institute for Land Management Planning
UzIstiqbolStat	Regional Statistical Authority
UZS	Uzbek Soum (currency)
WP	Water productivity
WSF	Water service fees
WUA	Water Users Association
ZEF	Center for Development Research, University of Bonn
ZT	Zero tillage

Part I
Introduction

Chapter 1
Cotton, Water, Salts and *Soums* – Research and Capacity Building for Decision-Making in Khorezm, Uzbekistan

Paul L.G. Vlek, John P.A. Lamers, Christopher Martius, Inna Rudenko, Ahmad Manschadi, and Ruzumbay Eshchanov

Abstract Khorezm, a district of Uzbekistan, is a textbook example of irrigated agriculture in the Aral Sea Basin causing the "Aral Sea Syndrome". It offers an opportunity to study the complex human-environment relations in the context of strong government control in transformation economies. Agricultural production and rural livelihood in Khorezm rely entirely on irrigation water supply based on a dense network of irrigation channels and drainage collectors. Inefficient use of land and water resources, inadequate institutions and policies, and underdeveloped agro-processing and service sectors are among the key issues threatening the economic and ecological sustainability of the region. An interdisciplinary research project was initiated by the Center for Development Research (ZEF) of the University of Bonn, Germany, together with the University of Urgench to develop science-based concepts and tools for the restructuring of land use and agricultural production in order to make more efficient use of natural resources. Simultaneously, the project aimed to develop recommendations for policies and institutions to enable economic development and environmental sustainability. As Khorezm is representative for many irrigated lowlands in the Aral Sea Basin, innovative concepts and technologies developed

---

P.L.G. Vlek (✉) • C. Martius • A. Manschadi
Center for Development Research (ZEF), Walter-Flex-Str. 3, Bonn 53113, Germany
e-mail: p.vlek@uni-bonn.de; gcmartius@gmail.com; manschadi@uni-bonn.de

J.P.A. Lamers • I. Rudenko
ZEF/UNESCO Khorezm Project, Khamid Olimjan Str., 14, Urgench 220100, Uzbekistan
e-mail: j.lamers@uni-bonn.de; inna@zef.uznet.net

R. Eshchanov
Urgench State University, Khamid Olimjan Str., 14, Urgench 220100, Uzbekistan
e-mail: ruzimboy@mail.ru

C. Martius et al. (eds.), *Cotton, Water, Salts and Soums*: *Economic and Ecological Restructuring in Khorezm, Uzbekistan*, DOI 10.1007/978-94-007-1963-7_1,
© Springer Science+Business Media B.V. 2012

in the project are expected to provide useful information to similar environments in Central Asia and elsewhere. This chapter provides the problem setting in the Aral Sea basin and Khorezm. The "Khorezm Project's" purpose and structure are briefly described. An overview of the chapters in this book is then given. Conclusions are drawn and an outlook is presented. Technological, economic and institutional innovations have to go hand-in-hand if long-lasting sustainable development is aimed for with increased environmental and economic security of the region's people.

Keywords Aral Sea Basin • Central Asia • Semi-arid climate • Interdisciplinary research • Landscape restructuring • Irrigation • Afforestation • Climate change • GIS/RS

1.1 Introduction

The widespread, and ever increasing physical and economic water scarcity in the Aral Sea Basin, exacerbated by climate change and land degradation, is threatening ecological sustainability and economic development in Central Asia. The economic transition that Uzbekistan is undergoing further complicates the restructuring of resource management. These problems are exemplified in the Khorezm region, a district located in the northwest of Uzbekistan. Khorezm depends to a large extent on agriculture. Thus, development is essentially based on the economic, ecological and social sustainability of agricultural land use. To increase our understanding of the linkages between human and environmental security, an interdisciplinary project was developed between the Center for Development Research (ZEF) of the University of Bonn, Germany, and the University of Urgench, Khorezm. The "Khorezm Project" followed an interdisciplinary approach that unites natural and human sciences to analyze the problems from different perspectives.

In contrast, the Soviet approach emphasized production maximization without taking environmental and social issues into account. This may have brought temporary economic development but also created the unsustainable situation that today is known as the "Aral Sea Syndrome" (WBGU 1998). The transition countries that inherited this situation suddenly found themselves under pressure to deliver improvements. However, with other, more pressing problems on their agenda, and their science institutions poorly prepared for the task, they often could not generate the needed change to the underlying structural problems. A deeper understanding of the complexity of the underlying problems and the means to ameliorate them was lacking. There seemed to be a need for an integrated approach in order to find the true causes of stagnation in behavioral and technological change.

This chapter introduces the project and the book. The book's title ("Cotton, water, salts and Soums – economic and ecological restructuring in Khorezm, Uzbekistan") invokes the major components of the problem setting:

- *Cotton* is the main crop, currently planted on about half of the irrigated agricultural land and using about half of the water that flows into the region (cf. Djanibekov et al. 2011b). Other important crops are wheat and rice. In this book, conditions of crop production, irrigation and fertilization are analyzed and the introduction of conservation agriculture in the region is discussed. We also describe the physical environment to which all innovations must be adapted. This includes an analysis of environmental conditions and climate change.

- Irrigation *water* is used in great quantities, leading to an average water use of 1 600 mm (i.e.; 16,000 m^3 per hectare and year), one of the highest water use ratios in the world. The low rainfall of 90 mm sharply contrasts with the potential annual evapotranspiration of about 1,400 mm and annual crop requirements of about 700–800 mm. The resulting percolation leads to a very shallow water table for most of the year in the region (cf. Djanibekov et al. 2011a; Tischbein et al. 2011). This book analyzes water management from different viewpoints, be they hydrological, social-anthropological and economic in nature. The result is a more integrated view of water resource management.
- *Salts* are contained in all surface water and accumulate in the topsoil when groundwater is brought to the surface through capillary rise where the water evaporates and the salts remain. If salinity levels build up too high, crop growth is affected. Thus, soil salinity is a major indicator of sound resource management and is reflected in agricultural productivity. Direct and indirect options to regulate water management and soil salinity are discussed in several chapters of the book.
- Finally, wealth is potentially created by agricultural production, here expressed in Uzbek currrency, *Soums*. Farmers produce crops to ensure a livelihood. However, state regulations may cause farmers to be sidelined with wealth going to the state. This affects rural development, poverty, and food security. In several chapters, economic analysis is undertaken to elucidate economic mechanisms to improve the current resource governance, for the benefit of human livelihoods and ecological sustainability.

We analyzed these and many other elements of the problem, to come up with suggestions for restructuring land and water use at the field, farm, and regional management levels. We also recommend restructuring the economic framework conditions that govern natural resource use in Uzbekistan. The research and educational goals of the Khorezm Project are aligned with the Millennium Development Goals of eradicating poverty and hunger and achieving food and water security, and also with the United Nations conventions on desertification/land degradation and climate change. They are also providing support to the long-term strategic programs that the EU and Germany developed for Central Asia. These aim at regional security, environmental protection, sustainable development, and a higher quality of life for the present and future generations.

1.2 Khorezm

Khorezm, an administrative district of Uzbekistan, is located in the northwest of Uzbekistan in the lower reaches of the Amudarya River – formerly the largest tributary of the Aral Sea. As part of the Turan Lowland of the Aral Sea basin, Khorezm is situated about 250 km south of the present shores of the Aral Sea.

Khorezm is one of the oases of the great historic civilizations of Central Asia, fed by the ancient river Oxus, today the Amudarya. For at least 3,000 years, waters from the Amudarya and Syrdarya rivers supported thriving agricultural communities (Tolstov 1948) – and since Soviet times a flourishing fishing industry – in the Aral Sea Basin (Fig. 1.1). But during the Soviet era, and more so since the late 1950ies, the Amudarya and Syrdarya waters were abstracted excessively and used in greatly expanded irrigation systems to secure the production of cotton, the "white gold". The area of irrigated cropland in the Aral Sea basin almost doubled, from 4.5 million ha in the 1950ies to 7.9 million ha in 2006 (cf. Tischbein et al. 2011). This dramatic expansion of the irrigation system resulted in a substantial decrease in water inflow to the Aral Sea from 43 km^3 year^{-1} in the 1960ies to an average of 9 km^3 year^{-1} during 2001–2005. Consequently, within decades the Aral Sea shrunk from being the fourth largest freshwater lake in the world (6.8 million ha surface area) to less than 20% of its former size. By 2006, the sea's level had dropped 23 m, the volume decreased by 90%, and the salinity risen to more than 100 g L^{-1} (Micklin 2008). The term "Aral Sea Syndrome" was coined to denote problems "associated with centrally planned, large-scale projects involving water resource development" (WBGU 1998) which aim at providing secure supplies of irrigation water and thus, food security, but fail to address the ensuing environmental disaster.

Khorezm covers an area of about 680,000 ha of mostly arid deserts, of which roughly 270,000 ha have been developed for irrigated agriculture. With an average annual precipitation of ~90 mm only (cf. Conrad et al. 2011), agricultural production and rural livelihood in Khorezm rely entirely on irrigation water supply. For this purpose, a dense network of about 16,000 km irrigation canals and ca 8,000 km drainage water collectors has been developed, mainly over the last decades of the Soviet period (Fig. 1.2).

This complex network of irrigation and drainage canals was designed to deliver water to large-scale collective farm units based on a centrally organized irrigation water scheduling and delivery system. Land reforms initiated after independence in 1991 have, however, resulted in the disintegration of the large collective and state farms into numerous smallholder farms (in Khorezm, from 117 *shirkats* in 2001 to 18,381 private farms by the end of 2008). This has led to a serious mismatch between the irrigation water supply system and the actual demand by the new private farmers. The establishment of Water Users Associations (WUAs) largely along the administrative boundaries of the former collective farms has been an attempt to bridge the gap between the higher-level water providers and the farmers (Zavgorodnyaya 2006). Due to lack of human and financial resources, the WUAs, however, have so far not been effective in organizing farm-level water supply management.

Another farm optimization process initiated by the Uzbek government in November 2008 reversed the process of land fragmentation, and led to the re-creation of larger farm units of at least 80–100 ha in size. This is too recent to analyze the effects of this reversal, but despite the various reforms, inefficient use of land and water resources continue. Moreover, inappropriate institutional settings and support frameworks, and underdeveloped agro-processing and service sectors are among the key constrains of economic and ecological improvement in the region.

1 Cotton, Water, Salts and *Soums* – Research and Capacity Building...

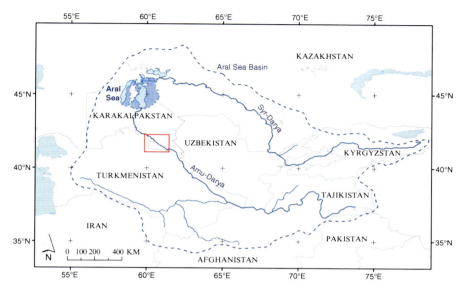

Fig. 1.1 Location of the Khorezm region (*square*) and the Aral Sea basin (*dashed line*)

In 2008, agriculture contributed 37% to the regional GDP. Agriculture is also important as a way of living and employment for the rural population. It is the main provider of raw materials for the agro-industrial sector. Of the 1.5 million people living in Khorezm, over 70% reside in rural areas and are mostly engaged in cotton, wheat and rice production (Djanibekov 2008). About 27.5% live below the poverty line of 1 US$ per day (Müller 2006).

Soils are generally low in fertility and organic matter content (cf. Akramkhanov et al. 2011), and substantial amounts of chemical fertilizers are used for the cultivation of the major crops (Kienzler 2010). Due to the application of large amounts of water during leaching and irrigation events, ground water tables reach critically high levels during the cropping season causing secondary soil salinization and land degradation – a widespread problem in Khorezm (Ibrakhimov et al. 2007).

1.3 The Khorezm Project

The overall goal of the Khorezm Project is to provide a comprehensive, science-based concept for restructuring of agricultural production systems in the Khorezm region. It was expected that – as Khorezm is representative for the irrigated systems in the lowland areas of Central Asia – innovative technologies and concepts developed here would have a potential for being up-scaled to the larger region and serve as examples for development in other, similar environments. To achieve this goal,

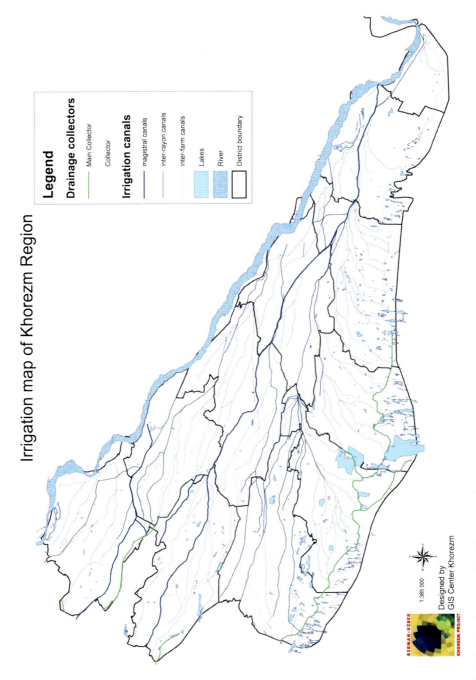

Fig. 1.2 Irrigation map of the Khorezm region

1 Cotton, Water, Salts and *Soums* – Research and Capacity Building...

Fig. 1.3 The thematic areas dealt with in the Khorezm project during 2001–2007

project research has been focused on various system components: land use, production systems, economy and society and institutions (Fig. 1.3). The core project research activities included monitoring and mapping of natural resource endowment and degradation, analyzing agrarian change and rural transformation with a special focus on gender perspectives, developing remote sensing and model-based technologies for improved irrigation water management and soil salinity control, analyzing socio-technical dynamics of irrigation water distribution and drainage management, value chain analysis of agrarian commodity sectors, such as cotton, winter wheat and vegetables, elaborating options for the phyto-remediation of degraded, saline land through set-aside programs for afforestation, adapting conservation agriculture equipment and practices, as well as crop rotation to sustainably improve the productivity of cropping systems. Implementing and adapting innovations with stakeholder groups, a process known as "Follow the Innovation" (Hornidge et al. 2011), and policy outreach and dissemination were an integral part of this project. These activities are grouped as shown in Fig. 1.3.

Simulation modeling has been an essential instrument of the project for integrating the disciplinary scientific findings, up-scaling the results, and predicting the long-term impacts of current and alternative land and water management options and policies. The overall restructuring concept is thus based on recommendations for (i) innovative technologies to enhance the economic viability and ecological sustainability of agricultural systems at the field, farm and WUA level; (ii) improved agricultural policies at regional and national levels and institutional restructuring for a more sustainable natural resource use.

1.4 A GIS Center in Urgench

The establishment and maintenance of a lab for Geographical Information Systems (Khorezm GIS Center) at the state University of Urgench (UrDU) has been one important component in the project. The GIS layers in the Central Data Base (CDB) comprise raster and vector data such as administrative boundaries (fields, WUAs, district, and region), infrastructure (irrigation and drainage network, roads, machinery and tractor parks), soil characteristics, groundwater information, and climate data. Regional GIS raster layers from satellite image analysis include among others land use, crop yield, and evapotranspiration. In addition socio-economic data (commodity prices, census data), and various reports are collected. As the only existing GIS lab in the Khorezm region, the project's GIS Centre has been fulfilling demands for spatial data, information and maps not only from the project staff but also from the local scientific community and regional authorities. Since the data for the CDB has been derived from different sources (field sampling, remote sensing, and secondary sources), the first task was a geometrical adjustment of these data sets to high-resolution satellite reference images. The database serves as the central system for the dynamic analysis of the resource endowment of the region, be it natural, human or economic, and provides data for the models developed in the project. The Khorezm GIS Center can be consulted for data by interested institutions and researchers.

1.5 Capacity Building

As one of the major internationally-funded long-term research projects in Uzbekistan, the Khorezm Project makes a significant contribution toward education and training of young local scientists. Since the onset of the project in 2002, 50 Ph.D. students, about half of them from Uzbekistan, have conducted their research in the framework of the Khorezm project and 28 have successfully defended, of which 13 are from Uzbekistan (August 2011). The Ph.D. candidates conducted their research under supervision of local and foreign experts while themselves supervising a large number of local M.Sc. (about 100) and numerous B.Sc. students. Furthermore, the capacity of University of Urgench staff was built as several of them obtained their Uzbek professorship in the project framework.

1.6 Structure of this Book and Overview of Contributions

This book is the second book based on this project. Wehrheim et al. (2008) analyzed legal, economic and social constraints for agricultural production and innovation in Khorezm. This book addresses the biophysical environment, describes cropping and irrigation management systems and takes an economic and anthropological look at the region's problems. The book summarizes work carried out during the first 6 years (2001–2007) of the project in Khorezm.

In the following sections we will give a brief overview of the contents of the book that is now in front of the reader.

1.6.1 Section II: The Physical Environment

Irrigated agriculture has been in this arid region for a long time. It was the basis for the blooming of the empire of the Khorezmshakh in the twelfth and thirteenth centuries, and of the seventeenth century, later the Khiva Khanate. Today this region is known as Khorezm (Al-Iqbal 1999).

Using mid to long-term weather and climate records, **Conrad et al. (Chap. 2)** suggest that the agro-climate condition in this region favour annual, warm-season crops such as cotton, wheat, rice and maize. Current air temperatures and growing degree-days are favorable for the main crops grown in the region. However, increasing winter temperatures have been observed, which improves the conditions for growing winter wheat. Soil temperatures, relevant for estimating the first planting days of crops such as cotton, have risen. Climate change is manifest in Khorezm by temperature increases, especially during the winter period. The 10-year average of summer temperatures in 1981–1990 was 0.2–0.5°C above the long-term average in 1930–1990. Thus the current state recommendations for planting cotton from mid-to-end April are becoming conservative; earlier planting dates may be adopted in the future to make use of the longer vegetation season. Spatial variability between the central and marginal parts of the irrigated "oasis" of Khorezm, and the desert itself appear to be small, but variability, and hence, risks increase towards the oasis margin.

The riverbed changes of the meandering Amudarya river over the millennia, as well as the most recent expansion in irrigated agriculture in Khorezm in Soviet Union times, have markedly influenced the development and formation of soils in the region (**Akramkhanov et al.; Chap. 3**). The result is a spatially diverse mosaic of soil conditions. While the irrigation and land use activities have made the top layer of soils, the agro-irrigation horizon, rather uniform, the underlying soils consist of a multi-layered and differentiated illuvium. The soil texture is dominated by silt loam, sandy loam and loams, characterized by rather low soil organic matter contents, and slightly- to medium- soil salinity. The top 60 cm are in general moderately to strongly saline reaching levels affecting crop performance. Soil microbiological activity is closely linked to soil organic carbon content whereas soil fauna reacts to land management. Thus, conservation agriculture combined with residue retention may help improve soil conditions in this region.

Cotton is the predominant crop in Khorezm. MODIS-satellite remote sensing data revealed the spatial distribution of cotton yields, which is not evident from district-aggregated official statistics. A regional crop yield model was developed by

Ruecker et al. (Chap. 4) relating the spatial distribution of cotton yields to management factors. This revealed the effect on cotton yields of soil texture, irrigation delivery, and water management. This knowledge can help target land use and infrastructure rehabilitation at regional and district levels, and at the level of water user organizations and individual farms. Being based on satellite data, the procedures developed by these authors is easily transferrable to other irrigation systems.

Large amounts of water are used for irrigation and soil salinity leaching in Khorezm, and this is a major cause of the observed shallow groundwater levels. These, in turn, lead to widespread soil salinity, which consequently demands more water for leaching (Akramkhanov et al. 2008). Currently, 25% of the total annual water supply to the region is used for leaching. This vicious cycle is of great concern in a situation of insufficient water supply and there is an urgent need for improved irrigation and cropping practices. It is not so much the increase in salinity of the irrigation water as the soil salinity caused by saline groundwater, which reduces agricultural productivity in the irrigated drylands of Uzbekistan. In a comprehensive analyses across scales from fields to water users association (WUA) to the district level (i.e., whole Khorezm), **Tischbein et al. (Chap. 5)** show that present irrigation water management leaves much room for improvement: overall technical irrigation efficiency is about only 27%. Drainage-water output reaches as high as 62–67% of the irrigation water input. The ill-functioning drains, the excessive leaching and the high water losses from the irrigation (conveyance and distribution) network are among the major causes of shallow groundwater tables. With gross water inputs of 2,600–2,800 mm, water supply regularly falls short of demand in tail-end locations. A series of irrigation assessment indicators (e.g. relative evapotranspiration (RET), depleted fraction (DF), drainage ratio (DR), or irrigation efficiency at field, network and scheme level) all indicate that, the region experiences an economic water deficiency due to weak infrastructure management and insufficient planning efficiency. For example, due to insecurity in water supply, farmers actively block the drainage system, to increase their water security. This factually improves the abovementioned efficiencies, but bears the strong risk of soil salinization. The authors argue that modernization should aim at optimizing the current system, introducing a flexible, model-based irrigation scheduling to replace the present rigid norms, and through improving infrastructure, maintenance and institutional water management procedures. Innovative technologies (e.g. drip and sprinkler irrigation) can then be considered as a next step. Likewise, improvements needed in the drainage network should first focus on major and local outlets before more costly interventions such as narrowing drainage ditch spacing and a more widespread introduction of tile drainage can be considered.

1.6.2 Section III: Society and Institutions

The project addressed the complex of land and water use planning in a confusing decision-making environment. The goal was to provide science-driven decision support. Both, the bio-physical as well as the institutional aspects were analyzed.

During the farm restructuring reform between 1992 and 2008, virtually all collective land was allotted to producers. This process resulted in a large number of small-scale farms that comprise multiple fields, scattered and located often at far distances from each other, and of various soil qualities and shapes. The original set-up of the infrastructure, suitable for a small number of large-scale farms, was not suitable for this array of many small-scale private farms, and agricultural efficiency and sustainability deteriorated. In response, the Government of Uzbekistan opted for a land consolidation to improve the overall efficiency in crop production. However, **Djanibekov et al. (Chap. 6)** convincingly argue that as long as the land consolidation is not accompanied by additional institutional reforms, the anticipated boost of resources use efficiency is unlikely to take place. Among these are easing the state procurement system, ensuring land stewardship by producers, and increasing the efficiency of the organizations providing agricultural services (ASPs).

Research by **Niyazmetov et al. (Chap. 7)** addresses the efficiency and functioning of agricultural service providers ASPs in Khorezm. ASPs suffer from serious flaws, preventing the efficient delivery of services to farmers. Property rights are poorly respected with local authorities having decision-making powers in ASPs and on farms. Corporate management is often poor because local authorities interfere in the election of ASP heads, making them more accountable to local authorities than to the business interests of the ASPs. Contractual arrangements are weak, affecting the reliability of interaction of the ASPs with their clients. Frequently, farmers cannot meet the demands for early payments for the services rendered. Commercial banks and their mini-bank branches in the villages face notorious cash-flow and liquidity problems whilst the farmers' direct access to their own bank accounts is constrained by regulations.

In **Chap. 8**, **Veldwisch et al.** describe socio-political aspects of the reforms introduced since independence in 1991, with a focus on agricultural water management. The reform has led to the establishment of basin boundaries and water user organizations (WUAs). These reforms were implemented by the national administration without much participation of farmers or water managers. In spite of the national administration's declared intention of introducing a certain level of decentralization, the authors conclude that "... *policies developed in 'society-centric policy processes' cannot easily be applied in countries with 'state-centric politics'*". In essence, water management remained virtually unchanged since independence, In fact, some aspects have worsened considerably.

The latter is illustrated in the quality of drinking water and the incidence of water-related diseases that have become alarming (**Herbst et al.; Chap. 9**). A large share of the rural population relies for their drinking water supply on unprotected groundwater dug or drilled wells. Both sources are frequently fecally contaminated and pose a risk. Particularly children are strongly affected by diarrheal diseases. A general lack of proper health-related behavior, food hygiene, sanitation facilities and their maintenance and sewage disposal increases the risk of these diseases. These are also some of the entry points for preventive measures to improve domestic drinking water management and use.

Affordable, secure and environmentally friendly energy is key to welfare, economic growth and sustainable development (**Eshchanov et al.; Chap. 10**). Suitable reform policies are needed taking into account the price and income elasticity of residential electricity consumption, which consumes 39% of all energy in Uzbekistan. This lies far above the world average of 28%. Electricity production is based o natural gas supplies, and the high share of domestic consumption reduces the countries' option to export fossil fuels. Using macroeconomic models that considered income, price of electricity, industrialization rate and an increase in residential space, the authors found that a price increase could result in an increased efficiency of residential electricity consumption.

1.6.3 Section IV: Land Management Improvement Options

Water and land in the agricultural landscape need to be considered together to increase resource productivity and achieve sustainable development.

After independence in 1991, the Uzbek government declared domestic wheat production a prime objective to achieve self-sufficiency in staple food. Wheat became the second strategic crop in the state order system after cotton. As irrigated wheat production has only a history of less than two decades, opportunities to improve productivity may still be available. The optimal combination of nitrogen (N) fertilization and irrigation, losses of the applied N and costs and benefits have therefore been studied by **Ibragimov et al. (Chap. 11)**. Given the high impact of a combination of N-fertilization and irrigation-water management on N_2O emissions (a greenhouse gas), they show that management practices should be modified to mitigate N_2O emissions and to sustain higher N-fertilizer use efficiency. The amount and timing of the N-fertilizer application and irrigation events can be manipulated to reduce N losses. Concomitant N fertilization and irrigation is to be avoided whenever possible. In general, management practices that have been shown to increase the N-fertilizer use efficiency in irrigated systems, such as sub-surface fertilizer application, fertigation and drip irrigation, will likely also reduce the N_2O emissions and thus are expected to lead to more sustainable agriculture.

Improved N-management provides stable crop yields of good quality and preserves the environment. Groundwater contributes considerably to satisfy crop water demand in the irrigated areas of Khorezm, and high nitrate levels in the same groundwater also represent a significant supplemental contribution to the soil-N balance and plant-N uptake as shown by **Kienzler et al. (Chap. 12).** If groundwater levels are reduced to avoid salinization, this forfeited N supply may need to be substituted to avoid N deficiency.

Conservation agriculture holds the promise to sustain soil fertility and increase water-use efficiency. These practices consist of a judicious mix of reduced tillage operations, crop residue management and retention, and introducing suitable crop rotations. Comparing permanent beds and zero tillage with conventional tillage methods under irrigated conditions, **Pulatov et al. (Chap. 13)** show that appropriate crop residue management and crop rotation can increase soil organic matter, improve soil structure, and enhance moisture holding capacities. While yield effects were not observed, operational costs were clearly reduced, fuel saved, and hazardous exhausts cut.

Proper crop rotation and the inclusion of more crops into the cropping systems is one of the three pillars of conservation agriculture. The studies reported by **Bobojonov et al. (Chap. 14)** revealed manifold ecological benefits gained from increased crop diversification, including the reduced probability of crop failure and an increased economic stability. Crop diversity is thus an important factor for risk management.

Degraded lands still being cultivated are a drain on the resource base of Khorezm. An alternative use of degraded or marginal land is the establishment of tree plantations. This is the topic of the contribution of **Khamzina et al. (Chap. 15)**. Trees provide ecosystem services such as enhanced soil fertility and carbon sequestration and concurrently provide timber, fruit, fodder and fuel-wood. The assortment of appropriate tree species and silvicultural techniques for converting degraded, salt-effected land into small-scale forest plantations are extensively evaluated by the authors. Bio-physical processes (soil-plant-atmosphere) following the establishment of the tree plantations, i.e. tree water use, N_2-fixation, litter fall turnover are described, as are the magnitude of ecosystem services generated by afforestation i.e. supply and quality of fuel-wood and fodder. Mixed-species tree plantations can thus exploit marginal, salt-affected land – about 15% of the area in the Khorezm region. This frees resources that can be directed to the productive croplands, increasing the overall land and water use efficiency.

Deforestation of the riparian *Tugai* forests has contributed to the emission of substantial amounts of carbon dioxide (CO_2) to the atmosphere and reduction in C storage in soil and vegetation (**Scheer et al.; Chap. 16**). The conversion of *Tugai* forests into irrigated croplands releases N_2O and CH_4 equivalent to 2.5 tons of CO_2 per hectare per year. The return of degraded unprofitable cropland into perennial plantations would compensate for these greenhouse gas emissions, and farmers could gain additional income if Uzbekistan would be ready to create or enter a carbon market.

1.6.4 Section V: Land and Water Management Tools

By linking models for crops, water, soil, and salinity with Geographical Information Systems (GIS), various tools were developed in the Khorezm Project to improve land and water management.

A farm-level economic ecological optimization model was developed as a land-use planning and decision-support tool that couples ecological and economic optimization of land allocation (**Sommer et al.; Chap. 17**). The model produces consistent and plausible outputs that can be used for quite complex scenario simulations. The simulation results enable a better understanding of the impacts of different cotton policies on the farm economy as well as on farmers' decisions with respect to land and water use in Khorezm. This has sparked a discussion on policy options that are available for promoting income and food security for rural producers in other areas with agronomic and economic conditions closely resembling those observed in Khorezm.

Oberkircher et al. (Chap. 18) investigated three WUAs and their member farmers to assess their actual water access and their perceptions on water-saving practices. The authors used a GIS to integrate social science with physiographic data. The farmers are aware that water access is strongly related to the proximity of WUAs to the river but could be improved with water-saving practices. Farm location within the WUA and land elevation also seems to influence access to water and shape water use behavior.

The vast infrastructure built for irrigation, together with an ill-functioning drainage network has lead to a build-up of very shallow groundwater, followed by water logging and salt accumulation in the soil profile. **Awan et al. (Chap. 19)** identify deficits in the management and maintenance institutions, inappropriate and inflexible irrigation strategies, and poor linkages between field level demands and the operation of the network. The groundwater contribution to crop water use is quantified and a model is developed that suggests mitigation strategies to raise irrigation efficiencies, reduce the impact of water stress on yield, thus improving water productivity. It also is the first managerial tool that integrates surface water from the irrigation system and groundwater and that therefore would be able to support managerial decisions on optimization conjunctive use of surface and groundwater under deficit water supply.

Bekchanov et al. (Chap. 20) argue that improving crop water productivity is a key to reduce food and water insecurity. To compensate for the general lack of information for farmers on soil salinity and available irrigation water, the authors used a mix of tools to estimate water allocation for cotton, wheat, rice, vegetables, and fruits. All these crops consume much higher amounts of irrigation water than presently recommended with cotton and rice as the highest water consumers. There is a large variability of crop-specific water productivity over the different regions and within the regions according to the location of farms and fields. Introducing water-wise options and less-water consuming crops remains a daunting challenge but would be beneficial particularly for downstream districts. A regionally differentiated cropping portfolio that accounts for variability in water availability and soil quality would go a long way to improve water productivity.

1.6.5   Section VI: Economic System-Management Reform

Economic, agriculture-based development of the country and Khorezm in particular has been analyzed with a general equilibrium model (CGE) (**Bekchanov et al.; Chap. 21**). The CGE model was developed for Khorezm's regional economy and the national economy of Uzbekistan, which permits the comparison of drivers and policies at both levels. It also allows comparison of the current state of affairs with various alternative scenarios. The national and regional databases include production, final demand, and input-output relations for 20 sectors of the economy, 7 of which relate to the agrarian sector. A liberalization of the present cotton production policy would not necessarily have immediate impacts on national and regional incomes, but policies aimed at increasing productivity of the main crops and of livestock production would raise private and government revenues. The authors concluded that regionalizing the development strategies in Uzbekistan would lead to better use of the comparative advantages of regions than the current central approach of a uniform nationwide development program.

This view is substantiated by the findings of a Value Chain Analysis of the cotton and wheat sectors (**Rudenko et al.; Chap. 22**). It reveals the potential of the agro-processing industry in Khorezm to impact both land and water use if the presently underused processing capacities would be better employed. Increased local processing of cotton fibre, for example, and export of textile products with higher value added could almost double regional export revenues. Developing this sector would maintain the targeted level of export revenue of the region with a lower rate of raw cotton production, which in turn could reduce land and water use. The freed-up land could be used for less intensive and resource-consuming crops such as tree plantations or pastures.

The introduction of water service fees has often been suggested as a 'silver bullet' solution for water use inefficiency. With a mathematical programming model that considers regional welfare, cropping patterns, export structure and the economic attractiveness of crops to agricultural producers, **Djanibekov et al. (Chap. 23)** analyzed the local potential for introducing water service fees. Introducing such fees would generate sufficient funds to cover operation and maintenance of the irrigation and drainage system only if set at very high levels, and introducing fees as a stand-alone measure would likely fail to achieve these targets unless supported by additional changes, such as the reduction of state production targets on crops.

1.7 Conclusions

The research conducted in the Khorezm Project between 2001 and 2007 in the irrigated areas of Khorezm, allows conclusions that may be applicable across Central Asia, in the Caucasus, and in other irrigated regions of the world that share similar characteristics (flat, irrigated drylands). The decades-long production of the "white gold", cotton, during the Soviet Union era allowed investments in schools, infrastructure, health provision and much more. However, it also has resulted in widespread land degradation and the demise of the Aral Sea. Continuing unsustainable agricultural practices threatens the environment and livelihoods alike. But irrigated agriculture, also offers the opportunities to address the challenges. Increasing land and water productivity is a major pathway to supporting and stabilizing the national and regional economy as well as providing options for individual prosperity and sustainable livelihoods.

At the field level, the project findings suggest that it would be beneficial to use marginal, strongly salt-affected and unproductive cropland for more suitable purposes that benefit the ecology and livelihoods of the population. When diverting the resources thus spared to more productive areas, overall resource conservation and increased water productivity will occur without compromising on agricultural productivity at the farm and regional level. Increased resource use efficiencies can be reached through state-of-the-art methods on more fertile land, with the use of GIS and remote sensing tools, rapid and near real-time soil salinity mapping, and the use of conservation agriculture principles and crop diversification. Innovations can minimize energy use, reduce operating costs and greenhouse gas emissions.

Progress at the field and farm level can only be made in this regard if the Government of Uzbekistan takes on the challenge of institutional and economic reform. The Uzbek government repeatedly has affirmed its willingness to deploy funds and resources to ensure water supply. While this may be the case, it may not be sufficient to address future exacerbation of water problems, due to increasing claims by upstream water users and by climate change. The supply of "low-hanging fruits" is dwindling, and more complex solutions may need to be created that look at efficiency and productivity increases. This will require further scientific support. Given the future role of agriculture in Uzbekistan, investments training of human resources that can solve some of these intricate problems should be given high priority.

Education is considered a key element to successful economic growth, because jobs of the future will require a higher level of skills. The Khorezm Project has strengthened research capacities at a higher-education establishment in Uzbekistan. This offered the advantage of integrating research with educational capacity building of local staff and students. The project was able to show that that this continuous effort has borne fruit. Yet, the success will ultimately depend on the partner country to take up the challenge, carry on with the capacity building and scale it out to other regions. An institutional reform of private and public agricultural extension is needed to be able to transfer knowledge to the inexperienced farmers, many of whom are former state-workers gone farming. They are novices in sustainable resource use and efficient enterprise development, and the investment will unlock improvements in the environment and rural livelihoods.

According to United Nations (2010), real GDP growth in Uzbekistan amounted to 8% in 2010, which was not only the highest in Central Asia but also has brought Uzbekistan to the top of the list with highest expected growth rates in 2011 (7%) and 2012 (8%). In some chapters of this book it is argued that the prosperity of Uzbekistan lies in the economic pursuit of a strategy of export-led-growth. Uzbekistan may prosper from manufacturing high-value products that can be exported and sold at competitive prices. This argues that farmers need better links to markets and trade, which may be achieved through the development of a thriving private sector. The alternative will be an endless continuation of subsidies. Cutting the present unsustainable subsidies and encourage home production may result in price increases of fuel and chemicals (fertilizers, pesticides) that need to be compensated by for instance higher farm-gate prices.

Recent experience worldwide has shown that emerging economies may be hit hard by booming oil prices since they use more oil per unit of output than rich countries do. The economy of Uzbekistan is energy-intensive and it profits from a highly subsidized petrol and gas price. But higher farm-gate prices are likely to be introduced when restructuring the commodity value chains, and higher prices would allow farmers to earn more income from their produce and increase their farm capital. This in turn could be used to invest in better, more efficient irrigation systems.

Cereal production in Central Asia needs to be boosted in both quantity and quality to ensure food security for the growing populations. At present, the producers in Uzbekistan are unable to react quickly to world market price movements, e.g. the food price hikes in 2008. Farmers in Uzbekistan are exposed to increased risks when producing alternative or additional crops. These risks stem from the needed up-front investments in all production factors for those new crops, including seeds, fertilizers and, eventually, machinery, with uncertain returns. The present unreliable water supply and imperfect market conditions often render investments unprofitable. The double goal of shielding farmers from risks and increasing security of returns on their investments can only be achieved by national strategic decisions. Minimum price guarantees or insurance schemes linked to water supply might overcome the risk aversion. Once such conditions are in place, sustainable development will be more easily achieved in the irrigated Central Asian drylands, to the benefit of people and nature.

Initiatives to address climate change in Central Asia are gaining ground. The present thinking by various international donor agencies is that not only Uzbekistan but also the other Central Asian countries should put further agricultural reforms ahead of other concern such as energy. But given the present inefficiency in the use of precious resources, energy management should remain a priority. Also the looming problem of rising food prices around the world should remain the focus of attention. Finally, much needs to be done to ensure water supply in downstream countries such as Uzbekistan. The five states in Central Asia including Uzbekistan must coordinate their efforts and work out a sustainable and transparent mechanism for cooperation on water and energy while committing to preserving the environment. Working under the principles of fair, sustainable and reasonable use of transnational water resources will be critical in this context.

1.8 Outlook

The interdisciplinary Khorezm Project addresses resource-use efficiency, economic viability and environmental sustainability of land use and agricultural production systems and aims to provide a science-based comprehensive restructuring concept for improved management of land and water resources in the region. Elements of such a concept emerge from the pages of this book, and a number of more general conclusions come into view.

Policies. It has become evident from several of the studies that Uzbekistan would greatly benefit from creating a more market-oriented policy environment that would lead to changes in land use and irrigation water management. However, this liberalization will benefit from careful studies of the options and consequences of judicious adjustment of the policy environment. In order to allow for better policy-making, the impacts of various scenarios and alternative options need to be investigated in an integrative manner. We now have models that allow optimization of the policy mix, and a forecast of the long-term effects of these policies. Tentative steps may be under way, as one ADB document (ADB 2006) concluded, *"the ongoing agriculture sector reform initiative in Uzbekistan ... seeks to reduce the mandatory state procurement targets for cotton and wheat"*. These processes need to be supported with sound recommendations based on scientific research, examples of which are provided in this book.

Efficient land use and water-saving technologies can only be introduced in an enabling policy environment. However, the right policy environment is of little consequence if the institutions are not in place to translate and implement these policies. In turn, institutional change would have little effect if there are no real options available to the farmers and decision-makers to gain efficiency, sustainability and profitability. The interplay of interventions at various spatial and temporal levels can be analyzed using computer simulation models, and support tools developed in the framework of this project.

Institutions. Poor performance of irrigation systems and low water productivity and crop yields in Khorezm points to severe institutional shortcomings that need to be overcome, e.g. with regard to water management institutions and service provision by agricultural organizations at the regional level. As all land management is related to agriculture, improvements in this sector will automatically increase the performance of natural resource management.

Institutions act at different levels. At the regional level, water distribution and government structures (such as the Basin Department of Irrigation Systems (BUIS) and *hokimiat*) decide on timely water distribution to the different parts of Khorezm. This level corresponds to the whole hydrographic basin of the lower Amudarya region, the 'basin' of Khorezm, and the smaller sub-basins of management (so-called TEZIMs of variable size) that are run by management sub-units of the BUIS. At the micro-catchment level of a WUA (typically of 2,000–3,000 ha size), water and land allocation are decided. At the high-resolution level of a farm (80 ha) or a single field (1–3 ha), the more specific problems of how land is prepared and used for crops need to be dealt with in an iterative manner. In the case of water management, it is here that farmers deal with irrigation management, fertilizer strategies, crop rotation, etc.

Land use is dealt with at all levels of intervention. Farm, land and resource use optimization require political decisions which, to be successful, will have to be supported institutionally. Technological decisions are chiefly dealt with at the farm-level scale. Information about successful land management alternatives can inform decision-making on policies, for example through farmers that have gained experience with these technologies. They can champion these technologies into the policy arena, so that adequate policies are created to enable the technological advancements. Training and capacity building and the creation of 'centers of excellence' that cover the region are urgently needed to support this process. The Khorezm Project has shown how this can be achieved.

Technological innovations and their adoption. Technologies that can help increase resource-use efficiencies and that can shift the current system toward more rational land and water use and sustainability of agriculture have been developed in the Khorezm Project. They have been tested under farm-level conditions in an integrated fashion to demonstrate their potential for the region. The next milestone will be the out-scaling through on-farm participatory approaches, preferably with governmental support providing the enabling policy and institutional environment.

However, as Vlek and Gatzweiler (2006) argued: (1) global environmental change is proceeding at a fast pace, (2) values and norms systems are necessary fundamental institutions for defining man-nature relations, and (3) institutional change at the level of embeddedness at which values are located, occurs slowly. Although at the national political level in Uzbekistan institutions change rather rapidly, there is little effort made to change the underlying value and norm system. As Diamond (2005) says, human beings "cling to those values which were the source of their greatest triumphs" but are now inappropriate. Overcoming reluctance and inability to change in the face of looming disaster are the key challenges for Central Asia.

References

ADB (Asian Development Bank) (2006) Poverty reduction: East and Central Asia: Uzbekistan. Annual report 2004, Part 2. http://www.adb.org/Documents/Reports/Annual_Report/2004/part050301_08_UZB.asp

Akramkhanov R, Sommer A, Martius C, Hendrickx JMH, Vlek PLG (2008) Comparison and sensitivity of measurement techniques for spatial distribution of soil salinity. Irrig Drain Syst 22:115–126

Akramkhanov A, Kuziev R, Sommer R, Martius C, Forkutsa O, Massucati L (2011) Soils and soil ecology in Khorezm. In: Martius C, Rudenko I, Lamers JPA, Vlek PLG (eds) Cotton, water, salts and soums: economic and ecological restructuring in Khorezm, Uzbekistan. Springer, Dordrecht

Al-Iqbal F (1999) History of Khorezm, by Shir Muhammab Mirab Munis and Muhammad Riza Mirab Agahi. Islamic Hist Civiliz 28:718. Brill, Leiden/Boston/Köln

Conrad C, Schorcht G, Tischbein B, Davletov S, Sultonov M, Lamers JPA (2011) Agro-meteorological trends of recent climate development in Khorezm and implications for crop production. In: Martius C, Rudenko I, Lamers JPA, Vlek PLG (eds) Cotton, water, salts and soums: economic and ecological restructuring in Khorezm, Uzbekistan. Springer, Dordrecht

Diamond J (2005) Collapse: how societies choose to fail or succeed. Viking, New York

Djanibekov N (2008) A micro-economic analysis of farm restructuring in Khorezm region, Uzbekistan. PhD dissertation, Bonn University, Bonn

Djanibekov N, Bobojonov I, Djanibekov U (2011a) Prospects of agricultural water service fees in the irrigated drylands, downstream of Amudarya. In: Martius C, Rudenko I, Lamers JPA, Vlek PLG (eds) Cotton, water, salts and soums: economic and ecological restructuring in Khorezm, Uzbekistan. Springer, Dordrecht

Djanibekov N, Bobojonov I, Lamers JPA (2011b) Farm reform in Uzbekistan. In: Martius C, Rudenko I, Lamers JPA, Vlek PLG (eds) Cotton, water, salts and soums: economic and ecological restructuring in Khorezm, Uzbekistan. Springer, Dordrecht

Hornidge A-K, Ul-Hassan M, Mollinga PP (2011) Transdisciplinary innovation research in Uzbekistan – 1 year of 'following the innovation'. Dev Pract 21(26):825–838

Ibrakhimov M, Khamzina A, Forkutsa I, Paluasheva G, Lamers JPA, Tischbein B, Vlek PLG, Martius C (2007) Groundwater table and salinity: spatial and temporal distribution and influence on soil salinisation in the Khorezm region (Uzbekistan, Aral Sea Basin). Irrig Drain Syst 21:219–236

Kienzler KM (2010) Improving the nitrogen use efficiency and crop quality in the Khorezm region, Uzbekistan, vol 72, Ecology and development series. ZEF/Rheinische Friedrich-Wilhelms-Universität, Bonn

Micklin P (2008) Using satellite remote sensing to study and monitor the Aral Sea and adjacent zone. In: Qi J, Evered KT (eds) Environmental problems of Central Asia and their economic, social, and security impacts, NATO science for peace and security series – C: environmental stability. Springer, Dordrecht

Müller M (2006) A general equilibrium approach to modeling water and land use reforms in Uzbekistan. PhD dissertation, Rheinischen Friedrich-Wilhelms-Universitat Bonn, Bonn

Tischbein B, Awan UK, Abdullaev I, Bobojonov I, Conrad C, Forkutsa I, Ibrakhimov M, Poluasheva G (2011) Water management in Khorezm: current situation and options for improvement (hydrological perspective). In: Martius C, Rudenko I, Lamers JPA, Vlek PLG (eds) Cotton, water, salts and soums: economic and ecological restructuring in Khorezm, Uzbekistan. Springer, Dordrecht

Tolstov SP (1948) Following the tracks of ancient Khorezmian civilization. Academy of Sciences of the USSR, Moscow – Leningrad

United Nations (2010) World economic situation and prospects 2011. http://www.un.org/en/development/desa/policy/wesp/wesp_current/2011wesp.pdf

Vlek PLG, Gatzweiler FW (2006) Can we walk the "Middle Path" in coping with global environmental change? In: Ecology and buddhism in the knowledge-based society. Dongguk University, Seoul

WBGU (1998) Wissenschaftlicher Beirat der Bundesregierung Globale Umweltveränderungen (Worlds in transition: ways towards sustainable management of freshwater resources). Springer, Berlin

Wehrheim P, Schoeller Schletter A, Martius C (2008) Continuity and change land and water use reforms in rural Uzbekistan Socio economic and legal analyses for the region Khorezm, vol 43, Studies on the agricultural and food sector in Central and Eastern Europe. Leibniz Institute of Agricultural Development in Central and Eastern Europe (IAMO), Halle, 203 p. http://www.iamo.de/dok/sr_vol243.pdf. ISBN 978-203-938584-938527-938589

Zavgorodnyaya D (2006) Water user associations in Uzbekistan: theory and practice. PhD dissertation, Bonn University, Bonn

Part II
The Background (Physical Setting)

Chapter 2
Agro-Meteorological Trends of Recent Climate Development in Khorezm and Implications for Crop Production

Christopher Conrad, Gunther Schorcht, Bernhard Tischbein, Sanjar Davletov, Murod Sultonov, and John P.A. Lamers

Abstract The extremely continental climate in the inner Aral Sea Basin is characterized by a high annual temperature amplitude, and constantly very low precipitation. Climate change in the Khorezm region, a district of Uzbekistan dominated by extensive irrigation agriculture, is manifest by temperature increases, especially during the winter period (40-year record). The 10-year average of summer temperatures in 1981–1990 was 0.2–0.5°C above the long-term average in 1930–1990. Time series of agro-meteorological parameters based on data (air and soil temperature, relative humidity, wind speed, precipitation, and radiation) of three meteorological stations in Khorezm revealed (a) air temperatures supporting good growing conditions for cotton, rice, wheat, sorghum, and maize; (b) comparatively good air and soil temperature conditions for cotton production; and (c) no negative climate impact on regional water demands for irrigation yet. Moderate spatial variability was observed between climate in the central irrigation system, at its southern margin to the desert, and in the desert itself (represented by the stations Urgench, Khiva, and Tuyamuyun, respectively).

C. Conrad (✉) • G. Schorcht
Remote Sensing Unit, University of Würzburg, Am Hubland, 97074 Würzburg, Germany
e-mail: christopher.conrad@uni-wuerzburg.de; gunther.schorcht@uni-wuerzburg.de

B. Tischbein
Center for Development Research (ZEF), Walter-Flex-Str. 3, 53113 Bonn, Germany
e-mail: tischbein@uni-bonn.de

S. Davletov
Urgench State University, Khamid Olimjan Str., 14, 220100 Urgench, Uzbekistan
e-mail: sanjar-22@mail.ru

M. Sultonov • J.P.A. Lamers
ZEF/UNESCO Khorezm Project, Khamid Olimjan Str., 14, 220100 Urgench, Uzbekistan
e-mail: murod@zef.uznet.net; j.lamers@uni-bonn.de

C. Martius et al. (eds.), *Cotton, Water, Salts and Soums: Economic and Ecological Restructuring in Khorezm, Uzbekistan*, DOI 10.1007/978-94-007-1963-7_2,
© Springer Science+Business Media B.V. 2012

For nearly all years between 1970 and 2007, Growing Degree-Days (GDD; based on crop specific baseline temperatures and maximum thresholds) indicated the suitability for all investigated crops. The conditions for winter wheat have over time improved in terms of GDD, especially in the southern margin of the irrigation system (Khiva). Absolute GDD minima, a parameter used for assessing the period of temperatures sufficient for crop growth, were found to be close to the lower boundaries for cotton. Water availability and fertilizer additions are other factors that determine cotton performance. Minimum GDDs only slightly reduced during an observation period of 21 years.

The earliest planting dates for cotton, derived from an analysis of soil temperature thresholds, range between 21 March and 18 April during the observation period at all three sites. In most years, the first suitable cotton planting dates occurred between April 1 and 5. The cotton planting dates at the southern margins of the irrigation system appear more variable, caused by an increased risk of late frosts. In light of this data, recommendations from the Ministry of Agriculture and Water Resources of Uzbekistan to plant cotton in Khorezm from mid- to end-April have become conservative and give room for site-specific adjustments.

The long-term (1970–2007) annual average potential evapotranspiration (PET) at Urgench station reached 1,378 mm year^{-1}. PET of the vegetation periods varied between 1,100 and 1,200 mm. Only a slight PET decline is detectable, i.e. crop water demand in Khorezm remains constant.

Keywords Central Asia • Temperature • Climate change • Cotton • Growing degree days

2.1 Introduction: Climatic Settings and Effects of Climate Change in Khorezm

Central Asia is located in the zone of the prevailing westerlies, but at a far distance from oceans. The Himalaya and Pamir mountain ranges hinder the monsoon from entering the sub continent (Glantz 2005). As a consequence, an extreme continental climate with short, very cold winters influenced by the Siberian High prevails in the inner Aral Sea Basin (ASB). In this region, global warming has caused a non-cyclical increase of air temperatures in the past 40 years (Chub 2000; Giese and Moßig 2004).

The Khorezm region is located in the inner Aral Sea basin, more precisely in the southern part of the Amudarya floodplain (Fig. 2.1, left part). Two Central Asian deserts partly from the border of the agricultural landscape in Khorezm: the Kyzylkum in the North, and the Karakum in the South. The regional climate is typified by a high annual temperature amplitude and very sparse precipitation (rain and snowfall) mainly occurring between November and March (Fig. 2.2). The Amudarya river is the only source of water for crop cultivation in Khorezm. Around 4.5 km^3 of

2 Agro-Meteorological Trends of Recent Climate Development...

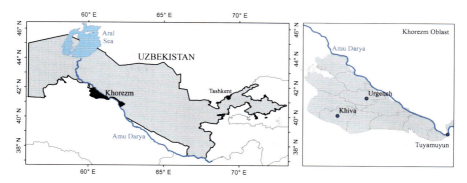

Fig. 2.1 Location of Khorezm in Central Asia (*left part*) and the location of the meteorological stations of Urgench, Khiva and Tuyamuyun within the Khorezm Oblast (*right part*)

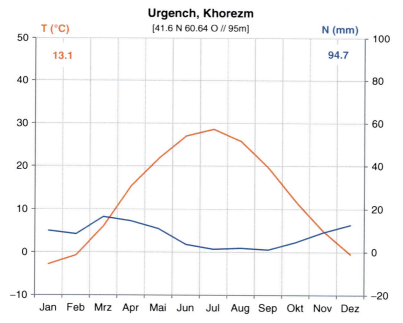

Fig. 2.2 Walther-Lieth diagram of the Khorezm region, based on the long-term data of the meteorological station in Urgench. *T* temperature, *N* precipitation

water are annually withdrawn from the Amudarya for irrigation and leaching of about 270 000 ha land throughout the Khorezm region (cf. Tischbein et al. 2011). The major crops are cotton, wheat and rice, which are complemented with maize, sunflower and sorghum, whilst in the many kitchen gardens a wide variety of crops and species is cultivated (cf. Djanibekov et al. 2011b).

A few indicators illustrate the impact of climate change in Khorezm. For instance, from 1970 onwards, average day temperatures (around 13–14°C) increased by 0.5–1.5°C. Especially during the past decade, the average temperature levels have significantly increased. The 10-year average of summer temperatures recorded between 1981 and 1990 was between 0.2°C and 0.5°C above the long-term average (1930–1990), which indicates the magnitude of climate change in the region. However, Chub (2000) highlighted that summer temperatures in the Amudarya delta increased only moderately compared to those at other stations of Uzbekistan. The temperature increase also varies spatially within the Khorezm region. The highest shift was found at the margins of the irrigation system in the vicinity of the Karakum desert.

A rise in mean temperatures has been observed mainly during winter months, whereas summer temperatures (April–October which corresponds to the vegetation season) seem to have been more stable. Accordingly, the length of the vegetation period (defined as the number of days with an average temperature higher than 10°C), remained nearly unaffected by climate changes in the past 40 years. Given that the current cropping plans are based on exploiting the upper limit of the vegetation period and that the economy in Khorezm strongly depends on agriculture (cf. Rudenko et al. 2011), the impacts of climate development on specific agro-meteorological parameters are still poorly understood in this region – despite the findings of several agro-climatic studies over entire Uzbekistan (e.g. Chub 2000).

The potential impact of various agro-meteorological parameters on the long-term growing conditions in Khorezm was analyzed using descriptive statistics and trend analysis. The parameter selected comprised growing degree-days, planting dates, and potential evapotranspiration, based on data for the past 40 years. The analysis focused on an impact of climate development on crop relevant details of the vegetation period and water demand. Temporal development of growing degree-days for cotton, rice, maize, wheat, and sorghum; planting dates of cotton (relevant for agricultural production in Uzbekistan); and potential evapotranspiration (PET) were assessed. The results were expected to allow understanding how sustainable the current cropping systems are under the predicted changes in climate, which, for being higher than global averages (IPCC 2007), are supposed to reduce future water availability in Central Asia.

2.2 Parameters and Methods

Long-term climate variables were available from three state-run meteorological stations in Khorezm (Fig. 2.1b) representing different climate conditions within the irrigation system of Khorezm. The stations in Urgench (data available between 1970 and 2007), Khiva (1973–2007), and Tuyamuyun (1980–2007) represent the situation of the central part of the irrigation system, its southern margin bordering the Karakum desert, and the transition zone between the deserts Karakum and Kyzylkum close to the Tuyamuyun reservoir, respectively. At all stations,

2 Agro-Meteorological Trends of Recent Climate Development...

Table 2.1 Baseline temperatures and duration of growing season (local agronomists, personal communication) used for the estimation of the crop specific growing degree-days

Crop	Growing season, day of year (Gregorian calendar)	Baseline temperatures (°C)
Wheat	263–196 (20.09–15.07)	4.5
Rice	130–283 (10.05–10.10)	10
Sorghum	94–263 (04.04–20.09)	10
Cotton	105–305 (15.04–31.10)	15.6
Maize	152–278 (01.06–05.10)	8

Source: Vaughn (2005)

temperatures, precipitation, relative humidity, wind speed and wind direction, and soil temperature were measured daily.

Weather data for the Khiva station was available for 1970–2007, but we used only data for 1986–2007 because in 1986 the location of the station had been changed, which had noticeable effects on the data record. At the meteorological stations of Khiva and Tuyamuyun wind speed measurements also appeared biased and were excluded from the study. Despite those artifacts and minor gaps in the data sets of the three stations, the high level of uncertainty attributed to meteorological data from stations in Central Asia by Giese and Moßig (2004) appears to be low for these three stations. In the following, the definitions of the selected parameters and the methods for analysis are outlined.

Growing degree-days (GDD) show to which extent the temperature conditions of a region are favorable for specific crops. GDDs are also used for predicting different stages of crop development or harvest time, and can indicate temperature stress factors (Yang et al. 1995; Snyder et al. 1999; Kadioglu and Saylan 2001).

GDDs were estimated according to the Baskerville-Emin method (Baskerville and Emin 1969). Depending on the crop, development of plants will occur only if the air temperature exceeds a minimum threshold for its development. Above this so-called baseline temperature, crop development increases linearly with temperature till an upper threshold. Here, a value of 30°C was selected, which was among the lowest values suggested by other authors. Rising the maximum threshold would lead to an increase in GDDs and hence to a less conservative assessment of growing conditions in Khorezm. Baseline temperatures and duration of growing seasons of the crops were based on official statistics and recommendations from local experts (Table 2.1). GDDs are particularly relevant between first planting to harvest.

In Uzbekistan, the **planting date for cotton** is relevant for assessing the duration of the vegetation period for this crop. Recommendations from the Ministry of Agriculture and Water Resources (MAWR) of Uzbekistan for Khorezm indicate to seed cotton each year between the middle and end of April. But the MAWR recommendations underline also that the progression of the soil temperature measured at a depth of 5 cm is essential for deciding exactly when to sow cotton. Once soil temperature reaches the threshold of 14°C and does not drop below this level for a period for 5 days in a row, the final, 5th day is regarded to be the suitable date for seeding cotton.

The **potential evapotranspiration (PET)** is an agro-meteorological parameter indicating expected water consumption by crops. Hence, PET is a basic parameter used to optimize irrigation scheduling, also in case of insufficient water supply. PET was calculated not only for the entire year, but also for the official vegetation period of cotton between April and the end of September. This period is relevant for estimating crop water demands and hence of high economic relevance for Khorezm.

PET was derived from temperature, relative humidity, net radiation, and wind speed according to the ASCE standardized reference evapotranspiration equation (Walter et al. 2005):

$$ET_{ref,short} = \left(\frac{0.408\Delta(R_n - G) + \gamma \dfrac{900}{T + 273} u_2 (e_s - e_a)}{\Delta + \gamma(1 + 0.34u_2)} \right)$$

where Δ is the slope of the vapor pressure-temperature curve (kPa°C^{-1}), R_n is net radiation (MJ m^{-2}), G is soil heat flux (MJ m^{-2} day^{-1}), γ is the psychrometric constant (kPa°C^{-1}), T = mean daily air temperature (°C), u_2 = daily mean wind speed at 2 m (ms^{-1}), e_s = mean saturation vapor pressure (kPa) and e_a = mean actual vapor pressure (kPa). Due to missing wind speed data, PET was only derived for the meteorological station of Urgench.

2.3 Long-Term Development of Agro-Meteorological Parameters and Its Implications

Table 2.2 summarizes the GDDs for the five crops, based on weather data collected at the three meteorological stations. Mean values and standard deviations suggest that Khorezm excellently meets the GDD requirements of all those crops. Compared to minimum requirements suggested in the literature, only minimum GDDs for cotton fell below critical minima in Urgench and Khiva. But minimum values for cotton given in the literature vary considerably (Table 2.2). Howell et al. (2004) argued for instance, that in northern Texas more than 1,130°C GDDs are required to reach full maturation of all cotton bolls.

Figure 2.3 shows the temporal development of GDDs for the major crops in Khorezm. Sorghum and maize (not depicted) show curves similar to rice. Despite data gaps in the time series, no clear statistical trend was observed during the observation periods at the three meteorological stations. Simple anomaly analysis suggests a hardly perceptible decrease of GDDs for cotton and rice, and an increase for wheat. At Khiva station, GDDs of wheat showed a weak increasing trend. However, due to limited time series of only 21 years for Khiva, this trend needs to be viewed with more caution than the findings of the other stations.

The range and standard deviations (Table 2.2) indicate a higher inter-annual variability at the margins of the irrigation system (Khiva) than in central Khorezm

2 Agro-Meteorological Trends of Recent Climate Development...

Table 2.2 Estimated growing degree-days (GDD) for different crops, based on 21 years (Khiva meteorological station), 27 years (Tuyamuyun meteorological station) 7 and 37 years (Urgench station) in different locations of Khorezm calculated according to the Baskerville-Emin method

	GDD	Cotton	Wheat	Maize	Rice	Sorghum
Urgench (37 years)	Mean	1,535	2,555	2,135	2,196	2,326
	Max	1,718	2,842	2,287	2,382	2,570
	Min	1,280	2,295	1,970	1,983	2,066
	STD	88	142	76	82	107
Khiva (21 years)	Mean	1,564	2,626	2,159	2,226	2,359
	Max	1,877	2,978	2,369	2,476	2,650
	Min	1,360	2,317	1,953	1,998	2,153
	STD	125	167	105	115	127
Tuyamuyun (27 years)	Mean	1,669	2,719	2,272	2,349	2,487
	Max	1,862	2,994	2,385	2,471	2,630
	Min	1,205	2,353	2,106	2,130	2,212
	STD	130	168	67	77	100
GDD minima ranges mentioned by secondary sources		1,055–1,389[a]	1,538–1,972[b]	1,222–1,555[a]	944–1,166[a]	N.a.

N.a. not available, *STD* standard deviation
[a]Ahrens (2008)
[b]Miller et al. (2001)

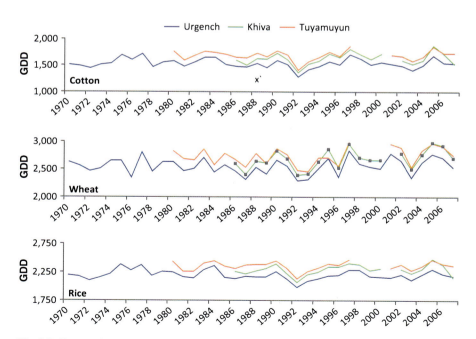

Fig. 2.3 Temporal development of growing degree-days (GDDs) for cotton, wheat, and rice at the stations Urgench, Khiva, and Tuyamuyun

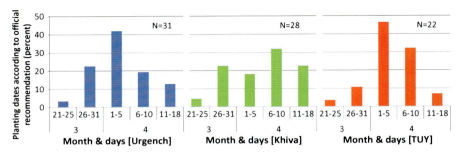

Fig. 2.4 Percentage of years which planting dates for cotton above the 14°C threshold, observed within 5-day intervals between 21–3 and 18–4 (Total number of observations in Urgench: 31, Khiva: 28, Tuyamuyun: 21). Months 3 (March) and 4 (April)

(Urgench). The higher variability would indicate a higher yield risk but this contrasts with the fact that GDDs in all years remained higher in Khiva than in Urgench (Fig. 2.3). On the other hand, tail-end located fields in the irrigation system experience water insecurity (Conrad et al. 2007) and the profitability of cotton production in the distal parts of Khorezm was postulated to be low (Bobojonov 2009). Hence, next to the probability to receive sufficient amounts of irrigation water (cf. Bekchanov et al. 2011), GDD could be used as an additional parameter to determine high (for instance tail-end locations) and low (e.g. top end locations) risk prone sites. In addition, even if plant growth components are sufficient to grow, say, cotton or wheat, the temperatures should not exceed certain maxima within sensitive plant developing phases if yield reductions are to be avoided (Mathews et al. 1995). For instance, GDDs do not reflect adverse effects on cotton growth caused by air temperatures above 39°C. These impacts were considered in the assessment of growing conditions in Uzbekistan by (Chub 2007) who calculated 13.6% losses of heat resources (the sum of air temperature conditions enabling crop growth, e.g. expressed by GDDs), due to unfavorable weather conditions and high rates of nitrogen fertilizers. Such parameters should be considered together with the temporal development of GDDs to provide for regional recommendations.

As a single parameter, however, GDDs far above critical minima showing no clear trends, do not indicate limited growing conditions. Based on the GDD estimates, the entire Khorezm appears suitable for cotton production, confirming previous conclusions by Ibragimov et al. (2007).

The **planting date for cotton** in Khorezm between 1977 and 2007 (meteorological station Urgench) varied with 25 days. Figure 2.4 shows the relative distribution of planting dates for the three meteorological stations grouped in 5-day periods.

In the central part of the irrigation system (Urgench), soil temperatures expectably achieved the 14°C baseline temperature earlier than at the desert margins of the irrigation system (Khiva). Urgench and Tuyamuyun exhibited a clear peak in the interval between April 1 and 5. In contrast, the distribution of observations in Khiva

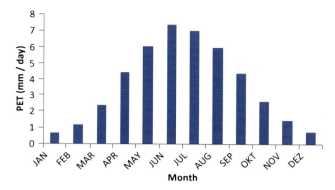

Fig. 2.5 Monthly averages of potential evapotranspiration (PET) based on the long-term data from the meteorological station in Urgench between 1970 and 2007

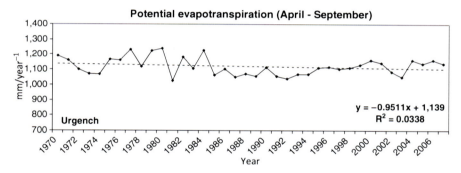

Fig. 2.6 Temporal (linear) trends of potential evapotranspiration (PET) summarized for the vegetation periods (April–September) between 1970 and 2007

appeared nearly equal, again indicating higher environmental variability and therefore higher planning insecurity. In all cases, the present blanket recommendation to sow cotton only after the middle of April has become conservative.

Based on the weather data (1970–2007) from Urgench station, the long-term annual average **potential evapotranspiration (PET)** amounted to 1 378 mm year^{-1}. Monthly averages from daily PET values (Fig. 2.5) were highest in June (with an average of 7.3 mm day^{-1}), followed by July (6.9 mm day^{-1}). Minima values for PET occurred in January and December (0.6 and 0.7 mm day^{-1}).

PET time series indicate constant water demands in the irrigation system in Khorezm (Fig. 2.6). Also, interannual variations of 200 mm PET, which can significantly affect water consumption by crops, occurred less frequently in the past 20 years in comparison to the 1975–1985 period.

The PET observations are partly in line with Jarsjö et al. (2007), who concluded for the entire ASB after 1950 that the increases in water losses through higher temperatures, and consequently higher PET, were counterbalanced by increasing precipitation levels during the same period. However, precipitation within the last 40 years as recorded at the meteorological stations of Urgench, Khiva, and Tuyamuyun remained stable.

Looking at the perspective of water distribution, Müller (2006) estimated that the probability of receiving sufficient amounts of irrigation water decreased in Khorezm during the last decades (cf. Djanibekov et al. 2011a, for details). Also a negative gradient of crop water availability in Water User Associations (WUAs) located at increasing distance from the water intake points of the irrigation system was found for the comparatively water-rich years 2004 and 2005 (Conrad 2006). Altogether, the observed PET developments may reduce the predicted acceleration of water supply problems in Khorezm.

2.4 Summary and Conclusions

This analysis of temperature and water related agro-meteorological parameters was based on middle- to long-term data of three meteorological stations representing different landscape conditions in the Khorezm region. In terms of growing degree-days (GDD), comparatively stable and good conditions for cotton, wheat, rice, maize, and sorghum in the entire region can be concluded. Improved growing conditions for wheat in the southern part of the region, as indicated by a positive trend in GDDs, coincided with the temperature increases at present mainly occurring during the winter months.

The analysis of soil temperatures, which according to official Uzbekistan recommendations is decisive for estimating first planting dates of cotton, generally confirmed the existing recommendations. All investigated datasets indicated suitable conditions for cotton planting between mid and end of April. However, the degree of uncertainty increases towards the Karakum desert, which borders Khorezm in the south.

Slightly decreasing PET developments imply that climate related crop water demands in the region are stable. This can be interpreted as a good sign in face of the expected increasing pressure on irrigation water resources in Khorezm.

The hypothesis of a negative influence of the desiccation of the Aral Sea on the agro-meteorological conditions in the Amudarya delta which has recurrently be pointed out (Giese 1997; Vinogradov and Langford 2001) can neither be confirmed nor rejected for the southern part of the Amudarya delta based on the 40-year data record. There is still a possibility that the desiccation of the Aral Sea reduces the temperature increase in Khorezm. To separate between regional and global change impacts on Khorezm, the inclusion of larger areas would be beneficial, and longer time series would be required to identify cyclic developments of climate variables. Simply as the total observation period is growing with time, agro-meteorological predictions in the inner Aral Sea Basin will improve, and the use of climate models will provide another option for estimating future scenarios.

References

Ahrens CD (2008) Essentials of meteorology: an invitation to the atmosphere. Brooks/Cole, Belmont, CA, USA

Baskerville GL, Emin P (1969) Rapid estimation of heat accumulation from maximum and minimum temperatures. Ecology 50:514–517

Bekchanov M, Lamers JPA, Karimov A, Müller M (2011) Estimation of spatial and temporal variability of crop water productivity with incomplete data. In: Martius C, Rudenko I, Lamers JPA, Vlek PLG (eds) Cotton, water, salts and *Soums*: economic and ecological restructuring in Khorezm, Uzbekistan. Springer, Dordrecht/Berlin/Heidelberg/New York

Bobojonov I (2009) Modeling crop and water allocation under uncertainty in irrigated agriculture: a case study on the Khorezm Region, Uzbekistan. PhD dissertation, Bonn University, Bonn

Chub EV (2000) Climate change and its impact on natural resources potential of the republic of Uzbekistan. Tashkent, Uzbekistan. Main Administration on Hydrometeorology under the cabinet of Ministers of the Republic of Uzbekistan. Central Asian Hydrometeorological Research Institute named after V. A. Bugayev

Chub EV (2007) Climate change and its impact on hydrometeorological processes, agro-climatic and water resources of the republic of Uzbekistan. Center for Hydro-meteorological Service under Cabinet of Ministers of the Republic of Uzbekistan (Uzhydromet). Scientific and Research Hydro-meteorological Institute (NIGMI)

Conrad C (2006) Remote sensing based modeling and hydrological measurements to assess the agricultural water use in the Khorezm region (Uzbekistan). PhD dissertation, University of Würzburg, Würzburg

Conrad C, Dech SW, Hafeez M, Lamers J, Martius C, Strunz G (2007) Mapping and assessing water use in a Central Asian irrigation system by utilizing MODIS remote sensing products. Irrig Drain Syst 21(3–4):197–218

Djanibekov N, Bobojonov I, Djanibekov U (2011a) Prospects of agricultural water service fees in the irrigated drylands, downstream of Amudarya. In: Martius C, Rudenko I, Lamers JPA, Vlek PLG (eds) Cotton, water, salts and *Soums*: economic and ecological restructuring in Khorezm, Uzbekistan. Springer, Dordrecht/Berlin/Heidelberg/New York

Djanibekov N, Bobojonov I, Lamers JPA (2011b) Farm reform in Uzbekistan. In: Martius C, Rudenko I, Lamers JPA, Vlek PLG (eds) Cotton, water, salts and *Soums*: economic and ecological restructuring in Khorezm, Uzbekistan. Springer, Dordrecht/Berlin/Heidelberg/New York

Giese E (1997) The ecological crisis of the Aral Sea region. Geogr Rundsch 49(5):293–299

Giese E, Moßig I (2004) Klimawandel in Zentralasien. Discussion Papers 17. Zentrum für internationale Entwicklungs- und Umweltforschung (ZEU), Gießen

Glantz MH (2005) Water, climate, and development issues in the Amudarya Basin. Mitig Adapt Strateg Glob Chang 10(1):23–50

Howell TA, Evett SR, Tolk JA, Schneider AD (2004) Evapotranspiration of full-, deficit-irrigated, and dryland cotton on the Northern Texas high plains. J Irrig Drain Eng 130:277–285

Ibragimov N, Evett SR, Esanbekov Y, Kamilov BS, Mirzaev L, Lamers JPA (2007) Water use efficiency of irrigated cotton in Uzbekistan under drip and furrow irrigation. Agric Water Manage 90:112–120

IPCC (2007) Intergovernmental panel on climate change. Climate Change 2007. Cambridge University Press. http://www.ipcc.ch/publications_and_data/publications_and_data_reports.shtml

Jarsjö J, Asokan SM, Shibou Y, Destouni G (2007) Water scarcity in the Aral Sea Drainage Basin: Contribution of agricultural irrigation and a changing climate. In: Qi J, Evered T (eds) Environmental problems of Central Asia and their economic, social and security impacts. Proceedings of the NATO advanced research workshop on environmental problems of Central Asia and their economic, social and security impacts. Springer, Dordrecht/ Tashkent

Kadioglu M, Saylan L (2001) Trends of growing degree-days in Turkey. Water Air Soil Pollut 126(1–2):83–96

Mathews RB, Horie T, Kropff MJ, Bachelet D, Centeno HG, Shin JC, Mohandass S, Singh S, Zhu D, Lee MH (1995) A regional evaluation of the effect of future climate change on rice production in Asia. In: Matthews RB, Kropff MJ, Bachelet D, van Laar HH (eds) Modeling the impact of climate change on rice production in Asia. CAB International, Wallingford, pp 95–139

Miller P, Lanier W, Brandt S (2001) Using growing degree days to predict plant stages. Montana State University Extension Service. SKU MT200103AG

Müller M (2006) A general equilibrium approach to modeling water and land use reforms in Uzbekistan. PhD dissertation, Rheinischen Friedrich-Wilhelms-Universitat Bonn, Bonn

Rudenko I, Nurmetov K, Lamers JPA (2011) State order and policy strategies in the cotton and wheat value chains. In: Martius C, Rudenko I, Lamers JPA, Vlek PLG (eds) Cotton, water, salts and Soums: economic and ecological restructuring in Khorezm, Uzbekistan. Springer, Dordrecht/Berlin/Heidelberg/New York

Snyder R, Spano D, Cesaraccio C, Duce P (1999) Determining degree-day thresholds from field observations. Int J Biometeorol 42(4):177–182

Tischbein B, Awan UK, Abdullaev I, Bobojonov I, Conrad C, Forkutsa I, Ibrakhimov M, Poluasheva G (2011) Water management in Khorezm: current situation and options for improvement (hydrological perspective). In: Martius C, Rudenko I, Lamers JPA, Vlek PLG (eds) Cotton, water, salts and *Soums*: economic and ecological restructuring in Khorezm, Uzbekistan. Springer, Dordrecht/Berlin/Heidelberg/New York

Vaughn D (2005) Degree days. In: Oliver JO (ed) Encyclopedia of world climatology, Encyclopedia of Earth sciences series. Springer, Dordrecht/Berlin/Heidelberg/New York

Vinogradov S, Langford VPE (2001) Managing transboundary water resources in the Aral Sea Basin: in search of a solution. Int J Global Environ Issues 1(3–4):345–362

Walter IA, Allen RG, Elliott R, Itenfisu D, Brown P, Jensen ME, Mecham B, Howell TA, Snyder R, Eching S, Spofford T, Hattendorf M, Martin D, Cuenca RH, Wright JL (2005) The ASCE standardized reference evapotranspiration equation. ASCEEWRI Task Committee Report 59

Yang S, Logan J, Coffey D (1995) Mathematical formulae for calculating the base temperature for growing degree days. Agric For Meteorol 74(1–2):61–74

Chapter 3
Soils and Soil Ecology in Khorezm

Akmal Akramkhanov, Ramazan Kuziev, Rolf Sommer, Christopher Martius, Oksana Forkutsa, and Luiz Massucati

Abstract The current status of the agricultural soils in Khorezm is closely linked to their development influenced by past river flows and more recent human-managed irrigation and drainage practices. The initial relief of the Amudarya river delta was formed by ancient channels that carried large amounts of sediments. When irrigation was introduced, sediments started to be deposited on croplands, a process which formed the now spatially distinct features in the topsoils. Today's landscape of the ancient river delta is thus significantly influenced by human interference. The areas presently used for irrigated agriculture are traversed by a network of irrigation canals, collectors and drains. Long-term irrigation formed a layer of uniform topsoil,

---

A. Akramkhanov (✉) • C. Martius • O. Forkutsa
Center for Development Research (ZEF), Walter-Flex-Str. 3, Bonn
53113, Germany
e-mail: akmal@zef.uznet.net; gcmartius@gmail.com; oksana.forkutsa@gmail.com

R. Kuziev
State Scientific Research Institute for Soil Science and Agrochemistry,
Kamarniso 3, Tashkent 100179, Uzbekistan
e-mail: gosniipa@rambler.ru

R. Sommer
International Center for Agricultural Research in the Dry Areas (ICARDA),
P.O. Box 5466, Aleppo, Syrian Arab Republic
e-mail: r.sommer@cgiar.org

L. Massucati
Institute of Organic Agriculture, University of Bonn, Katzenburgweg 3, Bonn
53115, Germany
e-mail: luiz.massucati@uni-bonn.de

C. Martius et al. (eds.), *Cotton, Water, Salts and Soums: Economic and Ecological Restructuring in Khorezm, Uzbekistan*, DOI 10.1007/978-94-007-1963-7_3,
© Springer Science+Business Media B.V. 2012

referred to as an agro-irrigation horizon, covering a multi-layered alluvium. This chapter provides an impression of the current state of land resources in the irrigated areas of the Khorezm region. We analyzed soil properties based on present and past data; the latter from a period before agriculture expanded and intensified. Soil texture is dominated by silt loam layers together with sandy loams and loams that constitute almost 80% of all soil layers. Organic matter in irrigated soils is low, constituting on average 7.5 $g\,kg^{-1}$ (0.75%) in the topsoil layers and decreasing in the deeper layers. Afforestation can increase SOC stocks. Soil microbiological activities reflected in the soil respiration rate were closely linked to soil organic carbon content and revealed considerable carbon accumulation under zero-tillage systems. Soil macrofauna density was highest under vegetation dominated by one tree species, Euphrates poplar (*Populus euphratica* Oliv.; Salicaceae), but its biodiversity was highest in cultivated sites (and dominated by small predatory arthropods). These results may serve as benchmarks in future monitoring of land use effects on soil quality. Most subsoils in Khorezm are slightly- to medium-saline, whereas the majority of topsoils above 60 cm are strongly saline. Between 1960 and 1990, moderately saline soils increased from 21% to 31%, and highly saline soils, from 6% to 21%.

Keywords Landscape • Salinity • Soil respiration • Micro- and macro-fauna • Texture

3.1 Soils Evolution and Land Resources

The development of soils in Khorezm has, over the millennia, been greatly and primarily influenced by water from a meandering Amudarya river. It therefore is relevant to understand the processes responsible for the spatial distribution of the different soil types. Central Asian oases, isolated areas of vegetation surrounded by deserts, are among the few regions in the world where irrigation resulted in the development of new anthropogenic soils overlaying the natural formations. Therefore, in particularly the upper layers of soils in the Khorezm region are relatively new formations compared to those soils that developed under natural conditions without human interference.

Irrigated agriculture in Khorezm has a long history (Tolstov 1948; Tsvetsinskaya et al. 2002). Soil evolution has been considerably faster under irrigation than in natural systems, and significant changes in soil properties can occur even within 10–100 years. Furthermore, intensive soil displacements occurred when the irrigation and drainage system was developed. Lithology, geomorphology, hydrochemistry and other local peculiarities such as the very flat terrain provide the background against which this intensive human interference has altered soil formation processes, including the mineralization of soil organic matter and microbiological activity, salt migration, illuviation and clay formation.

Features produced by water flows such as the riverbed, near-bed formations, and depressions can be distinguished and are described in detail by Felitciant (1964) and Tursunov (1981). The latter author explains that the initial relief of the Amudarya river delta in Khorezm was formed by two ancient channels, the so-called Daryalyk and Daudan. These transported large amounts of soil sediments, which were deposited and formed levees. Near-bed levees usually contain coarse-grained material, i.e. sand and silt. However, occasionally these are covered by finer grain material deposited during low-velocity floods. Areas further away from the riverbeds are characterized by layers of fine material that was deposited from suspended loads in slow-flowing waters, or even indicating areas of previously stagnant water that formed loamy and clayey soils. Also, since streams were constantly changing courses and braided the delta region, the vertical and lateral composition of the soils is very varied (Tursunov 1981).

Another factor that affected soil formation during the floodplain-alluvial periods consists in the marshy and meadow-marshy formations on flat areas between intra-bed depressions, which were covered by reeds and often flooded. The interaction of vegetation and floods formed the meadow soils, characterized by groundwater tables deeper than 3 m and a high organic matter accumulation on the surface. Elevated relief areas such as slopes and levees were also under meadow and meadow-marshy soil formation, but occasionally also transformed into *solonchaks*. These latter soils have developed mainly under a cover of grass vegetation and with a shallower groundwater table varying between 0.5 and 2.0 m (Kuziev 2006).

Large near-bed levees covered by bush and grass vegetation and experiencing groundwater table depths between 1 and 2.5 m formed meadow *Tugai* soils which morphologically resemble the features of the soils under *Tugai* forests (natural floodplain forests along the Amudarya, cf. Scheer et al. 2011). These soils received litter inputs from the vegetation, explaining why soil organic matter contents often surpass 3%.

Today's landscape in Khorezm, however, is significantly altered by human interference. Owing to the intensification of irrigated agriculture over the last 100 years or so, the areas converted into cropland have been traversed by a network of irrigation and drainage canals. Long-term, intensive irrigation formed a layer of uniform topsoil, referred to as the agro-irrigation horizon, which now covers the multi-layered alluvium deposits. Even ridged-hummocky sands, often present in the region, have been leveled and converted into farming land. A large part of the natural Khorezm landscape was transformed into cultured land.

Hydrogeology is another driver of soil formation. The initial hydrogeological conditions disfavored an intensive and widespread irrigated agriculture in Khorezm due to the flat topography, which limits lateral groundwater flow and results in shallow groundwater tables and increased soil salinity, despite the constructed collector-drainage network (cf. Tischbein et al. 2011). The regular occurrence of high groundwater tables, then, in turn, created hydromorphic conditions favorable to *solonchak* formation. The elevated groundwater table enhances vertical water movement, and considerable amounts of water are discharged via evaporation and transpiration rather than via the drainage system, leading to soil and groundwater

Table 3.1 Land resources of Khorezm, 1,000 ha (GKZGK 2009)

| | Agricultural land | | | | | | | | |
	Cropland	Trees	Marginal	Pasture	Household	Forest	Garden	Other	Total
Irrigated	208.6	13.6	4.5	6.4	43.0	0.4	0.1		276.6
Total	208.6	13.6	4.5	141.2	51.0	57.5	0.1	169.9	646.4

Table 3.2 Soil *bonitet* and area size (ha) according to land cadastre groups and soil classes of the agricultural land resources in Khorezm based on an assessment in 2005 (GKZGK 2009)

Land cadastre group	Bad		Below average		Average		Good		Very good	
Soil class	I	II	III	IV	V	VI	VII	VIII	IX	X
Bonitet	0–10	11–20	21–30	31–40	41–50	51–60	61–70	71–80	81–90	91–100
Hectares	n.a.	625	8,839	46,296	34,932	82,052	49,291	11,848	102	n.a.

n.a. not available

salinization. According to soil surveys conducted by the State Scientific Research Institute for Soil Science and Agrochemistry (SSRISSA) in 1996 (Kuziev 2006; Ibrakhimov et al. 2007), around 73% of the territory of Khorezm had groundwater tables at 1–2 m.

Khorezm occupies an area of 646,400 ha (GKZGK 2009), which includes a large desert area on the right bank of the Amudarya, which is often excluded from assessments of land resources; the irrigated areas cover only about 276,000 ha, the largest share of which is occupied by cropland (Table 3.1).

The following definitions (GKZGK 2009) apply to the categories in Table 3.1. Marginal lands include irrigated and rainfed areas that are out of production due to the deteriorated soil quality and irrigation management caused by inefficient use; or affected by erosion, strong salinization, gypsum, and poor land development. In the irrigated region, marginal lands are located in small patches at the border of cropped areas, and, provided that irrigation and drainage measures are implemented, these could eventually be recovered and used in agriculture (GKZGK 2009). Areas under trees are mainly fruit and mulberry plantings, and vineyards (cf. Khamzina et al. 2011). The household area can include buildings and kitchen garden plots (cf. Bobojonov et al. 2011).

The total irrigated area in 2009 was 1.5 times higher than that reported 30 years ago by SSRISSA (Kuziev 2006). More than 95% of the irrigated areas are meadow soils, whilst about 4% are marshy-meadow soils. Meadow soils, developed on alluvial deposits, are considered to be the most fertile land resources and have therefore been brought into cultivation. Soils developed on illuvium are considered less suitable for farming, explaining why only a fraction of these soils is currently cultivated (Kuziev 2006).

Soil quality conditions in Uzbekistan are reflected in the so-called soil *bonitet*. Soil *bonitet* is a relative score, ranging from 0 to 100, given to indicate the soil quality and natural fertility (GKZGK 2009). *Bonitet* is an aggregate of several parameters ranging from field characteristics (morphology, etc.) to results of laboratory analyses

3 Soils and Soil Ecology in Khorezm   41

for various soil properties (fertility, chemistry, etc.). Since cotton is the major crop in irrigated agriculture of Uzbekistan, it is used as the reference crop in soil *bonitet* assessments. Table 3.2 presents data on estimates of soil *bonitet* scores for the agricultural land resources of about 233,000 ha which includes cropland, trees, marginal land, and pasture. The average soil *bonitet* of these land resources is 53.

3.2 Soil Physical and Chemical Parameters

The description of the physical and chemical parameters of soils in Khorezm, and the following classification are largely based on a comprehensive data base of soils in the region collected by SSRISSA (Abdunabi Bairov, personal communication) organized into the Khorezm Soil Data Base by the ZEF-UNESCO Khorezm Project and, in the case of soil organic matter, supported by a comprehensive literature study carried out by Kuziev (2006).

3.2.1 Texture and Soil Horizons

The Khorezm Soil Data Base contains 511 soil profiles and altogether 2,157 soil layers. The Soviet Union system of defining textural fractions differs from the one adopted by FAO where fewer textural fractions are defined (Stolbovoi 2000). However, according to Stolbovoi (2000), who provides a correlation between the Soviet Union and FAO soil classification systems, these differences due to this partial incongruence of the classification systems are mostly irrelevant, and the generalized textural classes can be correlated adequately for practical purposes at a global scale (Table 3.3).

Out of all layers available in the Khorezm Soil Data Base, 1,884 had a complete set of information, allowing the conversion of soil texture classes into the FAO system (according to a formula provided in Shein 2009). Figure 3.1a shows the spatial distribution of the soil profiles, indicating a homogenous, representative sampling scheme. The available soil samples from the districts Shavat and Pitnyak (south-east of Khazarasp and not shown in this map) were not geo-referenced.

On average four generic soil horizons per profile are described in the database. For merely 35 soil profiles (~7%) more than 5, at maximum 7, horizons have been identified. Most soils in Khorezm are silt loams (USDA soil texture classification). Silt loam layers together with sandy loams and loams constitute almost 80% of all soil layers (Table 3.4). Clayey soils are hardly present in Khorezm (Fig. 3.1b). Heavy textured soils are frequent at deeper layers (Fig. 3.2).

This means that the impact of shallow groundwater and groundwater salinity on crop growth and secondary soil salinization, respectively, is potentially significant, because heavier soils better support capillary rise of groundwater (Brouwer et al. 1985).

Table 3.3 Correlation of particle size distribution between FAO and Soviet Union systems for soil textural classification (Stolbovoi 2000)

Name of texture fraction		Particle size (mm)	
		FAO system (1988)	Soviet Union system (1967)
Gravel, fine gravel		≥2	≥1
Sand	Coarse		–0.5
	Medium	–0.06	–0.25
	Fine		–0.05
Silt	Coarse		–0.01
	Medium	–0.002	–0.005
	Fine		–0.001
Clay		≤0.002	≤0.001
General classes	Coarse	–0.06	0.05
	Medium	–0.002	–0.001
	Fine	≤0.002	≤0.001

There is not much change along the soil profiles in the top 56 cm; profiles are rather homogenous. For instance, altogether 55% of the soils have identical first and second soil layer textural classes. Furthermore, another 33% of all soil profiles had closely matching first and second soil layers (i.e. neighboring soil textural classes in Fig. 3.1). This indicates that only about 12% of all soil profiles show a rather abrupt texture change in the top 56 cm soil. Considering the first four soil layers, this percentage increases to 25%, meaning that still 75% of the soils in the Khorezm Soil Data Base have a homogenous texture in the top ~129 cm. This is an important aspect, as it allows robust estimations of the vertical distribution of soil texture based only on topsoil data.

Soil parameters that highly correlate with clay content could be estimated with higher precision for monitoring purposes. The importance of monitoring the changes in agricultural land resources is acknowledged by the Government of Uzbekistan,[1] which has introduced guidelines to streamline monitoring of those resources throughout the country in three stages, namely, (1) preparatory works, (2) baseline field studies, and (3) long-term observations. By 2009, the second stage, baseline studies, was completed in most of the regions in Uzbekistan (GKZGK 2009).

3.2.2 Soil Organic Matter

The Khorezm Soil Data Base contains 1,806 entries for soil organic matter (SOM). Soil organic carbon (SOC) in Uzbekistan is conventionally chemically determined,

---

[1] Decree No. 496 of Cabinet of Ministers from 23.12.2000 on adoption of charter for agricultural lands monitoring.

3 Soils and Soil Ecology in Khorezm 43

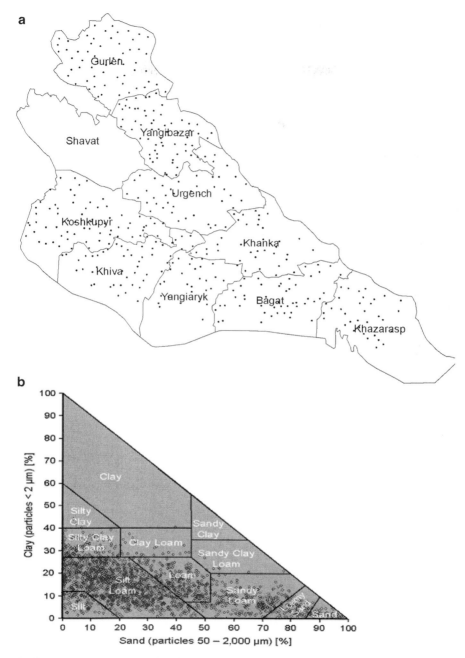

Fig. 3.1 (**a**) Approximate sampling locations of profiles, and (**b**) USDA soil texture classification of these soil samples and respective distribution for all soil layers described in the Khorezm Soil Data Base (Sommer et al. 2010)

Table 3.4 Frequency of soil texture classes of the irrigated areas in Khorezm (Khorezm Soil Data Base, Abdunabi Bairov, personal communication), according to USDA soil classification

Soil texture	N	%
Sand	59	3.1
Loamy sand	86	4.6
Sandy loam	229	12.2
Clay loam	25	1.3
Loam	241	12.8
Sandy clay loam	8	0.4
Silt loam	1,039	55.1
Silt	84	4.5
Silty clay loam	109	5.8
Silty clay	3	0.2
Clay	1	0.1
All	1,884	

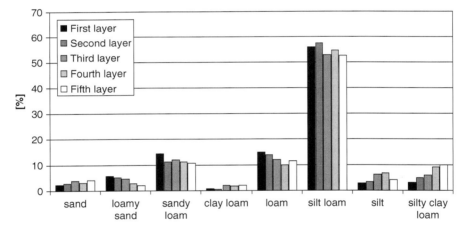

Fig. 3.2 Frequency distribution of eight soil texture classes in the upper five soil layers in Khorezm

using a slightly modified form of the Walkley-Black wet-oxidation (potassium dichromate) method (Nelson and Sommers 1982). For the conversion, it is assumed that SOM contains 58% SOC (Vorobyova 1998).

SOM contents turned out to be low in the soils of Khorezm (Khorezm Soil Data Base). On average, SOM is 7.5 g kg^{-1} (0.75%) in the topsoil layers, decreasing to 3.9 g kg^{-1} at 156–210 cm depth, with considerable natural variation. The lower quartile of SOM at 0–32 cm was 5.4 g kg^{-1} and the upper quartile 9.4 g kg^{-1} (Fig. 3.3). SOM contents of the data set (n = 148) described by Kuziev (2006) generally concur with those of the Khorezm Soil Data Base. SOM content of heavy- and medium-textured irrigated soils varied from 6–7 to 11–14.5 g kg^{-1}, gradually decreasing with

3 Soils and Soil Ecology in Khorezm

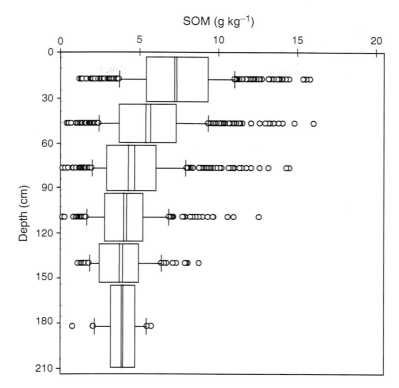

Fig. 3.3 Soil organic matter of soils in the Khorezm soil database; box-whisker plots, *bold lines* within the box indicate average values

depth. Light-textured soils contained 4.5–8.0 g kg^{-1} SOM, but this decreased to 0.2–0.3 g kg^{-1} at deeper depth. The overall SOM storage of the 0–50 cm layer thus amounts to 16–79 t ha^{-1}.

The subsoils of irrigated meadow soils are rich(er) in SOM as compared to the subsoils under natural formations where organic litter contributes to only topsoil (Kuziev 2006). Only due to irrigation, SOM can decrease from 12–20 g kg^{-1} to 7–12 g kg^{-1} (Kuziev 2006).

The comparison of the data from 1950 to 1959 for the 0–30 and 30–50 cm layers with those from the 1970s and 1990ies demonstrates a general trend in SOM reduction (Fig. 3.4). Meadow oasis soils located on ancient riverbeds and near-bed deposits had the highest SOM contents, followed by irrigated meadow soils. The lowest SOM content was found on meadow soils developed on recent Amudarya deposits.

While irrigation water tends to wash out SOM, it is also one often neglected source of SOM (Kuziev 2006), and deposits suspended sediments rich in nitrogen, phosphorus, potassium, OM and a number of microelements. Irrigation water is a significant source of mineral replenishment of irrigated soils, particularly for

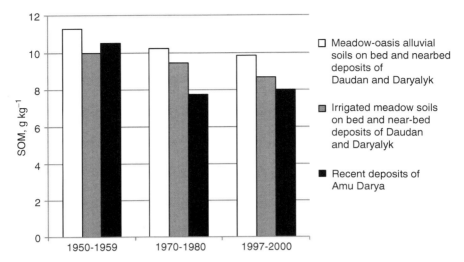

Fig. 3.4 Comparison of soil organic matter content of irrigated topsoil 30 cm from 1950 to 2000 (Adapted from Kuziev 2006)

automorphic soils of the desert zone, which are prone to enhanced SOM losses compared to soils with hydromorphic or semi-hydromorphic conditions. However, following the construction of the Tuyamuyun reservoir on the Amudarya, located immediately upstream of Khorezm, most suspended solids in the river water are now being retained in this reservoir (Tashkuziev 2003).

3.2.3 Soil NPK

Due to the low SOM content in Khorezm soils, total soil nitrogen (N) levels are also low, ranging from 0.2–0.8 g N kg^{-1} (0.02–0.08%), with storage of 1.6–4.9 t N ha^{-1} in the 0–50 cm layers. The total N in irrigated soils reaches 4–6 t ha^{-1}, out of which approximately 2 t ha^{-1} are in the topsoil.

Total phosphorus (P) ranges from 450 to 3,000 mg P kg^{-1}. Marshy-meadow soils have usually high (1,000–3,000 mg P kg^{-1}) total P contents, which convert into storage amounts of 8.7–15.2 t P ha^{-1} in 0–50 cm layer. Yet, only 10–20% of the stored amount is available to plants, 50–60% is less available and 20–40% is practically not available to plants (Kuziev 2006).

The soils in the arid region of Khorezm are naturally rich in potassium (K). The total potassium content in the upper horizons ranges from 3,000–31,000 mg K kg^{-1}, however, only 10% of the soils have a total K of over 2,000 mg kg^{-1}. Storage in 0–50 cm layer varies from 71 to 160 t K ha^{-1}.

The Khorezm Soil Data Base contains 1,869 entries for soil nitrate. The Soviet Union method to determine soil mineral N (NO$_3$ and NH$_4$), is for various reasons

3 Soils and Soil Ecology in Khorezm

Table 3.5 Classification of soil mineral N content with regard to optimal plant growth (WARMAP and EC-IFAS 1998) and its equivalent amounts considering exemplarily 0–30 cm soil depth (Soil bulk density assumed to equal as 1.5 $g\,cm^{-3}$)

Mineral N content, mg kg^{-1}	Classification	Mass equivalent, kg N ha^{-1} 30 cm^{-1}
<20	very low	<90
20–30	Low	90–135
30–50	Medium	135–225
50–60	High	225–270
>60	very high	>270

(e.g. air-dry soil samples, different chemical analysis) not 1:1 comparable with standard mineral N determinations in other countries. Therefore, also the classification of NO_3 contents with regard to optimal crop growth deviates from European norms (Table 3.5). It appears that the Soviet Union method of chemical analysis of mineral N gives consistently considerably higher concentrations of NO_3 and NH_4 than standard "Western" analysis. Hence the following classification was adjusted accordingly.

The NO_3 contents in the Khorezm topsoil alone, with on average 34.5 mg kg^{-1} (median: 19.6), would classify the soil to be medium-rich in mineral N (Fig. 3.5). The heterogeneity of NO_3 contents in 0–30 cm depth was, however, high as evidenced by the range from a lower quartile of 8.5 mg kg^{-1} to an upper quartile of 53.7 mg kg^{-1}. Median NO_3 content decreases from 12.0 mg kg^{-1} in 30–56 cm depth to 6.3 mg kg^{-1} below 156 cm. Interestingly, NO_3 contents below 156 cm vary more than in the three layers above. This may be an additional indication of topsoil nitrate leaching and accumulation in this sub-layer, at least in some soils.

3.2.4 Soil Salinity and C Stock

According to Kuziev (2006), the area of *solonchaks* had decreased from 10.8% to 1.6% between the 1960s and 1990s, due to ameliorative measures and after having been used for farming. Weakly saline or non-saline areas decreased from 74% in 1960 to 48% 1990. In the same period, moderately saline soils increased from 21% to 31%, and highly saline soils, from 6% to 21%.

The Khorezm Soil Data Base contains 2,053 entries for soil salinity expressed in total dissolved solids percentages (%TDS). These values were converted into electrical conductivity, EC_e, using the relationship suggested by Abrol et al. (1988): EC_e (dS m^{-1}) = %TDS / 0.064. Using this conversion, the findings showed that the median EC_e decreases from 10.1 dS m^{-1} in 0–32 cm to roughly half of this value (4.7 dS m^{-1}) in 32–60 cm depth. Below 60 cm depth, the median soil salinity is basically constant at around 3.8 dS m^{-1}. As with NO_3, heterogeneity of salinity is high in the topsoil, where EC_e ranged from a lower quartile of 3.3 dS m^{-1} to an upper quartile of 25.8 dS m^{-1} (Fig. 3.6). According to the classification of soil salinity by Abrol

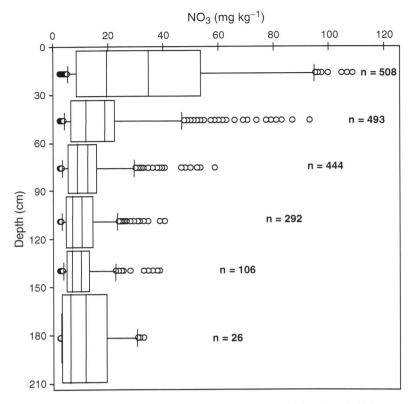

Fig. 3.5 NO$_3$ content of soils in the Khorezm soil database; box-whisker plots, *bold lines* indicate average values

et al. (1988), most subsoils in Khorezm are slightly- to medium-saline, whereas the majority of topsoils above 60 cm is strongly saline (Figs. 3.6 and 3.7).

Land resources in Khorezm can be classified by land use. Agricultural land use covers the largest share of the landscape, compared to natural land use systems such as *Tugai* forest (cf. Scheer et al. 2011) and desert. In the following we summarize findings of various studies (Forkutsa 2006; Massucati 2006; Hbirkou et al. 2011) of selected parameters in *Tugai* forest, desert, plantations of poplar (*Populus euphratica*) and elm (*Ulmus pumila*) trees, shelterbelts between fields, and cotton fields under conventional and conservation agriculture trials referred to as zero tillage on sandy (ZS) and loamy soils (ZL).

The snapshot survey of soil salinity conducted by Hbirkou et al. (2011) suggests that the EC$_e$ increased in the order *Tugai* forest ≤ desert site ≤ long-term afforestation sites < fallow land (Table 3.6). However, their data also shows that carbon stocks (total carbon (C$_t$) and SOC) were highest on long-term (80 years) afforestation sites, and lowest in the desert ecosystem, suggesting that long-term afforestation had led to an increase of the SOC stock in the topsoil (0–20 cm) to levels even above those in the native *Tugai* forest.

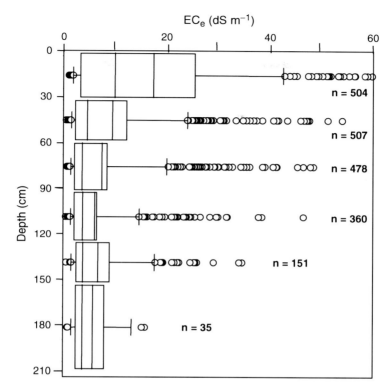

Fig. 3.6 Salinity of soils in the Khorezm soil database; box-whisker plots, *bold lines* indicate average values

Further analysis by Hbirkou et al. (2011) of SOC stocks of fallow land, a 4-years old afforested site, and an 80-years old afforested site suggests a rather rapid initial increase in SOC, which tapers off with time. For instance, fitting all three SOC datasets to a nonlinear regression reveals an increase rate of 0.15 t SOC ha^{-1} a^{-1} after 20 years of afforestation, but a lower increase rate of 0.09 t SOC ha^{-1} a^{-1} after another 60 years of afforestation (Hbirkou et al. 2011).

3.2.5 Soil Microbiological Activity

Soil microbiological activity (Forkutsa 2006) has a key role in the ecological function of soil (decomposition of organic residues, essential soil nutrients for plants, soil quality) and maintenance of soil fertility. Soil micro-organismic activity depends on the presence of moisture, nutrients and organic matter. Soil respiration is known as a biological indicator for microbial decomposition of organic matter in the soil. Mechanical disturbance causes excessive aeration of soil, while intensity and

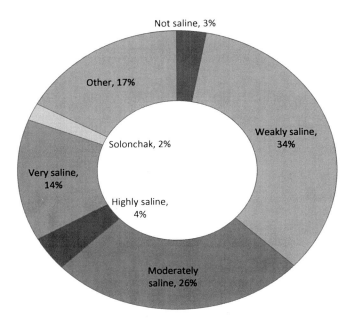

Fig. 3.7 Share of soil salinity classes in the Khorezm region according to the classification by Abrol et al. (1988)

Table 3.6 Selected soil ecology parameters of different land use systems in the Khorezm region

	C_{tot}, (t ha^{-1})[a]	SOC, (t ha^{-1})[a]	EC_e, (dS m^{-1})[a]	Soil fauna biodiversity (Hs')[b]	Soil fauna biodiversity (Hs'Max)[b]	Respiration, mg CO_2 m^{-2} h^{-1} [c]	Microbial biomass, μg C_{mic} g^{-1} dm[c]
Tugai	72.5	21.4	9.4	0.97	1.34	257.5	549.64
Desert	42.7	6.6	10.4	0.14	0.14	N/A	N/A
Tree strip	92.6	29.9	13.8	0.52	0.88	N/A	N/A
Conventional fields	77.8	21.5	17.1	1.12	1.25	419.2	654.17
NT loam				0.93	1.17	376.0	1001.04
NT sandy				0.72	0.78	241.0	216.95

Source: [a]Hbirkou et al. (2011), [b]Massucati (2006), [c]Forkutsa (2006)

frequency of disturbance results in a differential distribution and loss of organic matter (Gajri et al. 2002).

The effect of farming practices on soil quality was evaluated at three agricultural sites, which differed in soil management techniques (conventional tillage and zero tillage; with and without mulch), soil texture (loam and sand), and *Tugai* forest. The carbon dioxide output is consistent with the overall metabolic activity of the soil microflora. Although no pronounced difference between the sites and treatments was found, the respiration rates among the sites decreased in the order: Amir Temur Farm (Loamy) > Khiva Farm Loamy > *Tugai* forest > Khiva Farm Sandy; i.e. the

3 Soils and Soil Ecology in Khorezm

Table 3.7 Soil microbial biomass-C (C_{mic}), soil organic matter (SOM), microbial quotient and metabolic quotient on four sites in different districts of Khorezm (values followed by the same letter are not significantly different; Tukey test, HSD, P<0.05)

Site	Treatment*	C_{mic}, (μg C_{mic} g^{-1} dm)	SOM, (%)	Microbial Quotient, (C_{mic}/C_{org}, %)	Metabolic Quotient, (mg CO_2-C g^{-1} Cmic h^{-1})
Khiva Loamy	C	970.79 b	0.87 bc	13.8	1.9
	C+	765.49 a	0.69 bc	8.7	4
	Z	1430.48 c	0.79 bc	18.2	2.4
	Z+	837.39 ab	0.79 bc	10.6	3.1
Khiva Sandy	C	195.56 ab	0.37 a	5.3	3.9
	C+	77.92 a	0.35 a	2.2	7.3
	Z	312.82 b	0.33 a	9.4	3.9
	Z+	281.49 b	0.43 a	6.6	1.6
Amir Temur	C	654.17 b	0.99 c	6.6	8.1
Tugai forest	N	549.64 b	0.68 b	2.3	2.5

*C conventional, Z zero tillage, N natural ecosystem, + with a mulch of plant residues

agriculturally managed loamy soils have a higher microbial activity than the natural *Tugai* forests and the sandy soils.

Sandy textured soils (Khiva Sandy site) with poor SOC contents (1.7–2.8 g kg^{-1}) had a low range of respiration rates (75.0–390.8 mg CO_2 m^{-2} h^{-1}). In contrast, loamy soils with higher SOC (3.8–6.2 g kg^{-1}) showed respiration ranges from 130.83–949.2 mg CO_2 m^{-2} h^{-1}. Spatial differences across the sites were mostly explained by C % and soil moisture.

Soil microbial biomass-C (C_{mic}) resulted in similar patterns among all sites. The lowest C_{mic} was found at the Khiva Sandy site, in the range of 77.9–312.8 μg C g^{-1} dm, and the highest range at the Khiva Loamy site with 765.5–1,430.5 μg C g^{-1} dm. Yet, in spite of the similar soil texture in both Khiva Loamy and Amir Temur sites, the higher C_{mic} values (765.49 μg C g^{-1} dm or higher) in the Khiva Loamy site, where soil conservation techniques were applied (cf. Pulatov et al. 2011), indicate a better soil quality than in Amir Temur where only 654.17 μg C g^{-1} dm were measured.

In this study, the microbial quotient (C_{mic}/C_{org}) varied in the range of 2.2–18.2 (%) (Table 3.7). Often, this value lies in the range of 1–5% (Sparling 1997), however some studies conducted in similar environmental conditions on wheat crop fields have shown a similar wide range of soil microbial ratio values (6.2–2.6%; Insam 1990; Uçkan and Okur 1998). Higher values can be interpreted as a carbon accumulation, whilst lower ones stand for carbon loss. Yet, Insam and Öhlinger (1996) stated that the soil microbial quotient is highly influenced by climate and thus, arid areas produce higher values than humid (and less hot) environments. Therefore, the microbial quotients in this study indicate that significant carbon accumulation had occurred on sites under zero-tillage treatment. The metabolic quotient (qCO_2) (Table 3.7) is the rate of soil respiration in terms of the microbial biomass and indicates the qualitative influence on microbial biomass. Lower qCO_2 ratio values stand for more efficient microbial turnover. Based on microbial and metabolic quotients, our study shows a slight positive effect of zero tillage on soil quality.

Results of site comparisons of the mean difference in soil respiration and SOM were in consistency with studies of Conant et al. (2000) and Huxman et al. (2004), who suggest that soil respiration in semiarid ecosystems is mainly controlled by soil C content and moisture, which should also apply for arid regions as Khorezm. In the microbiological analysis, similar to the field measurements, the mulching contribution was not captured; possibly due to the short period of residue management in this field prior to the data collection, and also due to the chosen sampling depth (0–20 cm). However, the average values of soil microbial biomass carbon for conventional fields were lower than those for zero-tillage fields, which indicate that an initial, favorable SOM accumulation takes places under zero-tillage practice.

3.2.6 Soil Fauna Density and Diversity

For the screening of the soil fauna, Massucati (2006) collected macro- and mesofauna using soil cores of 20 cm diameter and 20 cm depth. Additionally, litter samples were collected from 80 cm diameter areas on the forest floor. The soil macro-invertebrates were then extracted in a heat-moisture gradient in a Berlese apparatus.

The density of the soil fauna (macro- and mesofauna) in Khorezm ranged between 29 individuals m^{-2} in the desert Kyzylkum[2] and 3,994 individuals m^{-2} in the *Tugai* forest (Fig. 3.8). The upper level fauna density value in the *Tugai* forest was caused mainly by the extreme abundance of ants (Hymnoptera: Formicidae), which made up 93% of the total individuals at this site. The abundance of this taxon is high in arid habitats, where soil fauna assemblages are mostly characterized by xerothermal species (Veile 1992). The soil from the *Tugai* forest served also as habitat for many larvae of flies (Diptera: Asilidae, Sarcophagidae) and beetles (Coleoptera: Chrysomelidae, Cryptophagidae), and a variety of spiders (Arachnida: Gnaphosidae, Lycosidae, Pseudoscorpionida). In tree strips[3] around a cotton field the density was 1,344 individuals m^{-2}, mainly characterized by Formicidae. Both tree strips and forest offer better habitats for social insects such as fire ants (Myrmicinae; *Solenopsis* sp), to build epigeic nests, since these areas are not used for agriculture. Termites are often found in the desert zones (Abdullaev et al. 2002) but they also affect buildings, often historical ones, e.g. in the ancient city of Khiva, an UNESCO heritage site.

The soil fauna density under conventional tillage was 267 individuals m^{-2}, showing no significant difference from the density in the *Tugai* forest, tree strips and in zero tillage fields. Analogous to the forest site and tree strips, ants were also the most

---

[2] Fixed and mobile dunes with open, psammophytic, shrub-like communities and semi-shrub, succulent plants, as well as tessellated *takyr* soils with a thin loam crust without vegetation.

[3] Planted trees (predominantly *Populus ariana* and *Elaeagnus angustifolia*) within a cotton field, for wind erosion protection.

3 Soils and Soil Ecology in Khorezm

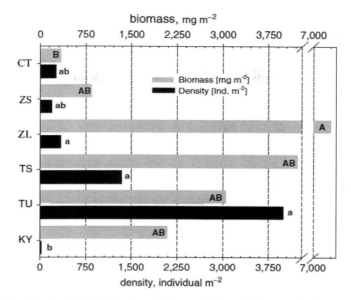

Fig. 3.8 Mean density (individual m^{-2}) and biomass (fresh weight; mg m^{-2}) of the soil fauna at different study sites in Khorezm (*CT* conventional tillage, *ZS* zero-tillage (sandy soil), *ZL* zero-tillage (loamy soil), *TS* tree strips, *TU Tugai* forest, *KY* Kyzylkum desert). Sample sites followed by the same letter are not significantly different (ANOVA, Scheffé, $\alpha = 0.05$)

important component of the soil fauna assemblage at this study site. This may primarily be explained by the favorable environmental conditions of adjacent tree strips, where social insects find refuge to build long-lasting nests. Secondly, ants have a wide diet spectrum (Martius et al. 2001), enabling them to colonize many different systems.

Under zero-tillage in the ZS and ZL sites (zero-tillage on sandy and loamy soil, respectively) (Fig. 3.8) isopods (woodlice; Isopoda: *Oniscidea*) were the most dominant taxon, accounting for 41% of the total soil fauna. Isopods (and myriapods) belong to a suitable indicator group for soil disturbance, since they are sensitive to soil tillage, and as a consequence are rare in cultivated soils (Paoletti 1999; Holland 2004). This group, which was observed across the tree strips too, enhances nutrient cycling by decomposing organic detritus and transporting it to moister microsites in the soil. Other two important taxa for soil structure and formation were observed under zero-tillage, earthworms (*Oligochaeta; Lumbricidae*) and myriapods (*Diplopoda;* millipedes). Earthworm and myriapod assemblages can positively be affected by conservation agriculture as observed in previous studies (e.g. Chan 2001; Holland 2004). For isopods and myriapods in particular, the retention of surface mulch modifies the physical soil properties (e.g. prevention of water evaporation), increasing the architectural complexity of the soil surface environment, and as consequence facilitating their movements (Holland 2004). However, the residues left on the soil surface did not affect the population of generalist predators, such as ground beetles and spiders, as previously concluded (e.g. Szendrei and Weber 2009). Thus, both conservation agriculture and tree strips may improve the environmental sustainability of irrigated arable lands in Khorezm, by offering refuge, shelter over the irrigation and cultivation season, overwintering grounds, and food reservoirs, especially in springtime, for many soil taxa.

Biodiversity, calculated by the Shannon-Wiener-Index, was the highest (1.3) under conventional tillage (CT), although it was not significantly different from *Tugai* and zero-tillage fields. The high diversity of invertebrates seems to be linked to the very specific soil fauna community under cultivated soils.

3.2.7 Soil Fauna Biomass

The soil fauna biomass, in contrast with soil fauna density, was the highest under conservation agriculture on the loamy soil (ZL) with 7,099 mg m^{-2} (all data in fresh mass) and the lowest under conventional tillage (CT) with 337 mg m^{-2}. In natural *Tugai* stands, biomass was in between, with 3,055 mg m^{-2}. Surprisingly, the biomass of the soil fauna in tree strips and even in the desert was higher than under conventional tillage fields (4,231 and 2,090 mg m^{-2}, respectively). Mole crickets (Ensifera; *Grylotalpa*) accounted for 49% and 94% of the total biomass in the tree strips and desert, respectively.

Soil fauna biomass was expected to be low in these very dry, hot, winter-cold regions, and the observed values (if divided by 3 to provide for a crude fresh-to-dry mass recalculation; Martius unpublished) are actually less than half those at the lower end of the biomass range in healthy, natural, temperate middle-European beech forests (which have a soil fauna dry mass of 5–15 g m^{-2}; Schaefer and Schauermann 1990) or in Amazonian tropical primary rainforests (approx. 3 g m^{-2} dry mass; Höfer et al. 2001). However, conservation agriculture was able to elevate the soil fauna biomass levels above those in the natural *Tugai* forests stands; in spite of the widespread soil salinization and widespread (past) pesticide application, but consistent with the higher moisture levels in the irrigated croplands. Under zero-tillage, the biomass of earthworms (*Oligochaeta; Lumbricidae*) accounted for 79% of the total biomass at this study site. Earthworms are widely known as ecosystem engineers, providing a wide variety of ecosystem services linked to soil formation and retention of organic matter.

Under conventional tillage, the lower biomass combined with a high number of individuals showed that the soil fauna assemblage at this study site is densely populated by small-sized taxa. This result is consistent with the idea that the anthropogenic condition influences the soil fauna community. The invertebrates under CT were on average smaller, but they are resilient to cultivated/irrigated soils. According to Kladivko (2001), larger species are more vulnerable to soil cultivation than smaller ones, due to the physical disruption, changes in soil environment conditions, and the burial of crop residues. The soil fauna community under CT was characterized by polyphagous prepadors, which comprised carabid (Coleptera: Carabidae; Bembidiinae) and rove beetles (Coleptera: Staphylylinidae; *Platystethus sp.*), ants (Formicidae: *Aphanogaster* sp.), and sheet web spiders (Linyphiidae). This fauna assemblage is expected to play a large role in regulating insect populations in this agrocenosis, which might contribute to the ecological services (i.e. pest control) of conservation agriculture. This finding is supported by previous studies showing a preference of carabid and rove beetles for cultivated fields (Paoletti 1999; Holland 2004) and a possible adaptation of sheet web spiders to arable fields (Krause 1987).

3.3 Summary and Conclusions

The upper horizons of the soils in Khorezm are new formations due to the combined effects of river flow dynamics and the introduced irrigation systems, when compared to soils developed under natural conditions, i.e. without human interference. More than 95% of the irrigated area consists of meadow soils; about 4% are marshy-meadow soils. Meadow soils, developed on alluvial deposits, are considered to be the best land resources and most of these soils have already been cultivated. Soils developed on illuvium are less suitable for farming with only a fraction of these soils currently being cultivated.

Most soils in Khorezm are silt loams, sandy loams, and loams (USDA soil texture classification) which constitute 80% of all soil layers. Heavier, clayey soils are largely absent in Khorezm. Since the texture of the top soil profiles turned out to be rather homogenous in the top 56 cm and even the top ~129 cm, it allows a robust estimation of vertical distribution of soil texture based only on topsoil data. This in turn allows estimating clay content and other highly correlated parameters, for example soil moisture holding capacity on large spatial scales for modeling purposes.

For this kind of dryland region with inherent low SOM and very few soil organisms, conservation agriculture seem to improve soil health (SOM) rapidly. The faunistic studies revealed the highest density of the soil macro- and mesofauna in the native riparian forest *Tugai*, mainly represented by ant assemblages (93%), compared with the habitats under agricultural use. In contrast, earthworm populations greatly contribute to biomass increase under conservation agriculture. The soil fauna biodiversity was greater under conventional tillage, but the assemblage here was mostly characterized by small predatory arthropods, pointing at disturbances from the tillage.

Summing up, the soils in Khorezm are mostly former illuvial and dryland soils heavily influenced by human interference; in some cases (soil near old settlements and those closer to the river) possible over millennia, but mostly as a result of the expansion of irrigated agriculture over roughly the last 80–100 years. The disturbances, drought, possibly the salinity and the typically high temperatures reduce soil biological processes to a high degree; but irrigated soils offer a certain degree of improvements in the living conditions of soil organisms for providing more steady moisture levels over the year. Soil life can be restored by conservation agriculture to a certain extent, which is consistent with findings on soil organic matter by Pulatov et al. (2011). Whether the resulting ecosystem services are of any importance for management purposes cannot be said at the present stage, but the large amount of predator feeding guilds in the soil fauna suggests that a certain contribution to crop pest control can be expected from those groups, which might be welcome under the currently often high pest incidence levels in cotton and wheat (Khamraev and Davenport 2004) (and possibly, it could be enhanced through adequate soil management). The study also indicates that earthworm monitoring might be a good way to assess improvements in soil biology, as earthworms represent a keystone species under the decomposers; their populations increase greatly under conservation agriculture, and they are easily monitored.

Most subsoils in Khorezm are slightly- to medium-saline, whereas the majority of topsoils above 60 cm are strongly saline. Areas covered by forests and deserts tend to have lower salinity levels. Due to poor soil organic matter (SOM) content, total nitrogen (N) levels are also low, this requires application of significant amounts of mineral fertilizers in order to grow crops, including nutrient intensive crop such as cotton. Despite the soils in the arid region of Khorezm being rich in potassium, to sustain and improve overall soil fertility there is a need for the balanced use of available resources and to avoid nutrient mining. Management options such as conservation agriculture and afforestation seem to bring fast benefits to these soils often impoverished by irrigation and land use and can to a certain extent mitigate (but not revert) soil salinity, currently the most limiting factor for agricultural productivity.

References

Abdullaev II, Khamraev AS, Martius C, Nurjanov AA, Eshchanov RA (2002) Termites (Isoptera) in irrigated and arid landscapes of Central Asia (Uzbekistan). Sociobiology 40(3):605–614

Abrol IP, Yadav JSP, Massoud FI (1988) Salt-affected soils and their management, vol 39, FAO Soils Bulletin. FAO, Rome, p 131

Bobojonov I, Lamers JPA, Djanibekov N, Ibragimov N, Begdullaeva T, Ergashev A, Kienzler K, Eshchanov R, Rakhimov A, Ruzimov J, Martius C (2011) Crop diversification in support of sustainable agriculture in Khorezm. In: Martius C, Rudenko I, Lamers JPA, Vlek PLG (eds) Cotton, water, salts and *Soums*: economic and ecological restructuring in Khorezm, Uzbekistan. Springer, Dordrecht/Berlin/Heidelberg/New York

Brouwer C, Goffeau A, Heibloem M (1985) Introduction to irrigation. Irrigation water management training manual, 1. Food and Agricultural Organization of the United Nations, Rome

Chan KY (2001) An overview of some tillage impacts on earthworm population abundance and diversity – implications for functioning in soils. Soil Tillage Res 57:179–191

Conant RT, Klopatek JM, Klopatek CC (2000) Environmental factors controlling soil respiration in three semiarid ecosystems. Soil Sci Soc Am J 64:383–390

Felitciant IN (1964) Soils of Khorezm province. Soils of Uzbekistan SSR, III. Tashkent, Uzbekistan, pp 133–211

Forkutsa O (2006) Assessing soil-borne CO_2 exchange in irrigated cropland of the Aral Sea Basin as affected by soil types and agricultural management. MA dissertation, Bonn University, Bonn

3 Soils and Soil Ecology in Khorezm

Gajri PR, Arora VK, Prihar SS (2002) Tillage for sustainable cropping. Food Products Press, New York

GKZGK (2009) National report on land resources of Uzbekistan. State Committee on Land Resources, Geodesy, Cartography and Cadastre, pp 95

Hbirkou C, Martius C, Khamzina A, Lamers JPA, Welp G, Amelung W (2011) Reducing topsoil salinity and raising carbon stocks through afforestation in Khorezm, Uzbekistan. J Arid Environ 75:146–155

Höfer H, Hanagarth W, Beck L, Garcia M, Martius C, Franklin E, Römbke J (2001) Structure and function of the soil fauna in Amazonian anthropogenic and natural ecosystems. Eur J Soil Biol 37:229–235

Holland JM (2004) The environmental consequences of adopting conservation tillage in Europe: reviewing the evidence. Agric Ecosyst Environ 103:1–25

Huxman TE, Snyder KA, Tissue D, Leffler AJ, Ogle K, Pockman WT, Sanquist DR, Potts DL, Schwining S (2004) Precipitation pulses and carbon fluxes in semiarid and arid ecosystems. Oecologia 141:254–268

Ibrakhimov M, Khamzina A, Forkutsa I, Paluasheva G, Lamers JPA, Tischbein B, Vlek PL, Martius C (2007) Groundwater table and salinity: spatial and temporal distribution and influence on soil salinization in Khorezm region (Uzbekistan, Aral Sea Basin). Irrig Drain Syst 21(3–4):219–236

Insam H (1990) Are the microbial biomass and basal respiration governed by the climatic regime? Soil Biol Biochem 22:525–532

Insam H, Öhlinger R (1996) Ecophysiological parameters. In: Schinner F et al (eds) Methods in soil biology. Springer, Berlin

Khamraev AS, Davenport CF (2004) Identification and control of agricultural plant pests and diseases in Khorezm and the Republic of Karakalpakstan, Uzbekistan. ZEF Work Papers for Sustainable Development in Central Asia, 8. http://www.khorezm.zef.de/fileadmin/webfiles/downloads/projects/khorezm/downloads/Publications/wps/ZEF-UZ-WP08-Khamraev-Davenport.pdf

Khamzina A, Lamers JPA, Vlek PLG (2011) Conversion of degraded cropland to tree plantations for ecosystem and livelihood benefits. In: Martius C, Rudenko I, Lamers JPA, Vlek PLG (eds) Cotton, water, salts and *Soums*: economic and ecological restructuring in Khorezm, Uzbekistan. Springer, Dordrecht/Berlin/Heidelberg/New York

Kladivko EJ (2001) Tillage systems and soil ecology. Soil Tillage Res 61:61–76

Krause A (1987) Untersuchungen zur Rolle von Spinnen in Agrarbiotopen. PhD dissertation, Rheinische Friedrich-Wilhelms-Universität Bonn, Bonn

Kuziev R (2006) Soil evolution in Khorezm region. Unpublished report, Urgench

Martius C, Römbke J, Verhaagh M, Höfer H, Beck L (2001) Termiten, Regenwürmer Und Ameisen: prägende Elemente der Bodenfauna tropischer Regenwälder. Karlsruhe, Staatliches Museum für Naturkunde Karlsruhe. Adrias 15:15–27

Massucati LFP (2006) Monitoring of soil macrofauna and soil moisture in a cotton field: an assessment of the ecological potential in irrigated agriculture in Central Asia (Khorezm province, Uzbekistan). MA dissertation, Bonn University, Bonn

Nelson DW, Sommers LE (1982) Total carbon, organic carbon and organic matter. In: Page AL, Miller RH, Keeney DR (eds) Methods of soil analysis. American Society of Agronomy, Madison

Paoletti GM (1999) Using bioindicators based on biodiversity to assess landscape sustainability. Agric Ecosyst Environ 74:1–18

Pulatov A, Egamberdiev O, Karimov A, Tursunov M, Kienzler S, Sayre K, Tursunov L, Lamers JPA, Martius C (2011) Introducing conservation agriculture on irrigated meadow alluvial soils (Arenosols) in Khorezm, Uzbekistan. In: Martius C, Rudenko I, Lamers JPA, Vlek PLG (eds) Cotton, water, salts and *Soums*: economic and ecological restructuring in Khorezm, Uzbekistan. Springer, Dordrecht/Berlin/Heidelberg/New York

Schaefer M, Schauermann J (1990) The soil fauna of beech forests: comparison between a mull and a moder soil. Pedobiologia 34(5):299–314

Scheer C, Tupitsa A, Botman E, Lamers JPA, Worbes M, Wassman R, Martius C, Vlek PLG (2011) Abundance of natural riparian forests and tree plantations in the Amudarya delta of Uzbekistan and their impact on emissions of soil-borne greenhouse gases. In: Martius C, Rudenko I, Lamers JPA, Vlek PLG (eds) Cotton, water, salts and *Soums*: economic and ecological restructuring in Khorezm, Uzbekistan. Springer, Dordrecht/Berlin/Heidelberg/New York

Shein EV (2009) The particle-size distribution in soils: problems of the methods of study, interpretation of the results, and classification. Eurasian Soil Sci 42:284–291

Sommer R, Djanibekov N, Salaev O (2010) Optimization of land and resource use at farm-aggregated level in the Aral Sea Basin of Uzbekistan with the integrated model FLEOM - model description and first application, ZEF- Discussion Papers On Development Policy No. 139, Center for Development Research, Bonn, July 2010, pp.102. http://papers.ssrn.com/sol3/papers.cfm?abstract_id=1650631

Sparling GP (1997) Soil microbial biomass, activity and nutrient cycling as indicators of soil health. In: Pankhurst CE et al (eds) Biological indicators of soil health. CAB International, New York

Stolbovoi V (2000) Soils of Russia: correlated with the revised legend of the FAO soil map of the world and world reference base for soil resources. Research Reports. IIASA, Vienna

Szendrei Z, Weber DC (2009) Response of predators to habitat manipulation in potato fields. Biol Control 50:123–128

Tashkuziev MM (2003) Soils of Khorezm province. Tashkent

Tischbein B, Awan UK, Abdullaev I, Bobojonov I, Conrad C, Forkutsa I, Ibrakhimov M, Poluasheva G (2011) Water management in Khorezm: current situation and options for improvement (hydrological perspective). In: Martius C, Rudenko I, Lamers JPA, Vlek PLG (eds) Cotton, water, salts and Soums: economic and ecological restructuring in Khorezm, Uzbekistan. Springer, Dordrecht/Berlin/Heidelberg/New York

Tolstov SP (1948) Following the tracks of ancient Khorezmian civilization. Academy of Sciences of the USSR, Moscow – Leningrad

Tsvetsinskaya EA, Vainberg BI, Glushko EV (2002) An integrated assessment of landscape evolution, long-term climate variability, and land use in the Amudarya Prisarykamysh delta. J Arid Environ 51:363–381

Tursunov L (1981) Soil conditions of irrigated lands of the western parts of Uzbekistan. Tashkent

Uçkan HS, Okur N (1998) Seasonal changes in soil microbial biomass and enzyme activity in arable and grassland soils. http://www.toprak.org.tr/isd/isd_77.htm

Veile D (1992) Ameisen-Grundzüge der Erfassung und Bewertung. In: Trauter J (ed) Arten- und Biotopschutz in der Planung: methodische Standards zur Erfassung von Tierartengruppen [BVDL- Tagung Bad Wurzach, 9. 10. Nov 1991]. Josef Margraf Verlag, Ökologie In Forschung Und Anwendung 5:177–188, Weikersheim

Vorobyova LA (1998) Khimicheskiy analiz pochv (chemical analysis of soils). Moscow University Press, Moscow

WARMAP, EC-IFAS (1998) Water use and farm management survey (WUFMAS) -agricultural year 1998 – Annual report. Water Resources Management and Agricultural Production in the Central Asian Republics (WARMAP-2); Executive Committee Interstate Fund for the Aral Sea (EC-IFAS); TACIS Project

Chapter 4
Spatial Distribution of Cotton Yield and its Relationship to Environmental, Irrigation Infrastructure and Water Management Factors on a Regional Scale in Khorezm, Uzbekistan

Gerd Ruecker, Christopher Conrad, Nazirbay Ibragimov, Kirsten Kienzler, Mirzahayot Ibrakhimov, Christopher Martius, and John P.A. Lamers

Abstract The spatial distribution of cotton yields in the Khorezm region exhibits larger differences than those indicated in statistics on a district scale. However, the yield distribution within districts and farms, and possible factors correlating with this pattern, are unclear. Here, we map and characterize the detailed spatial variation of cotton yield at a pixel size of 250 m and analyse relationships between cotton yield, environmental factors, hydrological infrastructure, and water management in Khorezm for the year 2002. A remote-sensing based yield modelling approach was employed using satellite data of the Landsat 7 Enhanced Thematic Mapper Plus (ETM+) and the MODerate resolution Imaging Spectroradiometer (MODIS). Regional GIS maps were developed for environmental factors such as soil texture and groundwater table, hydrological infrastructure (distance of water use associations to irrigation inlets, irrigation channel density, and seasonal actual evapotranspiration). Well-pronounced

---

G. Ruecker (✉) • K. Kienzler
International Bureau of the Federal Ministry of Education and Research at the Project Management Agency, German Aerospace Center (DLR), Heinrich-Konen-Str. 1, 53227 Bonn, Germany
e-mail: gerd.ruecker@dlr.de; kirsten.kienzler@dlr.de

C. Conrad
Remote Sensing Unit, University of Würzburg, Am Hubland, 97074 Würzburg, Germany
e-mail: christopher.conrad@uni-wuerzburg.de

N. Ibragimov • M. Ibrakhimov • J.P.A. Lamers
ZEF/UNESCO Khorezm Project, Khamid Olimjan Str., 14, 220100 Urgench, Uzbekistan
e-mail: nazar@zef.uznet.net; hayot_i@yahoo.com; j.lamers@uni-bonn.de

C. Martius
Center for Development Research (ZEF), Walter-Flex-Str. 3, 53113 Bonn, Germany
e-mail: gcmartius@gmail.com

C. Martius et al. (eds.), *Cotton, Water, Salts and Soums: Economic and Ecological Restructuring in Khorezm, Uzbekistan*, DOI 10.1007/978-94-007-1963-7_4,
© Springer Science+Business Media B.V. 2012

relationships were found between cotton yield and the factors soil texture, irrigation infrastructure and seasonal evapotranspiration, while the correlation was weaker between cotton yield and groundwater table. These correlations were spatially analyzed and interpreted to identify areas suitable for cotton cultivation. Soil zones with lower cotton yield and areas with an irrigation infrastructure less suitable for cotton were spatially demarcated; for these areas, alternative land use strategies are suggested. Overall, this study suggests that improved surface and groundwater management should be targeted to specific sites within certain soil zones, and needs to be delivered timely according to crop requirements. These are key regional management strategies for improving cotton yield on a regional scale in Khorezm. We demonstrated that information on where and when water management improvements should take place can be suitably provided for larger areas with a remote sensing approach. The remote sensing-based monitoring system allows evaluating area-wide indicators for irrigation performance on different scales. The information thus gained can then be delivered to local water users associations for their adaptation.

Keywords Environmental monitoring • Climate change • Irrigation system • GIS • Remote sensing • MODIS • Semi-arid climate • Central Asia

4.1 Introduction

Cotton is the major crop in the Khorezm region. For a site-specific improvement of the crop production it is important to evaluate the detailed spatial distribution of the crop yield in relation to relevant environmental, infrastructure, and water management factors. As the official yield information is aggregated to districts and regions, remote sensing is the premier technology for giving spatially explicit and unbiased yield information of large areas. This technology has thus been widely used to estimate crop yields on a regional scale (Quarmby et al. 1993; Baez-Gonzalez et al. 2002; Doraiswamy et al. 2003). Based on the development of an agro-meteorological model that integrates multi-temporal remote sensing and minimum field data, cotton yield was estimated in Khorezm in 2002 (Ruecker et al. 2007). The objective of this paper is (1) to map and characterize the detailed spatial patterns of cotton yield in Khorezm in 2002 by application of this model, (2) to indicate spatial relationships between cotton yield, hydrological infrastructure and environmental indicators.

4.2 Data and Methods

4.2.1 Remote Sensing Based Regional Crop Yield Model for Khorezm

The applied yield model is based on the biophysical principle that a strong relationship exists between the cumulative radiation quantity absorbed by the foliage during the crop growth and the biomass production (Monteith 1977). Based on this relationship, a model was developed for estimating spatially distributed cotton yields of the dominant local cotton variety (Khorezm-127) at 250 m resolution for Khorezm (Ruecker et al. 2007). This model considers air temperature stress and vapor pressure deficit stress on cotton development. Further parameters are local meteorological and crop specific data such as photosynthetic active radiation, light use efficiency and harvest index. The model relies on high spatial resolution Landsat 7 ETM+ satellite images to extract the area under cotton cultivation (Schweitzer et al. 2005). Time series of MODIS (Moderate Resolution Imaging Spectroradiometer) FPAR (MOD15) and NDVI (MOD13) satellite image products were used to estimate the seasonal biomass accumulation (Conrad et al. 2004).

4.2.2 Regional Scale Indicators of Environmental Characteristics, Hydrological Infrastructure and Water Management in Khorezm

Several environmental, infrastructure and water management indicators were considered, and their relationship with cotton yield on a regional scale was investigated. The selection of the indicators was based on literature review and consultations with local experts. Environmental indicators were parameterized by spatial patterns of soil texture at 0–30 cm depth and groundwater table data measured in wells every 8 days during the cotton growth period in 2002. Hydrological infrastructure indicators were based on an irrigation canal network layer of Khorezm. Spatial patterns of irrigation network density and channel distance between system intake and the water intake point of the water users association were generated by spatial analysis in a geographical information system (GIS). As water management indicator, the seasonal actual evapotranspiration, was derived using MODIS data in 2002 using the Surface Energy Balance Algorithm (Conrad et al. 2007). The GIS maps used in the correlation analysis with cotton yield are shown in Fig. 4.1.

4.3 Results and Discussion

4.3.1 Spatial Distribution of Cotton Yield on a Regional Scale in Khorezm

The yield model was run to calculate cotton yield on selected MODIS pixels that covered >80% of the cotton fields over Khorezm in 2002. The estimated yield of raw cotton within MODIS pixels ranged from 1.09 to 3.76 $t\,ha^{-1}$, with an average of 2.38 $t\,ha^{-1}$. A final regional cotton yield map was generated by spatial interpolation based on the cotton area retrieved from the Landsat 7 ETM+ land use classification (Fig. 4.2).

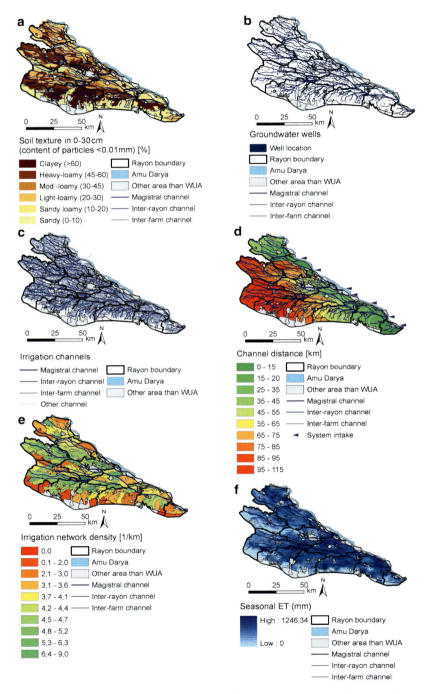

Fig. 4.1 Environmental and infrastructure factors of Khorezm including (**a**) soil texture, (**b**) ground water table, (**c**) irrigation network from which (**d**) channel distance and (**e**) irrigation network density were generated, and (**f**) seasonal evapotranspiration

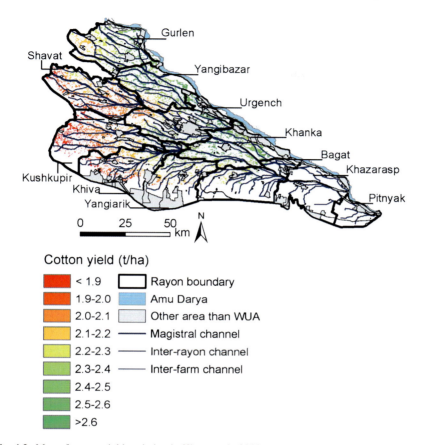

Fig. 4.2 Map of cotton yield variation in Khorezm in 2002

The map shows a highly heterogeneously distributed cotton yield within Khorezm with distinct, large spatial patterns with higher, lower and medium range at specific locations on a regional scale. The areas with higher cotton yields, depicted in bright and dark green signatures, occurred generally in the upstream areas of the irrigation system within the districts Bagat, Khanka, Yangiarik, Khiva, and Urgench. The other upstream areas within Yangibazar and Gurlen districts are mainly rice cultivation areas. The places with the highest cotton yields (>2.7 t ha^{-1}) were relatively small and located within the rice areas of the districts Urgench and Khanka which border the Amudarya river, and in areas north-east of the settlement area of Khiva.

In contrast to the upstream high cotton production areas, the production sites with low cotton yield (<2.1 t ha^{-1}) were mainly situated in downstream locations of the irrigation network. Larger spatial patterns of such low cotton production sites were found in southern Yangiarik, western, central and northern parts of Kushkupir, southern and north-eastern Shavat, and north-western Gurlen. The lowest cotton yields (<1.8 t ha^{-1}) occurred in north-west Kushkupir and south-west Shavat.

The cotton fields with yields at a medium range (2.1–2.4 t ha^{-1}) were mainly integrated in the central and downstream areas, but also in landscapes bordering the Amudarya river. It is recognizable that large parts of Kushkupir and Shavat, and smaller parts in Yangibazar, Khiva, Yangibazar, and Gurlen have cotton yields of such a medium range.

On a district scale, similar heterogeneous cotton yield patterns as on a regional scale could be found in many districts. Spatial gradients ranging from higher over medium to lower cotton yield patterns were found in up-, mid- and -downstream positions of the districts Yangiarik, Khiva, and Kushkupir. High to medium range cotton yields occurred in Bagat, Urgench and Yangibazar, while medium to low range yields occurred in Shavat and Gurlen. Only Khanka exhibited a nearly homogenous cotton yield distribution.

Depending on the number of cotton yield classes and their thresholds, further spatial gradients could be discerned on more detailed spatial scales of water users association or hydrological units comprising several fields. This is possible because the yield map is based on MODIS satellite images, which reflect the cotton yield in a spatially disaggregated pattern of 250 m pixels that match with fields of a medium size. In contrast, such insights into detailed spatial yield patterns and gradients would not be possible with the aggregated official information that is available on a district or regional scale. While the regional overview shows the main yield patterns, detailed yield information can be retrieved on district or WUA-scale. Thus, this remote sensing approach is able to provide important maps with crucial crop production information that can be used for both regional as well as district/ WUA-scale land and water use assessment and planning. The detailed yield information allows the site-specific analysis of underlying factors determining yield at regional scale.

4.3.2 Correlation of Cotton Yield with Environmental Conditions, Hydrological Infrastructure and Water Management in Khorezm

The results of the correlation analysis between environmental, infrastructure and water management indicators and cotton yield are shown in Fig. 4.3.

Considering soil texture in the first 30 cm, cotton yields were clearly higher on the moderately loamy, sandy-loamy, and light-loamy soils (Fig. 4.3a). The moderately and sandy-loamy soil textures occurred mainly in a broad zone along the western side of the Amudarya river stretching through the eastern parts of the districts Gurlen, Yangibazar, Urgench, Bagat, and Khanka (Fig. 4.1a). A zone with light-loamy soils connected directly to the south-west and represented the dominant soil texture in the central area of Khorezm, covering large parts of Khanka, Urgench, and Gurlen, while other districts had only smaller shares of this soil texture.

In contrast, lower cotton yields occurred generally in clayey, heavy loamy and sandy soils. These soil textures were found in areas bordering the desert margin and

4 Spatial Distribution of Cotton Yield and its Relationship to Environmental, Irrigation...

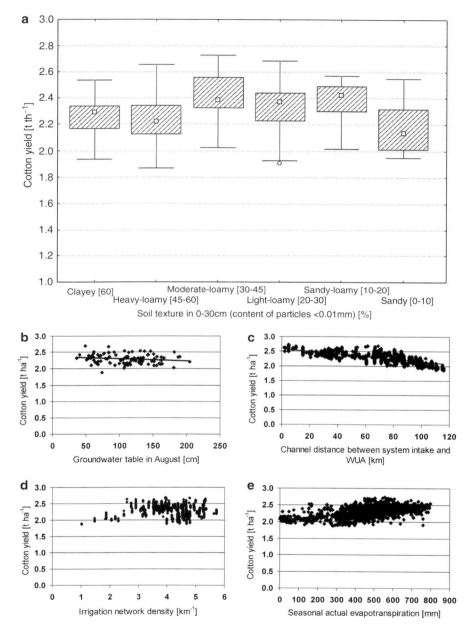

Fig. 4.3 Correlation of soil type, groundwater table, hydrological infrastructure and water management indicators with cotton yield in Khorezm in 2002. Numbers in brackets after soil type in (**a**) indicate % content of particles <0.01 mm

the connecting soil zones to the north-east. They occur in large shares specifically in Yangiarik, but also in Khiva, Kushkupir, and Shavat, and with smaller shares in Urgench, Bagat, and Khazarasp.

Overall, a clear zonal pattern can be discerned, with higher yields on soils along the Amudarya river and the central area, and lower yields in areas bordering the desert. In between and embedded there are smaller patches of soil types with yields that differ from this general pattern.

The correlation between groundwater table and cotton yields showed higher cotton on sites with shallower groundwater table, while lower yields correlated with deeper ground water table. However, the correlation was relatively weak in a regional scale perspective (Fig. 4.3b). This correlation pattern increased continuously from the early growth period (15 April–15 June), over the main growing stages (20 June–30 August) (correlations from both periods are not shown here), and was strongest during August. This pattern can be explained by the farmers' behaviour to try maintaining a relatively high groundwater table as a contribution to crop water supply by capillary rise: This is especially important to overcome crop water stress due to delayed, inefficient irrigation water supply, or to droughts. A similar correlation as in August appeared during the pre-season when leaching was introduced (10 January–10 April). A high groundwater table during the leaching period is practiced to wash the accumulated salts effectively out of the soil, as a strategy to prepare a suitable seedbed. Both strategies contribute to gain higher crop yields. In a spatial perspective, shallower groundwater tables were mainly found in upstream areas (e.g. Khanka, Gurlen, Pitnyak), while deeper groundwater tables occurred generally in tail end locations, such as in Khiva and Kushkupir.

The relationship between the infrastructure indicator "channel distance between system intake and WUA" and cotton yield was well pronounced. The linear correlation showed an R^2 of 0.5168 ($y = -0.0043x + 2.5849$). Higher cotton yields were found in WUAs that had shorter distances (0–40 km), medium yields had mid-range distances (40–80 km), while lower yields were in areas with far distances to the corresponding system intake (80–120 km). WUAs with a short canal distance to the system inlet where directly bordering the Amudarya (bright and dark green signatures). The location of these WUAs was rather identical with the spatial extent of the moderately-loamy and sandy-loamy zone (Fig. 4.1a, d). WUAs that were located in a far distance to the system inlet (bright to dark red signatures) corresponded generally with downstream areas, which again largely matched with clayey, heavy loamy and sandy soils. Finally WUAs that were in mid-range distances to the inlet (yellow to orange signatures) were mainly in the central area of Khorezm and coincided with light-loamy soils.

The investigation of another important infrastructure parameter, "irrigation network density", revealed that, in general, higher cotton yields corresponded well with areas that have a higher network density. However, this linear relationship is only valid for areas with network densities[1] up to ca. 3 km^{-1}. Such areas are mainly

---

[1] Channel density (km^{-1}) is calculated from channel length (km) divided by area (km^2).

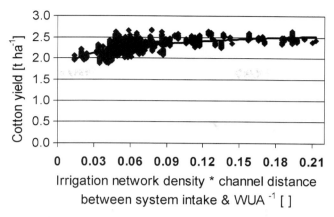

Fig. 4.4 Correlation between (irrigation network density * channel distance between system intake and WUA^{-1}) with cotton yield in Khorezm in 2002

located in downstream areas bordering the desert (compare red to dark orange signatures in Fig. 4.1e). Beyond this threshold, cotton yields are no longer increasing, but rather stagnating, exhibiting both higher ranges (approximately 0.8 t ha^{-1}) and fluctuations between lower and higher yields with a denser irrigation network (Fig. 4.3d). This saturation indicates that beyond ca 3 km^{-1} network density, other factors such as the previously mentioned WUA distance to system inlet have a strong impact to determine cotton yield.

The combination of the two infrastructure indicators (Fig. 4.4) shows a logarithmic type of relationship: e.g. (y = 0.2077Ln (x) + 2.8519; R^2 = 0.3912). Higher yields were in areas with higher network density and shorter canal distance, while lower yields were in areas with smaller network density and higher canal distance. Fluctuations can be understood by small-scale variability within districts and WUAs. In an assessment of a water management factor, a significant linear relationship between seasonal actual evapotranspiration (ET_{act}) and cotton yield was found for Khorezm in 2002 (Fig. 4.3e). Higher yields were found in areas with distinct higher ET_{act} (y = 0.0006x + 2.0425; R^2 = 0.296), pointing out that these areas gained high level of water inputs.

4.4 Conclusions and Outlook

Based on a regional crop yield estimation model that was largely driven by MODIS remote sensing data, spatial distribution of cotton yield was mapped for Khorezm and correlated with environmental, hydrological infrastructure, and water management factors. The map at 250 m pixel resolution clearly revealed a detailed differentiation of crop yield within the region and classified the region into areas with different yield production. The correlation of cotton yield with regional GIS-maps indicated the dependencies on soil texture, irrigation infrastructure, and water management.

Such detailed spatially distributed yield information and the dependencies on specific geo- and management factors are important to decision makers at regional, district, and WUA-scale for targeted land use planning/restructuring and infrastructure rehabilitation.

Spatially explicit information how crop yield is distributed and where management and infrastructure improvements should take place can be best provided for larger areas by remote sensing based monitoring systems that produce GIS-maps of key indicators such as crop yield. As this approach relies mainly on satellite data, it has a great potential for transfer to other irrigation systems.

References

Baez-Gonzalez AD, Chen PY, Tiscareno-Lopez M, and Srinivasan R (2002) Using satellite and field data with crop growth modeling to monitor and estimate corn yield in Mexico Crop Sci 42:1943–1949

Conrad C, Ruecker GR, Schweitzer C, Dech S, Hafeez M (2004) Modeling seasonal actual evapotranspiration with remote sensing and GIS in Khorezm region, Uzbekistan. In: Proceedings of the 11th SPIE international symposium on remote sensing, Maspalomas, Gran Canaria/Spain, 13–16 Sept 2004

Conrad C, Dech SW, Hafeez M, Lamers J, Martius C, Strunz G (2007) Mapping and assessing water use in a Central Asian irrigation system by utilizing MODIS remote sensing products. Irrig Drain Syst 21(3–4):197–218

Doraiswamy PC, Hatfield JL, Jackson TJ, Akhmedov B, Prueger J, Stern A (2003) Crop conditions and yield simulations using Landsat and MODIS. Remote Sens Environ 92:548–559

Monteith JL (1977) Climate and efficiency of crop production in Britain. Philos Trans R Soc B Biol Sci 281:277–294

Quarmby NA, Milnes M, Hindle TL, Silleos N (1993) The use of multi-temporal NDVI measurements from AVHRR data for crop yield estimation and prediction. Int J Rem Sens 14: 199–210

Ruecker GR, Shi Z, Mueller M, Conrad C, Ibragimov N, Lamers JPA, Martius C, Strunz G, Dech SW (2007) Regional scale estimation of cotton yield in Uzbekistan by integrating remote sensing and field data into an agrometeorological model. In: Proceedings of the international conference. Scientific and practical bases of improvement of soil fertility, Uzbekistan Cotton Research Institute, Tashkent

Schweitzer C, Ruecker GR, Conrad C, Bendix J, Strunz G (2005) Knowledge-based land use classification combining expert knowledge, GIS, multi-temporal Landsat 7 ETM+ and MODIS time series data in Khorezm, Uzbekistan. In: Erasmi S, Cyffka B, Kappas M (eds) Remote sensing and geographic information systems for environmental studies: applications in Geography. Göttinger Geographische Abhandlungen, Verlag Erich Goltze GmbH & Co. KG, Göttingen 113:116–123

Chapter 5
Water Management in Khorezm: Current Situation and Options for Improvement (Hydrological Perspective)

Bernhard Tischbein, Usman Khalid Awan, Iskandar Abdullaev, Ihtiyor Bobojonov, Christopher Conrad, Hujiyaz Jabborov, Irina Forkutsa, Mirzahayot Ibrakhimov, and Gavhar Poluasheva

Abstract The combination of hydrological research findings from field to regional level in the irrigated croplands of Khorezm, Uzbekistan, revealed that water availability at farm and field level fails to meet agricultural requirements in parts of the region, although water withdrawal from the Amudarya river is huge. In 2004 and 2005, the seasonal gross water input to a sub-unit of the Khorezm irrigation and drainage system – a hydro-unit investigated as a case study - constituted 2,630 mm and 2,810 mm, respectively, in which the share of pre-season leaching amounted to 700 mm each year. These findings correspond well with the 2,240 mm water input monitored in the vegetation period of 2005 at Khorezm-wide level. Reduction of

---

B. Tischbein (✉)
Center for Development Research (ZEF), Walter-Flex-Str. 3, 53113 Bonn, Germany
e-mail: tischbein@uni-bonn.de

U.K. Awan
Center for Development Research (ZEF), Walter-Flex-Str. 3, 53113 Bonn, Germany

University of Agriculture, Faisalabad, Pakistan
e-mail: ukawan@uni-bonn.de

I. Abdullaev
Deutsche Gesellschaft für Internationale Zusammenarbeit (GIZ) GmbH,
Abdullaev Str. 2 A, 100100 Tashkent, Uzbekistan
e-mail: iskandar.abdullaev@giz.de

I. Bobojonov
Department of Agricultural Economics Farm Management Group, Humboldt-Universität zu,
Berlin, Germany
e-mail: ihtiyorb@yahoo.com

C. Conrad
Remote Sensing Unit, Department of Geography, University of Würzburg,
Am Hubland, 97074 Würzburg, Germany
e-mail: Christopher.Conrad@uni-wuerzburg.de

C. Martius et al. (eds.), *Cotton, Water, Salts and Soums: Economic and Ecological Restructuring in Khorezm, Uzbekistan*, DOI 10.1007/978-94-007-1963-7_5,
© Springer Science+Business Media B.V. 2012

actual evapotranspiration (ET) was observed in the range of 5–10% at sub-unit level, more than 25% in tail-end locations in Water Users Associations and 30–40% in single fields. Technical overall irrigation efficiency in the vegetation periods 2004 and 2005 averaged 27.5% in the sub-unit and 26% for whole Khorezm. These low efficiencies in the sub-unit are plausible as also drainage water output reaches high shares, with 62% and 67% of the irrigation water input in 2004 and 2005, respectively. On the other hand, the local practice of allowing for shallow groundwater tables that feed the crop in a system of 'furrow and sub-irrigation via capillary rise' raises the *de-facto* efficiency to 38%, but entails the well-known (Section "Regional level of Khorezm") salinity problems of the region. The depleted fraction - the ratio between actual evapotranspiration and the sum of water inflow via irrigation and rainfall - in the peak irrigation season 2005 was around 0.3 at Khorezm level. With groundwater tables above 1.4 m and moderately saline groundwater, approx. 65–70% of the irrigated areas are at risk of waterlogging and salinization in April and July each year (major leaching and cropping periods, respectively). The dysfunctional drainage system, together with excessive leaching at field level and huge water losses from the irrigation (conveyance and distribution) network are major reasons for the observed shallow groundwater in the region.

With reference to technical aspects only, current problems with the water management system in the region are mainly caused by (1) inflexible irrigation scheduling, (2) low efficiencies of water application at field and network level, (3) inappropriate infrastructure and insufficient maintenance, (4) limited options for groundwater management and (5) a general lack of input data to support water management. These factors are therefore starting-points for interventions towards an improvement of water use. Such measures include among others the advanced determination of net irrigation needs taking the temporal behavior and site-specific dependencies of the field water balance into account, the optimization of the field water application process, and raising network efficiency. Together, these interventions have a water-saving potential of 50% relative to the current water withdrawal from the river.

Keywords Amudarya • Water and salt balancing • Irrigation scheduling • Irrigation performance • Water saving potential

---

H. Jabborov
Urgench State University, Khamid Olimjan Str., 14, 220100 Urgench, Uzbekistan
e-mail: h.jabbarov@mail.ru

I. Forkutsa
GWSP Program, Center for Development Research (ZEF),
Walter-Flex-Str. 3, 53113 Bonn, Germany
e-mail: forkutsa@gmail.com

M. Ibrakhimov • G. Poluasheva
ZEF/UNESCO Khorezm Project, Khamid Olimjan Str., 14, 220100 Urgench, Uzbekistan
e-mail: hayot_i@yahoo.com; gavhar2005@rambler.ru

5.1 Introduction

Due to the arid climate, agricultural systems in Khorezm depend entirely on irrigation and drainage; drainage is a prerequisite to control soil salinity. Huge water withdrawals (Conrad 2006), reduced actual ET (Conrad 2006; Forkutsa 2006), shallow groundwater levels (Ibrakhimov 2005) and widespread soil salinity (Forkutsa 2006) indicate deficiencies of the current water management system. This all leads to harmful implications for agricultural production, the environment and the living conditions of the population in Khorezm and in downstream areas.

Improving water management is a prerequisite for the sustainable restructuring of land and water use in Khorezm. But changes of water management need to be embedded in an overall conceptual framework to integrate agricultural, economic and institutional interventions.

The research described in this chapter aims at assessing current water management and suggesting ways towards a more efficient and effective use of water resources. The reasons for current deficits are embedded at all scales of technical water management (field level, irrigation network, drainage facilities), and thus, research activities addressed these levels.

For modeling and understanding the system, we considered the irrigated field as the basic spatial element of the irrigation and drainage schemes, and we started our analysis at this level. Recommendations aiming at improved water use efficiency directly address farmers with concepts for field water management. Improved scheme management is most promising as a bottom-up approach that starts from field level. In accordance with the basic procedures necessary to improve field water management, research at field level was divided into two components. First, we modeled field water and salt dynamics and established a tool to determine net irrigation data (irrigation timing and amount), taking into account site-specific conditions, and enabling a flexible reaction to the changing environment. Second, we improved water use efficiency by applying the double-side type of irrigation as a site-specific approach to achieve higher uniformity of water application along furrows.

We used the findings obtained at the field level as input for research at the next level: In a 850 ha sub-unit or micro-basin of the irrigation system we analyzed the irrigation and drainage system management and assessed irrigation efficiency and appropriateness of water supply at field level (by reduced actual ET) using a water balance approach. The sub-unit contains field-level supply and irrigation canals, but not the highest canal hierarchies (magistral canals and inter-rayon canals). Tackling the causes for limited irrigation performance enabled us to develop options to improve water management strategies.

In a larger sub-unit, corresponding to a Water Users Association (WUA), the current norm-based water distribution was analyzed and used as benchmark to assess the intended improvements through a combination of a flexible irrigation scheduling model and groundwater modeling. This research activity emphasized the modeling approach that is described in more detail by Awan et al. (2011).

Studies at the regional level of Khorezm aimed at analyzing the spatio-temporal dynamics of the groundwater situation in the region and assessing irrigation performance. Shallow groundwater levels, which prevail in Khorezm, are of high importance for water management. On one hand, shallow groundwater may contribute advantageously to meet crop water requirements (especially in parts of the system with reduced availability of conventional canal water), but on the other hand, capillary rise from the shallow groundwater table accelerates the processes of secondary soil salinization. An approach using remote sensing techniques, Geographical Information Systems (GIS) and hydrological modeling allowed us to develop the water balance and upscale it to the whole Khorezm region.

A performance assessment based on the depleted fraction, reduction of actual ET and drainage ratio allowed detecting problems of the current water allocation with respect to spatial and temporal occurrence. The results provided a promising starting-point for improvements of water allocation at larger scales.

Major features of Khorezm relevant to the irrigation and drainage in the region are pointed out in Sect. 5.2. The main results of research activities separate for the relevant scales are summarized in Sect. 5.3. Subsequently, findings across the scales are being applied to target the problems and causes (Sect. 5.4). Conclusions are then drawn with respect to water saving potential approaches and developing tools for further improvement in water management in dry irrigated lowland regions of the Aral Sea basin (Sect. 5.5).

5.2 Major Features of Khorezm Relevant to Irrigation and Drainage

The arid continental climate rises the annual potential evapotranspiration expressed by the reference evapotranspiration (ET_0) in Khorezm to 1,100–1,200 mm, which is more than one order of magnitude above the average annual precipitation of 90–100 mm (cf. Conrad et al. 2011). As a consequence, crop production relies completely on irrigation. Since historical times, irrigation systems were constructed and operated to provide water from the Amudarya to agricultural fields.

In 1930, about 3 km^3 of water were withdrawn from the Amudarya to feed 121,200 ha of irrigated land in Khorezm (Alimov et al. 1979). This situation changed dramatically when Soviet planners designated Central Asia as the area to grow the strategic crop cotton (*Gossypium hirsutum L.*). The designed irrigation network allowed increasing withdrawals of water from the river and expanding the cropland into the desert. Already in the 1970s, the water withdrawal into Khorezm was 4.1–4.3 km^3, while the irrigated area had increased to 151,400 ha (Mukhammadiev 1982). The withdrawal per unit area reached the highest values in the decade between 1970 and 1980 with peaks in 1978 (34,200 $m^3 ha^{-1}$) and 1980 (33,000 $m^3 ha^{-1}$) according to Jabborov (2005). In 1999, the irrigated areas reached 276,000 ha, which were irrigated with a total of ca. 4.5–5.0 km^3 of water (Conrad 2006). Since the drought years in 2000 and 2001, the irrigated area has stabilized at around 265,000 ha and the total annual withdrawal from the Amudarya amounted to 4.5 km^3

in water abundant years (Bekchanov et al. 2010). The resulting withdrawal per unit area is in the range of 17,000 m^3 ha^{-1}. The total length of the irrigation canals in Khorezm is 16,400 km (GIS-lab of ZEF/UNESCO project), leading to a network density of around 60 m ha^{-1}.

Next to cotton, winter wheat (*Triticum aestivum* L.), fodder crops and rice (*Oryza sativa* L.) are currently the dominating crops (Bekchanov et al. 2010; Veldwisch et al. 2011).

Water allocation for irrigation and leaching is based on rigid procedural prescriptions. Irrigation amounts and timing are determined according to hydrological zoning by norms elaborated in the 1960s (Khorst and Ikramov 1995). The norms consider the crop, climate, soil characteristics (mainly texture and stratification) and groundwater levels of each zone. Amounts of pre-seasonal leaching are based on a soil salinity appraisal, which usually takes place in November. According to recommendations issued by the Uzbek Ministry of Agriculture and Water Resources (MAWR) in 1975, low-saline fields should be leached with approx. 400 mm of water, moderately saline fields with 500 mm, and strongly saline areas with 600 mm.

To manage the groundwater level and enable leaching, a drainage network was developed. From 1942 to 1950, local main and inter-rayon drains were constructed which discharge drainage effluents into the local lakes and depressions. From 1950 to 1961, local lakes and depressions, which received drainage discharge, were linked to the main collector drain Ozerny (lake drain), with the aim to lower water levels and release water into the Sarykamish depression that lies outside the borders of Khorezm. After 1961, main drains were constructed, combining all local drains into one network and discharging the majority of the discharge into the Sarykamish depression. The total length of the drainage network reaches approximately 8,000 km (GIS Lab of Khorezm Project). The depths of field drains are in the range of 1.5–2.0 m and inter-farm but main collector drains rarely are deeper than 2.5 m.

Problems of the current water management are indicated by the huge withdrawals from the Amudarya per unit area of irrigated land in Khorezm and by wide-spread occurrence of soil salinization. Forkutsa et al. (2009) cite analyses by MAWR which classified 55% of the irrigated lands in Khorezm as slightly saline (2–4 dS m^{-1}), 33% as medium saline (4–8 dS m^{-1}) and 12% as highly saline (8–16 dS m^{-1}) at the end of the vegetation period 2004, while according to Akramkhanov et al. (2011), moderately saline soils were already 31%, and highly saline soils, 21%, in 1990.

In the Uzbekistan-wide context, huge water withdrawals are of particular concern, because the withdrawals for agriculture are in the range or even exceeding the renewable water resources. In 1998, for instance, the withdrawal for agriculture in Uzbekistan was 54.4 km^3 compared to 50.4 km^3 renewable resources (UNESCO 2003). As only around 30% of the renewable resources are generated on the territory of Uzbekistan (UNESCO 2003), the problem has two aspects: sustainability of water use and dependency on upstream countries.

5.3 Research Findings: From Field to the Regional Level

5.3.1 Field Level

Determination of (a) net irrigation data (i.e. timing and amount) needed to fulfill the actual crop requirements and (b) application of water with highest possible efficiency are two basic procedures of water management at field scale. Combining improvements of both procedures are imperative to raise water use efficiency.

5.3.1.1 Modeling Soil Water and Salt Dynamics at Field Level

The major objective was to assess and improve water management at field level using net irrigation data as the main entry point. As an alternative to the current rigid practice of applying irrigation norms (Sect. 5.2), the determination of irrigation data was based on a flexible model. This approach allowed for adequate responses to any changes in weather conditions, water availability or crop development and enabled simulating alternative scenarios. In particular, the use of flexible modeling approaches allows conceiving deficit irrigation strategies in case of non-avoidable under-supply, which in turn minimizes water stress impacts on yield. This objective cannot be achieved when using generalized norms. Furthermore, flexible modeling allows site-specific solutions that are increasingly becoming relevant since land use changed from large, uniform production units (for which the norms were originally developed) to a high number of smaller water consumers with increasingly diverse requirements (cf. Djanibekov et al. 2011).

The approach combined (1) monitoring in 2003 the relevant components of soil water and salt balances under the farmers' current water management practice at two fields with typical local characteristics (3–4 ha field size, sandy and sandy loamy soil texture, sub-division into micro-plots to compensate for the irregular micro-topography), and (2) modeling water and salt dynamics using the HYDRUS-1D model (Simunek et al. 1998).

Model-based analysis revealed that capillary rise in the sandy-loamy field was quite high, ranging from 12% to 47% of the actual ET in 2004 (Forkutsa 2006); capillary rise was strongly variable according to location in the field. Compared to empirical approaches of capillary rise estimation, the model enabled a sound evaluation of advantages, such as partial matching of crop water requirements, and disadvantages, such as accelerating soil salinization, of the groundwater contribution.

Combining the HYDRUS model and the advanced dual crop coefficient concept (Allen et al. 1998) allowed the comparison of actual and potential values of transpiration and evaporation (Forkutsa 2006). According to Fig. 5.1, the first irrigation event on a sandy-loam field in 2004 was delayed causing a reduction in the actual transpiration and, consequently, in yields. Furthermore, with the model, optimal irrigation dates could be determined based on site-specific requirements and changing conditions. These data provided the informal base to start flexible irrigation scheduling of the overall system. In addition, results of Forkutsa (2006) showed low

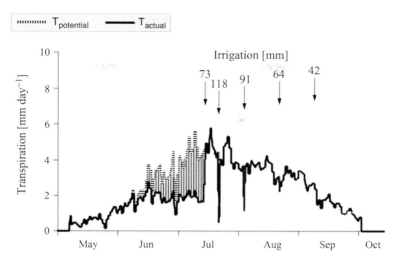

Fig. 5.1 Potential and actual transpiration in a sandy-loam field in 2003 (Forkutsa 2006)

(40–50%) application efficiencies. The reasons for such low values were irregular micro-topography, non-sufficient information on site-specific as well as temporal irrigation depths, missing infrastructure for proper dosage of the application discharge and non-reliable water availability in the irrigation system. In the context of interventions to address the causes behind the currently low irrigation efficiency, data on soil moisture provided by HYDRUS with high temporal and vertical resolution allowed determining proper net irrigation depths as a prerequisite to avoid over- as well as under-irrigation.

Modeling of the spatially and temporally distributed soil salinity in HYDRUS helped detecting those locations where salts accumulate, and also, determining appropriate leaching information. Considering the pre-leaching salinity levels, the target value of salinity after leaching (depending on the crop to be cultivated), introducing more salt-tolerant crops and improving leaching performance are among the promising measures to de-salinize soils more effectively. Besides, modeling salt dynamics with HYDRUS helped to analyze the leaching effects in the vegetation period and estimate the amounts of water losses due to deep percolation. Combining leaching and irrigation events could be considered as a basic alternative to the currently practiced pre-season leaching which requires huge amounts of water (700 mm gross input to sub-units; Sect. 5.3.2).

5.3.1.2 Improving Field-Level Water Application by Double-Side Irrigation

Improving water use efficiency at field level during furrow irrigation requires flexibility in net irrigation data and advanced handling of the water application process. Besides optimizing the application discharge, laser-guided field leveling and advanced methods of furrow irrigation such as surge flow and double-side irrigation[1] are an option in Khorezm, due to its flat topography. Double-side irrigation generally resulted in a more uniform water distribution along the furrows, improved application efficiency and saved around 20% of the seasonal gross irrigation water input to the field (Poluasheva 2005). Furthermore, the sharp increase in salt accumulation, which was observed at the end of conventionally irrigated furrows due to low irrigation water application and high capillary rise, could be halved. The higher uniformity of water application and a reduced peak in salt accumulation at the end of the furrows increased the cotton yield by 30–35% (Poluasheva 2005). In addition, the higher uniformity and reduction in salt accumulation after the vegetation season lowered leaching requirement and eased the application of leaching water.

5.3.2 Analyzing and Improving Water Management at Sub-unit Level

Water management strategies at sub-unit level were analyzed at two sites: (a) a hydrologically defined unit fed by the *Serchalli* canal and (b) the Water Users Association *Shomakhulum*.

5.3.2.1 Hydrological Unit Fed by the Serchalli Canal

The major objectives were to (1) establish a water balance for the sub-unit to contribute to the upscaling from field to the Khorezm level, (2) assess the current irrigation strategies in terms of efficiency and appropriateness, (3) develop integrated water management approaches towards improved water use, and (4) analyze the drainage system. A hydrologically delineated sub-unit covering 850 ha, with a net irrigated area of 606 ha, was selected to monitor the current water management. Two periods were covered: November 2003 to October 2004 and November 2004 to October 2005.

The current water management of the sub-unit can be characterized by excessive input of irrigation and leaching water as well as a high output via the drainage system. The yearly water input amounted to 2,630 mm in 2004 and 2,810 mm in 2005, of which around 25% was applied in the leaching period. Drainage water output reached 62% and 67% of the input in 2004 and 2005, accordingly.

A water balance was established and net irrigation requirements calculated based on soil characteristics, typical cropping patterns, land use and meteorological data. According to observations at field (Forkutsa 2006) and regional (Conrad 2006) level, the water balance of the sub-unit revealed a reduction of the actual ET in the range of 5–10%. Reduced actual ET indicates non-appropriate water availability at field level: either sufficient water is not available, or it is not provided at the right time. The latter aspect was monitored at field level (Sect. 5.3.1). As the gross water

---

[1] Double-side irrigation provides water to the furrow from both ends at the same time, as opposed to the conventional technique where water is supplied only from one end. Both methods were compared at the SANIIRI research farm in Khanka district in 300 m long furrows.

5 Water Management in Khorezm: Current Situation and Options...

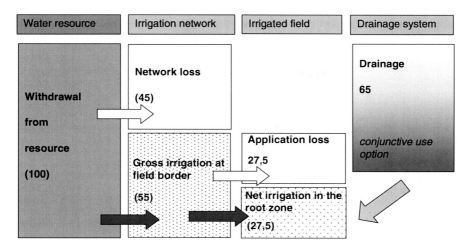

Fig. 5.2 Water flow and efficiency in a 850-ha sub-unit of the irrigation system in Khorezm. Average values for in 2004 and 2005

input was high, current shortcomings at field level were caused by inadequate and sub-optimal coordination of irrigation operational activities between field and network level as well as within the network system.

To estimate irrigation efficiency, net irrigation requirements of the current strategy were calculated and correlated with the measured water input. Direct measurement of spatially distributed soil moisture prior to and after irrigation events was not possible due to the sub-unit's size. The average technical irrigation efficiency of the sub-unit during the monitoring periods amounted to approximately 27.5%. In combination with observations at field level (Forkutsa 2006), an application efficiency of 50% and a network efficiency in the range of 55% could be derived (Fig. 5.2).

A FAO (2002) study estimated the current level of irrigation efficiency in 93 developing countries at 38%. To reach the 'more crop per drop strategy' until 2030 in regions where the share of withdrawals to renewable resources is already high, FAO states the need to raise irrigation efficiency to 53%. Comparing the 27.5% irrigation efficiency in the sub-unit with the target efficiency established by FAO demonstrates the urgent need to improve water use in Khorezm. Water saving potential is assessed in Sect. 5.5.

The overall efficiency drops predominantly in May (beginning of rice cultivation/ land preparation) and October (over-proportional network losses due to irrigation of spatially scattered fields under winter wheat, or locally performed leaching activities). The relatively high efficiencies in September were largely related to (i) the intention of farmers to avoid over-irrigation in the pre-harvest period, and (ii) the high irrigation depths in the late season leading to a relatively low share of percolation losses. The high irrigation efficiency as determined in September suggests that water use efficiency can be improved in Khorezm.

The irrigation system in Khorezm can be analyzed from different perspectives. Conventional analysis (and the one presented above) interprets the systems as a conventional furrow-irrigation system with an estimated overall irrigation efficiency of 27.5%.

If however, we interpret the system as a combined system of 'furrow and sub-irrigation via capillary rise', then the de-facto efficiency would be 38%, because the capillary rise, considered as a 'loss' in the conventional analysis, would then be interpreted as an irrigation input. In the Khorezm region, the latter approach is closer to reality insofar as the farmers perceive the groundwater contribution as an option to compensate for the unreliable water supply by the irrigation network. This is the reason for the blocking of drainage outlets often observed in the region (Sect. 5.3.3).

Both perceptions make sense depending on the purpose; if we try to improve the furrow system, the first analysis makes sense; if we would like to have an assessment that reflects the rationale of the Khorezmian farmer, the second analysis is more appropriate, in which at least a part of the 'losses' in the first approach is actually an input to the crop water requirements, and hence, a gain for the cropping system. Nevertheless, the second option entails increased soil salinity due to capillary rise, which further increases leaching requirements - a vicious cycle that is sustained also by the existence of the state order crop production system with its associated risks of losing land when farmers underperform (cf. Djanibekov et al. 2011; Veldwisch et al. 2011). Nevertheless, it is this perspective, which needs to be understood – the pressures under which local farmers operate and which determine their actions - if interactions with farmers on water management are to be successful.

The temporal behavior of the groundwater in the sub-unit matched the general trend observed at field and regional levels (Sect. 5.3.3) and was driven by the irrigation strategy. The impact of the regional groundwater flow on the water balance was small (at least in relation to the effect of the irrigation strategy), because the difference between water input and output to the sub-unit could be explained reasonably by the reduction of actual ET.

5.3.2.2 WUA *Shomakhulum*

A study was carried out in a 2,000 ha WUA, '*Shomakhulum*', in the vegetation periods 2006 and 2007 and the leaching period of 2008 (Awan 2010). In this study, the current water management based on norms was monitored. Results of the monitoring are used (Sect. 5.4.1) to analyze the plausibility of findings across the scales. The major objective was to develop a linked irrigation scheduling-groundwater model to work out integrated water management strategies (Awan 2010). With the model, optimal irrigation schedules can be derived, which lead to water saving and avoidance of water stress in comparison to norm-based schedules and in relation to schedules practiced by farmers (cf. Awan et al. 2011). Furthermore, the model supports assessing the long-term impact of water saving techniques on the groundwater level.

5.3.3 Regional Level of Khorezm

Research at the Khorezm-wide level aimed at (a) analysis of the groundwater situation and (b) upscaling of water balances and irrigation performance assessment.

5 Water Management in Khorezm: Current Situation and Options...

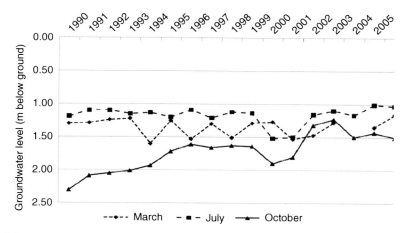

Fig. 5.3 Temporal changes of average groundwater table in Khorezm in March, July and October during 1990–2006

5.3.3.1 Spatio-Temporal Analysis of the Groundwater Situation

A continuous monitoring program of groundwater table and groundwater and soil salinity has been established by the regional authorities in Khorezm since the early 1970s. The Khorezmian Hydrogeological Melioration Expedition (OGME) of MAWR operates the network, which consists of around 2,000 observation wells in the Khorezm region alone.

Temporal Behavior of Groundwater Level and Salinity

The groundwater table was generally shallow throughout the region, averaging 1.61±0.51 m over the analysed period of 1990–2006. Average monthly regional groundwater levels in March (main leaching period) and July (peak irrigation season) amounted to 1.34 and 1.17 m, respectively. These levels by far exceed the threshold of 1.5 m, at which crop water supply is optimal (a blanket 'rule-of-thumb' for the local soils; Dukhovny 1996). Average groundwater readings of 1.77 m in October indicate that soils can be re-salinized from the still shallow groundwater levels under conditions of absent downward water percolation after the end of the irrigation season (except for the areas under winter wheat, which are still irrigated at that time). The analysis of the probability distribution of the 1990, 1994 and 2000 datasets showed that about two thirds of the April -the month immediately after leaching- and July data (on average 62 and 67%, respectively) had groundwater tables above the threshold level, versus only 19.7% of the October data (Ibrakhimov et al. 2007).

The average monthly March and July levels did not significantly fluctuate over the years (Fig. 5.3). In contrast, the groundwater table in October rose from ca.

2.3 m in 1990 to nearly 1.5 m in 1996, remained at that level until 1999 and rose to even shallower levels after the drought years 2000–2001, indicating a persisting trend to shallower groundwater tables.

Both intra-seasonal and over-seasonal (long-term) driving forces can be distinguished. The intra-seasonal variation of the groundwater table responds to leaching and irrigation strategies, because seepage and percolation from canals and during field water application are the only sources for groundwater recharge (with exception of a small area along the Amudarya, where the water level in the river influences the groundwater table). Analyses of monthly groundwater data show that the groundwater level becomes shallow after the leaching and the peak irrigation season, with only a brief delay.

The long-term behavior of the groundwater table is influenced by the annual withdrawals from the Amudarya. Deeper groundwater levels in the drought period 2000–2001 indicate reduced groundwater recharge in this period, due to lower water input to Khorezm and a tendency to use water more efficiently in drought periods (Bekchanov et al. 2010).

Winter wheat has been occupying increasing areas due to the government's policy of food independence (cf. Djanibekov et al. 2011; Rudenko et al. 2011). This entailed increased diversion and water use for the irrigation of winter wheat after the cotton harvest, which is the reason for the rising groundwater levels in October (and in general outside growing periods). As the wheat fields are spatially scattered throughout the region, losses from the irrigation network to groundwater are over-proportional during this time of the year.

The overall average groundwater salinity is 1.75 ± 0.99 g l^{-1}, falling into the FAO category of moderately saline waters (Rhoades et al. 1992). Locally, much higher groundwater salinity values were recorded, reaching as much as 14–15 g l^{-1}. At some locations, the only slightly saline groundwater of 0.5 g l^{-1} could have been re-used for irrigation through pumping, but this is currently not systematically practiced, in part due to the fact that farmers and water user organizations lack means of easily determining groundwater salinity levels.

The average values of groundwater salinity in the October, April and July measurement periods did not differ significantly. The spread of the salinity values was generally skewed towards the lower ranges, and 50–60% of the values in each period indicated moderately saline groundwater. Less than 1% of the values in each period exceeded 7 g l^{-1}, the threshold for high salinity defined by Rhoades et al. (1992), a level at which salinity starts affecting crop performance. The temporal changes of the groundwater salinity during the study period were also insignificant ($P > 0.05$), ranging from 1.6 to 2.0 g l^{-1} in all 3 measurement periods. Thus, it is not the groundwater salinity so much that damages crop performance directly, as it is the secondary soil salinity level build-up due to the salinization process that occurs with shallow groundwater levels, when water evaporates while the salts accumulate in the topsoils (cf. also Akramkhanov et al. 2011).

Spatial Distribution of Groundwater Table and Salinity

Because shallow, moderately saline groundwater tables are likely to increase soil salinization, it is important to identify those areas in the region where groundwater is particularly shallow and saline. The maps of the average April, July and October measurements of the groundwater table (Fig. 5.4) revealed areas with significantly shallower groundwater ($P = 0.05$) in the southern, western and North-western districts of the region. Striking in these maps is the similarity of the spatial patterns of the groundwater table in each measurement period: the groundwater table was shallower in the southern but deeper in the central parts of the region even in October when no, or rare irrigation events occur. Interpolation and comparison of the groundwater table maps in each individual measurement period showed that the spatial distribution of the groundwater tables was very similar to those of the averaged April, July and October measurements.

Similar to the spatial distribution pattern of the groundwater table, salinity was more pronounced in the southern and western parts of the region ($P < 0.05$, Fig. 5.4) and did not differ spatially across the measurement periods.

The maps in Fig. 5.4 show that the majority of the cropland experienced shallow groundwater tables (0.4–1.1 m, far above threshold levels) at salinity levels ranging between 1.5 and 4.6 g$\,$l^{-1}. Substantial soil salinization takes place in these areas. These observations indicate that the drainage system seems to function poorly due to: (1) existing outlet problems due to the widespread occurrence of hydraulic bottlenecks in the collectors and in the main collector and (2) active blocking of drainage outlets by farmers to increase the groundwater contribution to crop water supply in these downstream areas as a strategy to cope with water shortage. Improvements of the drainage system should start with these bottlenecks.

With groundwater tables above 1.4 m and moderately saline groundwater, approximately 65–70% of the irrigated areas are at risk of waterlogging and salinization in April and July. The position and extent of these areas remained unchanged during the whole study period. Only in July, these areas reduced to ca. 47% in the drought year 2000, down from ca. 66% in 1990 and 1994.

To assess the impact of shallow groundwater on topsoil salinization, Ibrakhimov et al. (2007) compared salt input by capillary rise and irrigation water using hydrological measurements in two case studies. Capillary rise from the shallow groundwater appeared to be the major factor contributing to topsoil salinity, since the salt input from groundwater exceeded the salt input from irrigation water by around 40%.

5.3.3.2 Up-Scaling Water Balances and Assessing Irrigation Performance

To upscale the water balance from field level to the entire Khorezm region and assess the irrigation performance, remote sensing techniques were combined with hydrological monitoring and the result was analyzed within a GIS (Conrad 2006).

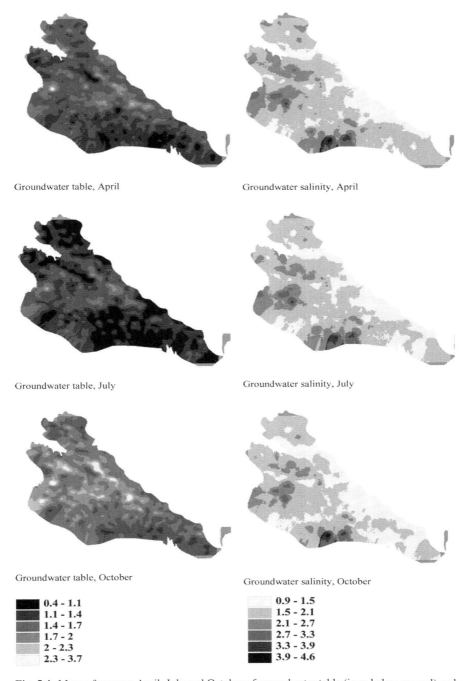

Fig. 5.4 Maps of average April, July and October of groundwater table (in m below ground) and salinity (in g l⁻¹) in Khorezm during 1990–2000, based on data from approx. 2000 observation wells (Ibrakhimov 2005)

GIS was applied to map the existing network of irrigation and drainage canals and to calculate distances between the irrigation intake points and canal entry points to the WUAs. Although actual evapotranspiration (ET) is fundamentally important for establishing water balance and assessing irrigation performance, reliable information on this parameter and especially its spatial distribution was lacking, and an algorithm for large area land use classification through remote sensing was developed (Conrad 2006). Furthermore, remote sensing data were used to compile the spatio-temporal distribution of actual ET in the whole of Khorezm with the SEBAL approach (Bastiaanssen et al. 1998). Water in- and output during the vegetation period (April 1 to September 30) of 2005 was monitored with gauges covering 82% of Khorezm's irrigation and drainage networks. Irrigation performance assessment was based on the water accounting approach (Molden 1997) and indicators suggested by Bos et al. (2005).

The water balance for the entire Khorezm region shows that 5.38 km^3 of water were directed to Khorezm for agricultural use in the vegetation period 2005. Actual ET accounted for 2.64 km^3 and precipitation was rather small (0.09 km^3). The balance could be closed by a realistic storage change of −0.13 km^3 (i.e. less water being stored in soil moisture and the aquifer at the end of the vegetation season than at its start, immediately after leaching).

Given the total irrigated area in Khorezm of 240,000 ha in 2005, the gross irrigation depth amounted to an average of 2,240 mm. Spatially averaged actual ET of cotton over the whole season was estimated to 768 mm in 2004 and 774 mm in 2005, whereas actual ET of rice amounted to 798 mm and 824 mm in 2004 and 2005, respectively. The water balance could be closed with exception of minor changes in storage. These could be explained by the variation between the emptying soil moisture and groundwater storages at the end of the vegetation period in September compared to the filled storages in April, immediately after leaching. The only minor storage change supports the conclusion that the impact of regional groundwater in- or outflow on the water balance is small.

Spatial variation in gross irrigation depths ranges between 1,750 and 2,920 mm, excluding a sub-unit of the Palvan-Gazavat sub-system which is taking 5,240 mm. Plausible reasons for this very high intake are a very large fish farm in this sub-system and high irrigation network losses in canals designed to convey water to the Dashauz region in Turkmenistan. These canals are now oversized, since irrigation water to Dashauz is supplied by a new canal constructed in Turkmenistan.

With high water availability in the vegetation period 2005, the expected tail-end problem was less severe in this year. But a comparison of actual and potential ET indicates an under-supply in WUAs located more than 85 km away from the main inlet at Amudarya (these suffered over 25% reduction in actual ET). The drainage water output in Khorezm was generally high, amounting to 55% of the water input. As the numbers refer to the vegetation period and exclude the leaching phase in spring, the drainage ratio indicates low irrigation efficiency.

The depleted fraction indicates the ratio between actual evapotranspiration and the sum of water inflow via irrigation and rainfall. Depleted fraction in Khorezm over the monitoring period was estimated at 0.48, which is below the target value of

0.6 and the critical limit of 0.5 given by Bos (2004). The measured values of the depleted fraction (0.48) and drainage ratio (0.55) add up to 1.03 and are thus plausible (for longer periods (no influence of storage change) and larger areas, the sum should approach 1).

The depleted fraction of 0.48 indicates that 52% of the water withdrawals do not contribute to meet actual evapotranspiration, thus revealing underperformance of the Khorezm irrigation system.

Considering the temporal behavior and spatial distribution of the depleted fraction indicates critical periods and problematic areas, which both need to be prioritized when designing improved management strategies. The depleted fraction in Khorezm reaches its minimum in June (0.3), one reason for which is that the preparation of rice fields in June requiring huge water inputs. Depleted fraction remains below 0.4 during the peak irrigation season in July, which indicates high water losses; these could be improved with better irrigation scheduling.

Analyzing the spatial variability of the depleted fraction allows identifying areas with severe water management deficits. For example, in the already mentioned Palvan-Gazavat sub-system, the depleted fraction throughout the main irrigation season was lower than the Khorezm-wide average and dropped below 0.25 in June and July. Thus, the irrigation performance assessment provides valuable information for water managers revealing those parts of the irrigation system and periods, which need implementation of improving measures with a high priority. Concentrating on critical periods and areas might help to increase the cost-effectiveness of interventions.

5.4 Integration of Findings Across the Scales

Before integrating the research findings at various levels to identify the problems (Sect. 5.4.2) and their roots (Sect. 5.4.3), the consistency of the results across the scales is analyzed (Sect. 5.4.1).

5.4.1 Plausibility/Consistency of Results Achieved at Different Scales

The findings on gross irrigation input at medium-scale are in line with those at regional scale. In the vegetation period 2005, the spatially averaged gross irrigation depth amounted to 2,110 mm in the 850 ha sub-unit, while the same variable amounted to 2,240 mm for the whole of Khorezm. The slightly lower number in the sub-unit is consistent with the fact that the losses in the magistral and main canals (efficiency 90–95%; Jabborov 2005) conveying the water to the sub-unit system are not included in the number of the sub-unit. The gross irrigation depths in the sub-unit in Khiva amounting to 1,930 mm in 2004 and 2,110 mm in 2005 are higher than the 1,590 mm monitored in the *Shomakhulum* WUA in 2007 (cf. Awan et al. 2011). This difference is due to the small share of rice cultivation in *Shomakhulum* (1.5% of the irrigated area) in comparison to the 10% of area cropped to rice in the sub-unit in Khiva. Furthermore, overall water availability in 2007 was lower in Khorezm, which increased pressure to irrigate more efficiently, leading to a lower gross irrigation depth. This is reflected by the overall efficiency in the *Shomakhulum* WUA in the range of 33% (cf. Awan et al. 2011).

The irrigation efficiency was not directly determined on the basis of the Khorezm-wide measurements. The efficiencies were up-scaled from the findings at medium-scale (~28%), taking into account the 90–95% efficiency of the main canals (Jabborov 2005). This allowed approximating the technical efficiency of the overall irrigation scheme in Khorezm, which was estimated by this procedure to be at 26%. Relating the actual ET to the water input in the region led to an approximation of the depleted fraction in the range of 48%.

To compare the depleted fraction with the overall irrigation efficiency, the contribution by capillary rise from the groundwater needs to be considered. The procedure of estimating the depleted fraction does not differentiate between the contribution of irrigation water to actual ET and capillary rise, whereas technical irrigation efficiency considers the soil water deficit resulting from the difference between ET and capillary rise (and rainfall). Taking into account that 30% of the actual ET in Khorezm stemmed from water supplied by capillary rise, the overall irrigation efficiency equivalent to the above-depleted fraction was estimated to be 34%. The discrepancy between this value and the 26% obtained by extrapolation from the medium-scale assessment can be explained by: (a) the common practice of water re-use at the large scale, (b) the fact that large-scale ET determined by remote sensing includes components besides agricultural crops (it includes evaporation from irrigation and drainage canals and in tendency leads to an overestimation of ETa from cropped fields, and in turn to an overestimation of the depleted fraction), and (c) the contribution of soil moisture storage in April (as the starting date of the large-scale analysis). The latter is supported by comparatively high actual ET values for this period in which most of the crops are not yet planted, resulting in an increased depleted fraction.

The drainage ratio (drainage water output related to irrigation water input) at Khorezm level was 55% in 2005, whereas the analysis of the medium-size area led to a value of 67% for the same year. The difference is consistent with the fact that the monitoring period in the latter case covered the whole year, including leaching, while the Khorezm-wide scale assessment regarded the vegetation period only and did not include the leaching period. Furthermore, the drainage water re-use practice that becomes evident at the large-scale lowered the drainage portion.

Analyses on medium- and large-scale were in agreement with the option to close the balances using realistic values of the storage change (in case of the large-scale assessment) and applying a reasonable reduction of actual ET (in case of the sub- unit).

5.4.2 Problems of Water Management

Basically, the problem of the current water management in Khorezm is conditional on discrepancies between huge water withdrawals from the resource and yet insufficient water supply at farm level, at least in various regions (cf. Bekchanov et al. 2011). Although monitoring in 2004 and 2005 revealed that annual gross water withdrawal was in the range of 2,630–2,810 mm including a 25% share for pre-seasonal leaching, in considerable parts of Khorezm and during relevant periods of crop development, the crop water requirements at farm level were not fulfilled.

The mismatch between the site-specific and time-variable requirements was often due to improper irrigation timing and inadequate amounts of irrigation water administered, leading to over- or under-supply. All this brought about a reduction of actual ET, up to 40% at single field and in downstream locations. Furthermore, the positive and expected effects from leaching were limited by the poorly functioning drainage infrastructure (for example, insufficient depth, outlet problems, etc.) and the frequently observed habit of farmers blocking drainage water collectors, as well as a non-appropriate consideration of spatial (i.e., poor knowledge by farmers of within-field soil salinity variability) and temporal (initial salt content, salt tolerance of the intended crop) leaching requirements.

The inadequate status of the irrigation infrastructure and the missing coordination of operational activities between field and irrigation network and within the irrigation network all contributed to the low overall irrigation efficiency. Results of ponding tests in canal sections (Awan 2010) showed that high losses (in the range of 5% of the operational discharge per 100 m canal length) occurred in rather small canals of low hierarchy, whereas the canals of higher hierarchy had much lower losses. This was due to the fact that the water level difference between the canals and the groundwater is small or reversed, leading to low losses and in some cases even reversed flows; i.e., when the groundwater level is above the canal water level no water losses from the canal occur. Furthermore, siltation creates nearly sealed canal bottoms.

5.4.3 Causes of the Existing Problems

Analyzing the technical dimensions of the irrigation system revealed deficiencies that were caused mainly by the inflexible irrigation scheduling, non-appropriate irrigation and drainage infrastructure, insufficient maintenance of this infrastructure, limited options for groundwater management and general lack of input data to support water management with timely and spatially detailed information. Due to the socio-technical nature of irrigation systems, technical and institutional mismanagement were closely interlinked, showing a mix of reasons for the current problems (cf. Oberkircher et al. 2011; Veldwisch et al. 2011).

A relevant example of a link between technical and institutional factors is the occurrence of operational losses in Khorezm as a consequence of insufficient coordination between irrigation activities at field/farm and network level. Even though exact quantifications of operational water losses are hard to achieve, inappropriate and careless water use very likely had a considerable impact on the losses in the network. This is mirrored in the direct linkages between irrigation and drainage. For example, the wide-spread use of pumps with fixed discharge parameters does not allow regulating the water flow, and the provided discharge typically exceeds the discharge required at field level, and a share of the discharge directly ends up in the drainage system (Awan 2010). Furthermore, although many farmers out of own initiative increasingly use pumps for irrigation to bridge water insecurity, such uncontrolled actions of individuals or smaller groups of farmers entail the danger not only of over-extraction but also sharpening the unequal use of water and land resources. Hence, technical measures for improvements in the performance of irrigation and drainage systems need to be embedded in institutional concepts and assessed against their economic effects and environmental impact.

5.5 Conclusions

Conclusions refer to the assessment of the water saving potential (Sect. 5.5.1), the drainage situation (Sect. 5.5.2) and the developing of tools to support water management (Sect. 5.5.3).

5.5.1 Water Saving Potential

To simulate the water saving potential, three scenarios were combined: proper timing of irrigation at field-level, raising application efficiency from the current 50 to 65% by advanced handling of the furrow irrigation technique (discharge control, double-side irrigation, applying surge flow), and improving network efficiency by better coordination of operational activities and by appropriate network maintenance (from current 55 to 65%). Combining these approaches would lead to a reduction in cotton gross irrigation requirements from 1,750 to 850 mm. According to the simulation findings, the achieved water saving does not change the degree of reduced evapotranspiration, which is considered an indicator for the appropriateness of water supply at field level.

This indicates the upper limit of water saving in Khorezm that can be achieved by farmers and water managers. The introduction of measures such as drip and sprinkler irrigation or the lining of canals is often suggested, but these are constrained by the required relative high initial investments (Bekchanov et al. 2010). Therefore, we did not assess these options as they are not realistic to implement in the short- or mid-term perspective. It seems better to give priority to measures making optimal use of the current system. Furthermore, although advanced technologies have a high potential to raise irrigation efficiency, they do not represent a solution to the current problem of spatial and temporal unreliability of water supply. Therefore, the introduction of new technologies needs to be embedded into those concepts that improve the reliability of water supply in the first place. This may be achieved for example by re-structuring the socio-technical system of water management or by realizing measures for de-central water storage (small reservoirs; conjunctive use of surface and groundwater).

Bekchanov et al. (2010) showed that overall efficiency can be increased considerably (compared to the current level) by using an additive combination of water saving interventions. This approach takes into account the interventions for water saving (for example: laser-guided land leveling), the range of water saving to be achieved by each intervention (saving potential in relation to current practice), and a realistic area on which the intervention could be introduced within a short time horizon (such as the size of the area to which laser-guided leveling realistically could be applied under today's conditions).

5.5.2 Drainage

Shallow groundwater, especially in July and August (1.2–1.15 m below surface), is caused by a combination of factors including the losses from the irrigation system, which recharge the groundwater, and the limited drainage network capacities, which function poorly in the periods of peak requirements of the system. Applying a cause-oriented approach to the functional components of the drainage system- which includes direct drainage, conveyance, ensuring outlet functioning -and considering the phases of interventions- i.e. operation, maintenance, re-design-, leads to various suggestions. First, major outlet problems should be eliminated, which currently are caused by insufficient depths at special cross-sections or hydraulic bottlenecks in the main collector. This could be achieved by deepening in particular those channel tracts, which have problematic cross-sections. Second, collectors and drainage ditches should be deepened where possible and feasible, and their maintenance should be intensified. Solving major and local outlet problems is the prerequisite to improve functioning of field drainage systems: more narrow spacing of field drainage ditches and eventually the introduction of tile drainage can be considered.

The approaches to improve the drainage system need to be embedded in an integrated framework. Improving the irrigation efficiencies will lower the discharge loads to the drainage system, because the percolation and seepage losses are presently the major contributors to groundwater recharge. As raising the groundwater level by blocking the drainage system is sometimes practiced by farmers to mitigate unreliable water availability in the irrigation network, the reliability of water provision in the irrigation system needs to be improved. These measures will allow farmers to discontinue the current understandable practice of drainage blocking, which would decrease soil salinization due to reduced capillary rise from the less shallow groundwater. Besides, sub-regions of Khorezm can be selected that offer reasonable conditions for conjunctive use. This approach will need to be based on information on groundwater salt content and will utilize surface runoff with low salt content from rice fields as well as overflow from irrigation to the drainage system.

5.5.3 Developing Tools for Water Management

Water management needs tools to support irrigation scheduling and drainage activities. Modeling relevant processes of the hydrological system under the influence of interventions by water management allows establishing and selecting optimal irrigation and drainage strategies. These should preferably be determined in close cooperation with water users at all relevant levels. The tools, based on flexible models, should cover the overall irrigation and drainage system and account for components and different levels, using the water requirements at farmer's level as a starting-point. This would also allow fulfilling the farmer's requirements considering the availability of resources and the capacities of the existing hydraulic networks as well as the needed reduction in impact on resources. This concept allows assisting the water user at different scales. The tools would provide stakeholders at field/farm, irrigation and drainage network level with water management information (water distribution plans; drainage activities), which are appropriate to elaborate strategies on these levels and consider links in the overall system.

The ultimate objective at the field level must aim at optimizing the water use efficiency in relation to yield and gross water input. Ensuring maximum water use efficiency can be obtained by two sets of interventions: adequate input of water to meet crop water requirements and to control soil salinity, and optimization of the application efficiency. Should the available water supply still be insufficient to meet the gross water requirement at farm level, then the irrigation strategy needs to be adapted. These adaptations must improve the application processes by advanced handling of the furrow method (intermittent flow; alternate furrow) and by introducing, for example, new application techniques, which are options to close the gap between demand and resources at farm level. Yet these sets of options need to be cross-checked with the present mainstream perception of farmers and water managers, as this allows determining also future training and educational programs.

The water needed at field level is supplied through the irrigation network. For that purpose, the discharge in the network needs to be regulated. According to the bottom-up structure of the suggested approach, the discharges in the network are based on the results at field level (irrigation time and amount) and take into account the network losses. It needs to be checked whether the current hydraulic capacity of the canals and of the hydraulic structures allows realizing the discharges needed to allow performance of irrigation at field level. The overall sum of discharge representing the gross requirement of the sub-unit needs to be realized by water allocated to Khorezm at regional level. In case of discrepancies (gross demand exceeding the available supply), strategy changes and intensified maintenance and re-design options can be derived.

In a similar way, water management at regional level aims at fulfilling the gross irrigation requirements resulting from sub-system level (WUA-level or areas fed by main canals). In case of Khorezm, the operation of the Tuyamuyun reservoir situated upstream immediately above the region (Froebrich et al. 2006) is a further option to adapt resources to requirements (at least regarding the temporal distribution). Between the regional and sub-system scales, remote sensing can be a viable tool to support strategic decision-making because of its ability to continuously gaining spatially distributed information on demand and consumption of water (land use and evapotranspiration).

At the different scales analyzed, the impact of water distribution on the groundwater level and the resulting requirements on the drainage system are considered in a linked irrigation scheduling-groundwater model. Capillary rise from the groundwater and drainage water outflow are the major outputs of the groundwater system. Groundwater level and discharge of drainage water (and salt contents) can be modelled at farm, sub-unit and unit level, taking eventual conjunctive use options into account. Also, re-design requirements will be accounted for, starting with improving the outlet situation and considering modifications of the drainage system, i.e. deepening the drains, closer spacing, and introduction of tile drainage.

References

Akramkhanov A, Kuziev R, Sommer R, Martius C, Forkutsa O, Massucati L (2011) Soils and soil ecology in Khorezm. In: Martius C, Rudenko I, Lamers JPA, Vlek PLG (eds) Cotton, water, salts and *Soums*: economic and ecological restructuring in Khorezm, Uzbekistan. Springer, Dordrecht

Alimov RA, Bedrinsev KN, Benyaminovich EM, Gavrilenko DM, Gulyamov YG, Dukhovny VA, Kadirov BN, Korjavin BD, Mershenskiy SI, Ozerskiy EI, Rachinskiy AA, Tersitskiy DK (1979) Irrigation of Uzbekistan, contemporary state and perspectives of irrigation development in the Amu Darya River Basin I. Fan, Tashkent, pp 217–286

Allen RG, Pereira LS, Raes D, Smith M (1998) Crop evapotranspiration – guidelines for computing crop water requirements. FAO Irrigation and Drainage Paper 56,300 p

Awan UK (2010) Coupling hydrological and irrigation scheduling models for the management of surface and groundwater resources in Khorezm, Uzbekistan. ZEF Ser Ecol Dev 73:155

Awan UK, Tischbein B, Kamalov P, Martius C, Hafeez M (2011) Modeling irrigation scheduling under shallow groundwater conditions as a tool for an integrated management of surface and groundwater resources. In: Martius C, Rudenko I, Lamers JPA, Vlek PLG (eds) Cotton, water, salts and *Soums*: economic and ecological restructuring in Khorezm, Uzbekistan. Springer, Dordrecht

Bastiaanssen WGM, Menenti M, Feddes RA, Holtslag AAM (1998) A remote sensing surface energy balance algorithm for land (SEBAL) 1. Formulation. J Hydrol 212–213:198–212

Bekchanov M, Lamers JPA, Martius C (2010) Pros and cons of adopting water-wise approaches in the lower reaches of the Amu Darya: a socio-economic view. Water 2:200–216

Bekchanov M, Lamers JPA, Karimov A, Müller M (2011) Estimation of spatial and temporal variability of crop water productivity with incomplete data. In: Martius C, Rudenko I, Lamers JPA, Vlek PLG (eds) Cotton, water, salts and *Soums*: economic and ecological restructuring in Khorezm, Uzbekistan. Springer, Dordrecht

5 Water Management in Khorezm: Current Situation and Options...

Bos MG (2004) Using the depleted fraction to manage the groundwater table in irrigated areas. Irrig Drain Syst 18:201–209

Bos MG, Burton MA, Molden DJ (2005) Irrigation and drainage performance assessment. Cromwell Press, Trowbridge

Conrad C (2006) Remote sensing based modeling and hydrological measurements to assess the agricultural water use in the Khorezm region (Uzbekistan). PhD dissertation, University of Würzburg, Würzburg

Conrad C, Schorcht G, Tischbein B, Davletov S, Sultonov M, Lamers JPA (2011) Agrometeorological trends of recent climate development in Khorezm and implications for crop production. In: Martius C, Rudenko I, Lamers JPA, Vlek PLG (eds) Cotton, water, salts and *Soums*: economic and ecological restructuring in Khorezm, Uzbekistan. Springer, Dordrecht

Djanibekov N, Bobojonov I, Lamers JPA (2011) Farm reform in Uzbekistan. In: Martius C, Rudenko I, Lamers JPA, Vlek PLG (eds) Cotton, water, salts and *Soums*: economic and ecological restructuring in Khorezm, Uzbekistan. Springer, Dordrecht

Dukhovny VA (1996) The problem of water resources management in Central Asia with regard to the Aral Sea situation. In: Micklin P, Williams DWH (eds) The Aral Sea Basin. Springer, Berlin/Heidelberg

FAO (2002) World agriculture: towards 2015/2030. A FAO study, Rome, p 106

Forkutsa I (2006) Modeling water and salt dynamics under irrigated cotton with shallow groundwater in the Khorezm region of Uzbekistan. ZEF Ser Ecol Dev 37:158

Forkutsa I, Sommer R, Shirokova Y, Lamers JPA, Kienzler K, Tischbein B, Martius C, Vlek LG (2009) Modeling irrigated cotton with shallow groundwater in the Aral Sea Basin of Uzbekistan: II. Soil salinity dynamics. Irrig Sci 27(4):19–30, Springer

Froebrich J, Bauer M, Ikramova M, Olsson O (2006) Water quantity and quality dynamics of the THC - Tuyamuyun Hydroengineering Complex - and implications for reservoir operation. Environ Sci Pollut Res 14(6):435–442

Ibrakhimov M (2005) Spatial and temporal dynamics of groundwater table and salinity in Khorezm (Aral Sea Basin), Uzbekistan. ZEF Ser Ecol Dev 23:175

Ibrakhimov M, Khamzina A, Forkutsa I, Paluasheva G, Lamers JPA, Tischbein B, Vlek PL, Martius C (2007) Groundwater table and salinity: spatial and temporal distribution and influence on soil salinization in Khorezm region (Uzbekistan, Aral Sea Basin). Irrig Drain Syst 21(3–4):219–236

Jabborov H (2005) Irrigation canals and collector-drainage systems in Khorezm. Internal report to the ZEF/UNESCO project

Khorst M, Ikramov R (1995) Main principals of zoning of the irrigated land in Uzbekistan on suitability for drip irrigation. In: Middle Asian Irrigation Research Institute named after Jurin VD (ed). Summary of research on drip irrigation, Tashkent, Tashkent, pp 13-24 (in Russian).

Molden DJ (1997) Accounting for water use and productivity. SWIM Paper 1. International Irrigation Management Institute, Colombo

Mukhammadiev UK (1982) Water resources use. Uzbekistan Tashkent, Uzbekistan, 87 p

Oberkircher L, Haubold A, Martius C, Buttschardt T (2011) Water patterns in the landscape of Khorezm, Uzbekistan. A GIS approach to socio-physical research. In: Martius C, Rudenko I, Lamers JPA, Vlek PLG (eds) Cotton, water, salts and *Soums*: economic and ecological restructuring in Khorezm. Springer, Dordrecht

Poluasheva G (2005) Dynamics of soil saline regime depending on irrigation technology in conditions of Khorezm oasis. In: Proceedings of the international scientific conference: science support as a factor for sustainable development of water management, Taraz/Khazakhstan, 2005

Rhoades JD, Kandiah A, Mashali AM (1992) The use of saline waters for crop production. FAO irrigation and drainage paper 48, Rome

Rudenko I, Nurmetov K, Lamers JPA (2011) State order and policy strategies in the cotton and wheat value chains. In: Martius C, Rudenko I, Lamers JPA, Vlek PLG (eds) Cotton, water, salts and *Soums*: economic and ecological restructuring in Khorezm, Uzbekistan. Springer, Dordrecht

Simunek J, Sejna M, Van Genuchten MT (1998) The HYDRUS-1D software package for simulating the one-dimensional movement of water, heat and multiple solutes in variably-saturated media, Version 2.0. US salinity laboratory, agricultural research service, US department of agriculture. Riverside California, p 177

UNESCO (2003) Water for people, water for life. The 1st United Nations World Water Development Report

Veldwisch GJ, Mollinga P, Zavgorodnyaya D, Yalcin R (2011) Politics of agricultural water management in Khorezm, Uzbekistan. In: Martius C, Rudenko I, Lamers JPA, Vlek PLG (eds) Cotton, water, salts and *Soums*: economic and ecological restructuring in Khorezm, Uzbekistan. Springer, Dordrecht

Part III
The System Setting
(Society and Institution)

Chapter 6
Farm Reform in Uzbekistan

Nodir Djanibekov, Ihtiyor Bobojonov, and John P.A. Lamers

Abstract Land-reform processes are catalyzing agricultural transition in the Commonwealth of Independent States (CIS), and this is well illustrated by the process of farm restructuring in Uzbekistan. In Khorezm, a region in the northwest of the country, this process is mirrored in the nationwide reforms: state-induced farm restructuring, state ownership of land, land reform to transfer land from collective to individual use, and continuation of area-based state targets for cotton at fixed prices. This study has the following two key objectives: (1) to explain how the land reform has changed the production structure in private farms, and (2) to describe the main changes in private farming during the reform process. The development of private farming since independence in 1991 has taken place in four phases, which are distinguished by the speed of reform, number and average size of farms, their structural specialization, and changes in the farms' cropping pattern. In the first two phases, farm restructuring led to a downsizing of producing units and to many independent farmers in coexistence with the old farming system. In the following period, the old system was completely dismantled. The final and most recent phase, imposed by national policy, has reversed this trend, and farms were increased to sizes similar to those in Soviet times; many farmers had to give back their long-term lease contract. Although it was expected that sooner or later

---

N. Djanibekov (✉) • J.P.A. Lamers
ZEF/UNESCO Khorezm Project, Khamid Olimjan str., 14, 220100 Urgench, Uzbekistan
e-mail: nodir79@gmail.com; j.lamers@uni-bonn.de

I. Bobojonov
Department of Agricultural Economics, Farm Management Group,
Humboldt-Universität zu Berlin, Philippstr. 13, Building 12 A, D-10115 Berlin, Germany
e-mail: ihtiyorb@yahoo.com

C. Martius et al. (eds.), *Cotton, Water, Salts and Soums: Economic and Ecological Restructuring in Khorezm, Uzbekistan*, DOI 10.1007/978-94-007-1963-7_6,
© Springer Science+Business Media B.V. 2012

farms would have to be larger to obtain viable and sustainable production units, it is argued that changing farm size alone, without any other supportive structural measures, will not provide sufficient incentives for reaching economically efficient farm enterprises. Along all stages of the farm restructuring process, the setup of agricultural infrastructure did not change much, or if it did change, then by state-imposed initiatives, which is constraining the development of farms and sustainable farming systems and practices.

Keywords Farm restructuring • Farm optimization • State procurement • Khorezm region • Aral Sea Basin

6.1 Introduction

For almost seven decades, decisions on agriculture in the former Soviet Union were made by the leaders of the Communist Party of the Soviet Union and by the leaders of the 15 socialistic republics in the Union (Suleimenov 2000). Following the dissolution of the Soviet Union in 1991, 12 of the former republics joined in the Commonwealth of Independent States (CIS), which is characterized by several features (Lerman 2009). For instance, people's livelihoods in the CIS depend significantly on the transformation of the agricultural sector, which was largely subsidized during the Soviet period (Rozelle and Swinnen 2009). At the onset of the transition from a centrally planned to a market-driven economy, agricultural production in these countries was based on levels and prices determined by state targets (Lerman 2009). The state owned the land and distributed it among the large collectives (*kolkhozes*) and state farms (*sovkhozes*). These were the core of the Soviet agricultural production, while subsistence production by small, rural households supplemented the national food supply (Lerman 1999). The economies of the Central Asian Countries (CAC) have distinctive features compared to the other transition countries (Bloch 2002). While being endowed with a relatively small arable land area per capita, CAC have a large share of rural inhabitants and show high demographic growth rates. Furthermore, the share of the rural population is continuously increasing, and as its income depends largely on agriculture, changes in this sector have an important impact on overall economic performance (Lerman 2009). For instance, in Uzbekistan, 64% of the population lives in rural areas, and 32% of the labor force is employed in the agricultural sector. Almost 25% of the gross domestic product of Uzbekistan is generated in the agricultural sector (Government of Uzbekistan 2007). As rangelands comprise almost 66% of the total land in the CAC (Sanginov et al. 2004), there is a general lack of arable land, and the expansion of irrigated cropland is limited due to its dependence on the decreasing amount of irrigation water from trans-boundary rivers. Cotton production dominates the agricultural production in the CAC, except for Kazakhstan (Pomfret 2007), but wheat production is on the rise.

The reforms implemented in the former Soviet Union countries, including the CAC, have been quite similar. However, there have been large differences in the

6 Farm Reform in Uzbekistan

Table 6.1 Number and average size of individual farms in Central Asia

Country	1992	1993	1994	1995	1996	1997	1998	1999
Number of farms								
Kazakhstan	3,300	9,300	16,300	22,500	30,800	42,500	51,300	58,400
Kyrgyzstan	4,100	8,600	12,800	17,300	23,200	31,000	38,700	49,300
Tajikistan	0	0	0	200	1,800	2,300	8,000	10,200
Turkmenistan	100	100	300	1,000	1,000	1,400	1,800	–
Uzbekistan	1,900	5,900	7,500	14,200	18,100	18,800	21,400	23,000
Average size of farms, ha								
Kazakhstan	238	533	406	348	412	452	542	386
Kyrgyzstan	25	44	67	43	63	48	25	20
Tajikistan	0	0	0	9	18	64	136	286
Turkmenistan	10	11	8	6	6	9	8	–
Uzbekistan	7	8	9	14	15	15	16	19

Source: Modified after Spoor and Visser (2001)

speed and manner of implementation of these reforms (Spoor and Visser 2001). In all these countries, the transition from the planned to a market economy has been driven by a different mixture of political and economic objectives (Csaki and Nucifora 2005). The discrepancies between centrally set plans and actually realizable farm output in the late 1980s led to continuous, substantial losses in agricultural production in the entire former Soviet Union (Wegren 1989). Therefore, one of the critical issues in reforming the agricultural sector in the aftermath of independence in 1991 has been the land reforms (Lerman 2009). The land reform consisted of distributing the land to individual producers in combination with farm restructuring programs whereby not only the land, but all state-owned and collectively operated production factors (except water) were distributed to the new market-oriented agricultural producers (Lerman 1999). The method of land distribution differed among countries (Spoor and Visser 2001). In most countries, except for Belarus, Uzbekistan, and Tajikistan, land was privatized, and land owners could decide whether to operate on it, sell surplus or buy additional land. In Belarus, Uzbekistan, and Tajikistan, the state retained the exclusive land ownership (Lerman 2009).

Furthermore, speed and degree of land reform differed (Spoor and Visser 2001). Different production unit sizes were considered to be optimal. By 1999, Armenia had the fastest rate of transferring arable land to individual farming businesses, and Uzbekistan and Belarus the slowest. Also the size of the created individual farms differed between countries. By 1999, in most cases the average size of individual farms had increased, except for Kyrgyzstan and Azerbaijan. Among the CAC, the largest relative increase in farm sizes was observed in Tajikistan and Uzbekistan, where land is owned by the state (Table 6.1). In some countries, the farm restructuring process was initiated to replace large production units by smaller ones (Lerman et al. 2002). This resulted in a progressive increase in the number of small farms without a comprehensive adjustment of the farm infrastructure (Spoor and Visser 2001). Hence, the farms may have become too small to be economically efficient using the agricultural infrastructure inherited from Soviet times (Spoor and Visser 2001) that was especially tailored to serving the large farms (Csaki and Nucifora 2005).

The reform processes in many former Soviet countries have been extensively described and compared (e.g. Lerman 1998; Spoor and Visser 2001; Lerman et al. 2004), but the changes in Uzbekistan have been so rapid and radical that an update is needed, particularly with regard to the agricultural sector. The reform process is illustrated in this paper using the case of the Khorezm region in northwest Uzbekistan, which we see as being representative for the irrigated lowlands along the Amudarya river, one of the two major water supply systems in the CAC. The 270,000 ha of irrigated land in Khorezm are, therefore, representative of a total of 1,060,000 ha of land between the Tuyamuyun water reservoir in the south and the Aral Sea shores in the north.

6.2 Farm Reform Process

After independence in Uzbekistan, the farm types and structures were modified through the land reform and geared towards market–orientation (Lerman 2008a). There were four critical phases of the farm restructuring process. Each phase was characterized by a different speed and level of regulations accompanying the transfer of state and collective property to private ownership. During all phases, the total cropland area in Khorezm remained the same as in 1991, and so the change in the number of private farms[1] was achieved through a change in size and number of production units. The farm restructuring process in general only affected large farms, which were on land leased by the state to the newly evolving private farms. In the 15-year study period, the total land leased to private farms increased 170 times. Rural households (*dehqon* farms) were not part of the farm restructuring process, and their land share increased only as a result of the regional population growth.

In the **first phase** of farm restructuring (1991–1998), the property rights of products produced by the agricultural production units were changed from state to collective ownership. In this phase, most *sovkhozes* were divided and transformed into *kolkhozes*[2] with the aim of reducing the government's financial responsibility for on-farm *sovkhoz* operations in favor of the self-supported *kolkhoz* system, and in turn relieved the state budget from expenditures (Bloch 2002; Guadagni et al. 2005). The procurement obligations for agricultural products were removed with the exception of cotton and grains such as rice and wheat (Pomfret 2000).

---

[1] The term 'private farm' is used in this paper, as it is widely used among scientists for this type of farm in Uzbekistan. Lerman (2008b) refers to them as 'peasant farms'.

[2] The main differences between *sovkhozes* and *kolkhozes* were the larger size of the *sovkhoz* and the source of finance, which in the case of *sovkhozes* came directly from the state budget, while *kolkhozes* were self-financed.

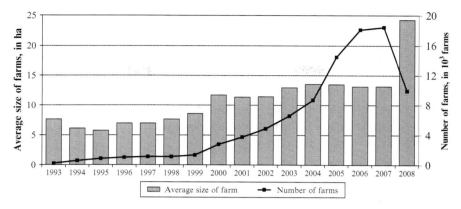

Fig. 6.1 Dynamics in establishment of private farms

The individual farming activities during this period resembled the lessor-lessee relationship between farm enterprise and an individual, as initiated under the leadership of Gorbachev. According to these arrangements, *kolkhozes* and *sovkhozes* leased abandoned and unused land to teams, cooperatives within farms, individual families, *kolkhoz* members or *sovkhoz* workers for periods of up to 50 years (Wegren 1989). However, the freedom to choose which crop to grow remained restricted, as the lessee had to grow crops that matched the *kolkhoz* or *sovkhoz* infrastructure. In this system of leased farming, the destination of the harvest was prescribed by the contract with the lessor, i.e., *sovkhoz* or *kolkhoz*, which was also the only source of farm inputs such as fertilizers, fuel (diesel) and machinery. Hence, *sovkhozes* and *kolkhozes*, but not lease farms, sold their harvest to the procurement agencies (Wegren 1989). At this stage, number and average size of private farms were relatively too small to contribute significantly to agricultural output (Fig. 6.1).

The **second phase** (1998–2003) started with the adoption of laws on agricultural cooperatives (*shirkats*) and private farms. This period is characterized by transformation of *kolkhozes* into *shirkats* (Uzbek word for shareholding) and intensification of private farming. In Khorezm for instance, all 132 *kolkhozes* (as of 1996) were involved in this transformation process: 123 were converted into *shirkats* and the remaining 9 were split into private farm unions (Kandiyoti 2002). Where *kolkhozes* were transformed into *shirkats*, the reorganization did not alter the system established during the period of collective farms. In the 'stay as is' approach of farm restructuring, *kolkhozes* were *de facto* just re-registered under the new category of *shirkats* (Lerman et al. 2002). Without a meaningful organizational restructuring, the efficiency of the *shirkats* remained low, e.g., only six *shirkats* in Khorezm made profits in 2002, while the rest suffered losses (Rudenko and Lamers 2006; Niyazmetov 2008). However, in the case of private farming, the introduction of the law on private farms in 1998 meant fundamental changes for Uzbekistan. According to this new legislation, private farms became entirely independent from *kolkhozes* and *shirkats* in their production decisions in terms of both input and output allocation (Bloch and Kutuzov 2001; Bloch 2002; Khan 2005). Three types of farms were distinguished in this legislation (Box 6.1). Farmers could purchase inputs independently and decide on cultivation plans and destination of their harvest. However, from the very beginning, real land privatization in Uzbekistan was never intended, but rather a change in size and number of agricultural producers through land-lease practices, although, in fact, the only payments for land to the state budget were imposed in the form of a unified land tax. Typically and importantly, low-quality lands were allotted to private farms, i.e., lands that had previously been operated by *shirkats* at a net loss (Khan 2005). Furthermore, private farms still had to fulfill certain state procurement quotas (Müller 2006). With poor land, small farm size and lack of accompanying reforms, these farms could hardly compete under the new harsh market environment due to high production costs and low crop yields.

Box 6.1. Typology of Private Farms in Uzbekistan

Three different types of private farms are recognized in the Uzbek legislation according to their production specialization. The largest and dominant farm type consists of **private cotton and wheat farms** specialized in the production of these two crops. These accounted for 55% of all farms in Khorezm in 2006. This type of enterprise must lease at least 10 ha of land designated by the government for the production of cotton and cereals. Within the state procurement system, these farms were heavily subsidized for producing these two strategic crops (Müller 2006). It was in particular this farm type that was subject of the farm optimization imposed in 2008, when their number was reduced while their size was substantially increased.

The second type is the **horticultural private farm**. In 2006, 30% of the private farms in Khorezm specialized in gardening, and grape and vegetable growing. According to the legislation, this type of farm must be at least one hectare in size but it is beyond the procurement system. The main crops produced by these farms are vegetables, fruits and grapes; however, horticultural farms produce other crops, e.g., wheat, potatoes, melons, and fodder crops basically for own consumption or sale at local markets. Additionally, horticultural private farms are allowed to keep a small stock of animals. These farms do not have to sign the procurement contracts for cotton and wheat.

In 2006, about 15% of the farms in Khorezm were specialized in rearing livestock including poultry. According to the legislation, the area leased to **livestock rearing farms** is directly related to the size of their animal stock in the ratio of 0.33 ha per cattle equivalent, and there should be at least 30 heads of cattle equivalents corresponding to 10 ha land. In addition to animal rearing, livestock farms can produce cash crops. Similar to the horticultural private farms, animal-rearing private farms enjoy more decision-making freedom than crop-growing private farms. At the end of the third phase of the farm reforms in 2006, there was no evident shift in the structure of farm types compared to 2005.

6 Farm Reform in Uzbekistan

This second phase can be considered as the first attempt towards an extensive downsizing of agricultural production units from large agricultural cooperatives into private farms (Bloch 2002; Khan 2005). At this stage, the number of *shirkats* in the Khorezm region reduced from 123 to 102. The state still prioritized the establishment of private farms, and as their number increased, the average size remained 7.5 ha. Nevertheless, although the land was redistributed, the *shirkats* still dominated in the resource use and total production. The share of sown area in private farms as a percentage of total area sown in Khorezm increased from 3% in 1998 to 30% in 2003 (Table 6.2).

The **third phase** (2003–2008) was characterized by an intensified fragmentation of *shirkats* into private farms. The nationwide privatization was first introduced in the non-profitable *shirkats* in 1998 and then gradually implemented in the remaining *shirkats* till 2006. This period has been defined as "decollectivization" (Swinnen and Mathijs 1997), and was characterized by accelerated expansion of private farm land and transfer of land to individuals under lease agreements for agricultural production. From this moment onwards, private farms became the core of agricultural production. On average, about 130 new private farms emerged per *shirkat*. The state preferred establishing private farms with larger areas; for instance, during this reform phase, the average size of a private farm in the Khorezm region increased from 7.5 ha in 1998 to 13 ha by the end of 2003. In 2006, about 15 ha were the preferred farm size.

In the last 2 years of this third phase, the expansion of private farmlands slowed down, as all *shirkat* lands had become redistributed to private farms. In 2007, the number of private farms in Khorezm increased by only 320 farms with an average size of 15.4 ha (Fig. 6.1). With the absence of land-lease payments, and as all land had been distributed for production, a potential newcomer to farming could only be allotted land that the state had taken from other private farms or land where farmers had returned their lease contract. By the end of this phase, about 18,000 private farms were registered in Khorezm (OblStat 2008).

Changes in cropping patterns in Khorezm initiated by the national reforms aimed at maintaining export revenues and achieving grain self-sufficiency (Djanibekov 2008). From this perspective, the regional cropping patterns in Khorezm show four major trends: (1) as the regional land is allocated primarily by the state procurement quota, the cotton area has remained the same, (2) the area with food crops, especially winter wheat, increased drastically in accordance with the declared national objectives, (3) the increase in wheat area was achieved at the expense of annual fodder crops such as fodder maize, and (4) at the expense of perennial crops, in particular alfalfa. As farm restructuring progressed, all these trends were also projected in the cropping pattern of the private farms.

In the first phase of the farm restructuring process, private farming in Khorezm had gained a more commercial orientation. For instance, the area of land for rice production by private farms comprised the largest share of cropland (Fig. 6.2), which explains the creation of a specific "rice farmer" category in the subjective classification of crop farms by Veldwisch et al. (2011). In fact, the first two phases of land reform saw the leasing out of land with low productivity to private farms,

Table 6.2 Four phases of farm restructuring process

	First phase	Second phase	Third phase	Fourth phase
Period	1992–1998	1998–2003	2003–2008	2008–present
General objective of farm restructuring	Collectivization of state owned farms	Partial decollectivization	Complete decollectivization	Farm optimization, leading to a consolidation of cotton and wheat producers
Main transformations in farm types and structures	Transformation of *sovkhozes* into *kolkhozes*	Transformation of *kolkhozes* into *shirkats*; partial transformation of *shirkats* into private farms	Complete transformation of *shirkats* into private farms	Private farms merged into larger units through reallocation of land from farms with size <30 ha to larger ones
Dominant type of agricultural producer	*Kolkhozes*, *sovkhozes* and rural households (*dehqon* farms)	*Shirkats*, private farms and *dehqon* farms	Private farms and *dehqon* farms	Private farms and *dehqon* farms
State procurement system	Cotton, wheat and most agricultural products	Cotton and wheat	Cotton and wheat	Cotton and wheat
Additionally created service agents in agriculture	No	Water users associations; machinery and tractor parks	Water users associations; machinery and tractor parks	–
Dominant form of labor management in agriculture	Production links (*zveno*) and brigades	Family contracting (*pudrats*)	Permanent and seasonal employment	Permanent and seasonal employment
Share in regional cropland and average size at beginning	0.5% 7.5 ha	3% 7.5 ha	30% 13 ha	82% 24 ha

Source: Updated from Djanibekov (2008)

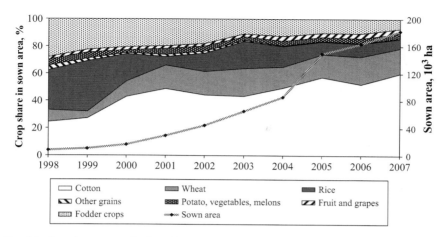

Fig. 6.2 Cropping pattern of private farms 1998–2007

while the cropland areas of higher productivity were kept for cotton *kolkhozes* and *shirkats*. When private farms later started replacing *shirkats*, they became responsible for producing cotton and wheat, and so the cropping pattern of private farms shifted towards these two crops (Djanibekov 2008). For example, in 2007 these two crops occupied almost 80% of the farms' sown area.

Agriculture in Uzbekistan is organized in a dual system of symbiotic relationship between commercial farms and subsistence (*dehqon*) farms (Djanibekov 2008). As each private farm also has a small household plot to operate along with its farm fields, these small private farms have a dual residence-business objective where the decisions of the actual farm business are interlinked with the decisions on the operation of the household plot, which have an average size of 0.2 ha in Khorezm (Lerman 2008a). In 2003, these *dehqon* farms cultivated about 17% of the total cropland, but produced almost 61% of the gross agricultural product of Khorezm, mainly due to almost the entire livestock production (Djanibekov 2008). This is interesting insofar as livestock is rarely seen as an important farm business factor in this cotton-dominated area. In addition, household farms contributed the largest share to potato, vegetables and grape production. As farm restructuring progressed, the contribution of private farms to the regional production became significant. For instance, in 2006, private farms produced almost all cotton in Khorezm. After 2006, wheat, the main staple commodity, was mostly produced on private farm fields. Furthermore, private farms produced most of the regional rice, fodder, melons and fruits. In contrast, the production levels of the livestock and poultry farms remained far behind those of rural households. In 2006, while private farms cultivated about 81% of the farmland in the region, rural households produced almost 98% of the milk and meat.

The **fourth phase** of farm restructuring started in autumn 2008, when crop-growing farms with areas of less than 30 ha were requested to return the land they were leasing from the state. In a very short period, lease contracts with these cotton- and wheat-producing farms were cancelled by the state, and the land became part of larger production units, resulting in the concentration of agricultural production in fewer, larger farms. Typically, one large farm thus replaced 5–6 smaller farms. Private farms involved in other production activities, e.g., gardening, horticulture and livestock, were not subject to this farm optimization plan.

The general layout of the farm optimization process is demonstrated here by the example of the Gurlen district of Khorezm. In this district, the optimization process affected 1,365 private farms with an average size of 15.1 ha. From this group, 1,135 farms (average size 11.5 ha) closed their business and their lands were leased to 229 large-scale farms, which further expanded from an average of 32.6–89.8 ha. The land, however, was not necessarily leased to larger farms only. A closer look at the data shows that about 5% of the initial farms smaller than 5 ha and about 30% of the newly expanded farms had initial lease contracts for less than 20 ha. Furthermore, although the majority of the closed-down farms had comprised cotton- and grain-producing enterprises, in some cases farms specializing in horticulture and livestock rearing were also included in this farm optimization process. As a result of this process, the number of cotton- and wheat-producing private farms in Khorezm decreased by about 85% to 2,400 farms, and the total number of farms decreased by 50% to about 9,800 farms. One argument guiding the optimization process was to merge smaller private farms into more solid farms without large public expenditures on infrastructure. It was expected that larger farms would more effectively produce cotton and grain with the infrastructure built to serve *sovkhozes* and *kolkhozes* during Soviet times.

The distribution of farms became more skewed with the progress of the farm restructuring reforms. For instance, farmland availability in the years 2002, 2005 and 2009, which mirrors the end of the second phase, middle of the third phase and beginning of the fourth farm restructuring phase shows that private farms were mainly comprised of two groups, (i) farms with an area of less than 5 ha, which specialized in gardening and horticulture, and (ii) large farms. With the progressing farm reforms, establishment of private farms in Khorezm with a larger area was prioritized, which in the end has led to a considerable inequality in land distribution for private farms in this region. For instance, whilst in 2005, 80% of the farms leased 45% of the farmland, after farm optimization in 2008, 83% of the farmland was leased by only 20% of the farms. The newly established private farms still consisted of scattered fields. At a first look, farm optimization had only partly led to an amalgamation of fragmented small fields into larger ones, and thus this did not solve the problem of scattered fields.

6.3 Constraints of Agricultural Production in Uzbekistan

The following sections describe the current situation of agricultural production in Uzbekistan and the various constraints under which a farmer has to operate. These include the policy on cotton production, land tenure, design of irrigation infrastructure, and lack of agricultural service organizations. The number of private farms in

6   Farm Reform in Uzbekistan   105

Khorezm increased largely due to the mandatory disbandment of *shirkats* and not so much through improvement of the infrastructure for private entrepreneurship in the agricultural sector. The development of farming activities and farm incomes is thus bound by several constraints inherited from the previous set up of agricultural production (Niyazmetov 2008).

6.3.1   Cotton Production Policy

The specificity of the cotton procurement policy has affected private farm development in Uzbekistan. Many previous reports (e.g. Pomfret 2000; Guadagni et al. 2005) underline that the present cotton procurement policy in Uzbekistan is quantity based. However it is, in fact, area based. The policy is predominantly export oriented and aims at generating national export revenues (Guadagni et al. 2005). In order to achieve these policy objectives, the state defines and fixes the size and location of fields on which farmers have to grow cotton. The system affects farmers' production decisions and leaves 40–50% of the land for other cropping activities (Djanibekov 2008).

The government procures the cotton harvest from the farms at a state procurement price below export parity under market conditions (Rudenko 2008). To support cotton (and winter wheat) production, the state provides farmers with selective benefits in the form of subsidies for fertilizers, fuel and machinery services (Bobojonov 2009). Furthermore, subsidies are also provided in the form of price differentials for inputs and cotton byproducts, arbitrary allocation and reallocation of financial resources, maintenance and operating costs of irrigation system, and as credit postponements, tax remissions and preferable credits (Djanibekov 2008). In spite of these arrangements, if the cotton yield is low, the procurement prices for cotton may lead to farm losses (Djanibekov 2008).

In order to increase the incentives for farmers in cotton production, the government narrowed the difference between the world market and virtual[3] farm-gate prices for cotton fiber by increasing the farm-gate prices for raw cotton. If the procurement prices are based on the official exchange rate of US$ to the UZS, in 1998, the world market price for cotton fiber was 67% higher than the virtual farm-gate price. Whilst between 1998 and 2007 the world market price for cotton fiber dropped, the virtual farm-gate price increased 1.15 times. As a result, in 2007 the gap between the virtual farm-gate and the world market price dropped by 17%.

With the removal of input subsidies, input prices during the same period tended to grow faster than the farm-gate price for raw cotton. Between 1998 and 2007, for instance, the price of nitrogen fertilizer, in terms of the US$ exchange rate

---

[3]In Uzbekistan, according to the procurement system, farmers are paid for raw cotton, while the cotton is exported as fiber. On the world market, there is in fact no price for raw cotton but only for cotton fiber, and it depends on quality.

increased 1.49 times and fuel prices increased 3.25 times. In the same period, output prices for other commodities grew fast. As a result, in Khorezm farmers responded by diverting inputs foreseen for cotton to other crops with higher market value (Veldwisch 2008), e.g., rice and vegetables. The state soon became aware of these practices and, to ensure that farmers allocated sufficient inputs and labor to fulfilling procurement quotas on cotton and wheat, started to regulate farm production of commercial crops. The state restricted rice production through direct measures such as control of the farmers' cropping pattern, and indirectly by the reallocation of timing and amount of irrigation water (Veldwisch 2008). As cotton cultivation is strategic, the water allocation for this crop is prioritized, which negatively affects the yields of other crops (Veldwisch 2008). As a result, the mechanism of producing cotton also defines the current organization of infrastructure in which farms can quasi-privately select the allocation of all production inputs except land and water.

6.3.2 Land Tenure

The exclusive ownership of land by the state is another obstacle for developing private farms. The land-lease right of farmers in Uzbekistan is limited to nontransferable, usufruct rights. The users are not allowed to sell, mortgage, or exchange the land (Lerman 2008a). Only *dehqon* farms, or rural households, are granted the land for lifetime use and have inheritable land rights. Any change in the area under cotton can be a reason for forfeiture of a farm's land lease. Furthermore, the state as the single landowner can expropriate land from a farmer if necessary. This makes private farming only quasi-private, as the main productive asset is not owned privately (Lerman 2008b). Land tenure as motivation and incentive for long-term investments that preserve the land has been controversially debated in literature. On the one hand, some studies on land tenure underline that due to an insecurity of tenancy rights, tenants are reluctant to invest in soil quality improvements (Besley 1995), e.g., soil conservation practices, and particularly to embark on long-term investments, e.g., tree windbreaks (Kerkhof 1990). Hence, it was argued that prior to the implementation of large-scale soil improvement activities, land rights should be secured to farmers (Jacoby et al. 2002). On the other hand, research in four countries of West Africa and Southeast Asia showed that the tenure insecurity does not necessarily reduce investments in land quality, and that it can even favor long-term investments if they extend the long-term tenure security (Neef 2001). Next to land ownership, numerous other factors such as low social status of engagement in agricultural activities, low returns to invested labour, and more profitable off-farm activities demonstrated to be of equal importance when analysing the motivation for investments. Whereas there are many studies in other countries, this controversial aspect remains insufficiently clear in transition countries such as Uzbekistan.

As farm restructuring progressed, fields (parcels) of large farms were leased out to private farms. This process did not take into account the general idea of an economically efficient farm as one production unit, where parcels should be located

6 Farm Reform in Uzbekistan

next to each other and form a single territory of a farm unit. As a result, private farmland in Uzbekistan is scattered over different fields, sometimes at quite long distances from each other (e.g., in 2004 some fields were about 10 km apart; Djanibekov, unpublished observations). In the current absence of land markets, it is difficult for private farms to increase production efficiency by adjusting and improving their land holdings without approval from the state.

The activities of livestock rearing farms are constrained in a different way. The land these farms are entitled to is insufficient for fodder production, and as arable land cannot be declared as pastures, livestock farms face a notorious lack of feed/ pastures and have to rely mostly on costly fodder in form of byproducts of grain and cotton (Djanibekov 2008). The same applies to the extension of tree plantations. The state legislation prohibits the change in land use from crop growing to gardening (Kan et al. 2008), which is a major factor constraining the introduction of agroforestry in Khorezm.

6.3.3 Access to Service Organizations

Another constraint resulting from the state procurement policy is that distribution and sales outlets for mineral fertilizers and fuel are often far from the farmers' fields. Having access to efficient agricultural service organizations in the rural areas is another important factor for the economic efficiency of many private farms in Khorezm (cf. Niyazmetov et al. 2011). While there were around 19,000 private farms in Khorezm in 2007, the number of servicing organizations was far too small to provide adequate services (Table 6.3). The larger the farm, the more machinery-dependent the production becomes. Since most agricultural equipment is old and unreliable for timely operations, vast investments are needed to entirely renew the machinery parks as well as long-term credits for farms to purchase new machinery.

The input supply channels are represented by the state agencies, which sell their commodities at state-determined prices and in quantities to cover mainly the input demand for cotton and wheat production. Almost no official input markets are available for other crops. Due to the absence of formal supplementary input markets,

Table 6.3 Availability of agricultural service organizations in Khorezm

Agricultural service organization	Number of service organizations		Growth in number, %	Number of farms served in 2007
	2006	2007		
Machine and tractor stations	130	127	−2	150
Fertilizer and selling outlets	72	85	18	224
Fuel selling outlets	105	109	4	174
Extension offices	46	35	−24	543

Source: Niyazmetov (2008)

the inputs necessary for the production of crops other than cotton and wheat can be obtained through parallel markets (Djanibekov 2008). The demand for informal input markets is primarily created by farms producing crops beyond the state procurement system, which usually suffer from a lack of such inputs at crucial moments in the production season.

The general lack of cash in the rural area is generated by a reduced access to credit and cash flow. The farms can only obtain cash through sale of their harvest at the market. However, the marketing infrastructure for agricultural commodities is chronically underdeveloped in entire Uzbekistan, including in the case study region Khorezm (Niyazmetov 2008). For instance, most farmers in Khorezm reported that they were not able to produce vegetables due to a lack of marketing outlets and processing capacities (Djanibekov 2008). Only a few farmers who had special contracts and ties with traders or processors coming to their fields at harvest could reap profits from these free markets.

6.3.4 Infrastructure Design

As was the case with the institutional infrastructure, the technical agricultural infrastructure from Soviet times was not affected by the land reform, but rather farm production costs increased. Agricultural production in the region entirely depends on irrigation water. Thus, the largest problem is the irrigation infrastructure and management system, which was originally designed during Soviet times to serve large farm units such as *kolkhozes* and *sovkhozes*. As research in China showed, an over fragmentation of fields hampered the functioning of irrigation and drainage systems and aggravated the impact of natural disasters (Hu 1997). Following the downsizing of production units, many activities aimed at adjusting the available irrigation system to serve the numerous and scattered small farms. However, it is recurrently argued that not only should the technical challenges be addressed but also the social challenges in the water distribution process need to be dealt with to ensure efficiency under the radical changes, e.g., the planned delivery of water to the tail-end users of the irrigation system (Abdullaev et al. 2008). Similarly, the use of farm or rented machinery on such scattered small plots is costly for a farm, since it generates additional transportation costs (Dixon et al. 2001).

6.4 Discussion

The gradual farm restructuring process in Uzbekistan consisted of the liquidation of *sovkhozes*, transformation of *kolkhozes* into *shirkats*, and then a fragmentation of the latter into many private farms. The institutional changes that constituted farm restructuring in Uzbekistan were mainly the transfer of land to individuals, which led to the establishment of a new form of farm enterprises. These new private farms remained state dependent on inputs such as fuel and fertilizer and, most importantly, water, and had to meet state-determined production quotas. With this dependency and the absence of a suitable farm-serving infrastructure, the economic efficiency of private farmers remains low. Furthermore, as the state owns the land, it has the power to change number, size and structure of land users when necessary. Before 2008, such adjustments were directed towards the transfer of collective land to private land through lease agreements and not in the reverse direction.

It is commonly argued that land should be distributed to the rural population to help the people to escape from poverty and that it should not be concentrated in large farms (Lerman 2005). However, the farm-serving infrastructure is outdated and can be an obstacle for gaining economic efficiency of small farms (Dixon et al. 2001; Niyazmetov et al. 2011). Additionally, in Uzbekistan, the problem of poorly developed farm infrastructure (Niyazmetov 2008) is exacerbated by the persistent state procurement system (Rudenko 2008), which also predetermined the current trend of land reform. Along all the stages of farm restructuring, private farms have to follow production boundaries assigned by the state, which constrains farm production and profit potential (Djanibekov 2008), but rather delivers export revenues to the state budget (Müller 2006).

As the private farm units became the dominant form of agricultural production, how they were perceived by the state changed. In phase three, when all large state and collective farms were transferred into numerous private farms, it became evident that due to the setup of infrastructure, the stability of cotton production was undermined by a vast number of private farms. In this situation, the state had to cope with the dilemma of how to promote private farming in the current infrastructure without limiting the export revenues from cotton. In this context, an increase in farm size, i.e., land consolidation, can contribute to an increase in farm productivity by creating farms with fewer parcels that are larger and better shaped and thus allow introducing better farming techniques (FAO 2003). If private ownership for land were to exist, farms would adjust to the production environment by exchanging, renting or purchasing land (Dixon et al. 2001). Since land is in state ownership, the modification or optimization of the farm size and number have been initiated by the government. Hence, at the end of 2008, farm sizes were optimized through a process of confiscating and merging farmlands into larger farm units.

Although the economies of scale allow more agricultural products to be produced at lower per-unit costs and due to increases in productivity, changing the farm size alone without improving the farm infrastructure is not sufficient to provide incentives for increasing the economic efficiency of producer units. The process of farm optimization and land consolidation at the end of 2008 also revealed the insecurity of the land tenure system and sent wrong signals to farmers with respect to investments in land improvements. The development of the entire farm-serving infrastructure is required. Relaxing an individual factor constraining farm activities will result in improved production, but the effects will be limited by other obstacles. For instance, the removal of the current prohibition on subleasing of land would allow more efficient producers to lease additional land from the less efficient ones. As long as farmlands are mostly allocated to cotton cultivation under low procurement prices, lease of extra land will be limited. In the same manner, the development of the food-processing sector would promote production of fruits and vegetables to the level that is limited by the rules implied by the state procurement policy and water-distributing infrastructure. The modification or abolishment of the state procurement for cotton can be an option to improve the economic efficiency of the farmers. If the farmers are released from the state procurement targets, the gross agricultural income will increase, but it will also lead a to higher demand for water as more farmers will likely go for the most profitable and water-intensive cropping activity, i.e., rice cultivation (Djanibekov 2008), unless incentives for crop diversification are introduced (cf. Bobojonov et al. 2011). To avoid such deflections from sustainable farm development, the state needs to pay more attention to adjusting the farm infrastructure in a way that if, for instance, the state procurement system continues, profits generated from other activities are high enough to overcome farm losses due to the endured obstacles. In case of a persisting cotton policy, the new expanded farms may not gain an increase in productivity and will continue to fail to meet the state target.

6.5 Conclusions

Land consolidation is particularly important for Uzbekistan and other former Soviet Union countries, where the infrastructure was created to serve large-scale farms. In almost all these countries, the farm restructuring process downsized agricultural producers. The main drawback of the approach of producer downsizing was that the infrastructure previously maintained by and serving only a few large farms had deteriorated and was not adjusted to serve the many newly emerged land-leasing farms. Hence, the observed return in 2008 to the previous production structure where land is concentrated in a smaller number of larger crop-growing farms seems viable under the current infrastructure setup. A package of policies needs to be promoted to increase agricultural productivity and efficiency of the farms. The initiated land consolidation will demonstrate in the next few years whether a return to larger farms is feasible and what adjustments in infrastructure would be required to turn them into sustainable production units.

References

Abdullaev I, Nurmetova F, Abdullaeva F, Lamers JPA (2008) Socio-technical aspects of water management in Uzbekistan: emerging water governance issues at the grass root level. In: Rahaman M, Varis O (eds) Central Asian water. Water & development publications. Helsinki University of Technology, Helsinki, pp 89–103

Besley T (1995) Property rights and investment incentives: theory and evidence from Ghana. J Polit Econ 103(5):903–937

Bloch P (2002) Agrarian reform in Uzbekistan and other Central Asian countries. Working paper no. 49, Land Tenure Center, University of Wisconsin, Madison

Bloch P, Kutuzov A (2001) Rural factor market issues in the context of agrarian reform. Land Tenure Center, BASIS project, Statistical Compendium

Bobojonov I (2009) Modeling crop and water allocation under uncertainty in irrigated agriculture: a case study on the Khorezm region, Uzbekistan. PhD dissertation, Bonn University, Bonn

Bobojonov I, Lamers JPA, Djanibekov N, Ibragimov N, Begdullaeva T, Ergashev A, Kienzler K, Eshchanov R, Rakhimov A, Ruzimov J, Martius C (2011) Crop diversification in support of sustainable agriculture in Khorezm. In: Martius C, Rudenko I, Lamers JPA, Vlek PLG (eds) Cotton, water, salts and *Soums*: economic and ecological restructuring in Khorezm, Uzbekistan. Springer, Dordrecht

Csaki C, Nucifora A (2005) Ten years of transition in the agricultural sector: analysis and lessons from Eastern Europe and the former soviet union. In: Holt M, Chavas JP (eds) Essays in honor of Stanley R. Johnson. Berkeley Electronic Press, Berkeley

Dixon J, Gulliver A, Gibbon D (2001) Farming systems and poverty: improving farmers' livelihoods in a changing world. FAO and World Bank, Rome and Washington, DC, p 412

Djanibekov N (2008) A micro-economic analysis of farm restructuring in Khorezm region, Uzbekistan. PhD dissertation, Bonn University, Bonn

FAO (2003) The design of land consolidation pilot projects in Central and Eastern Europe. FAO Land Tenure Stud 6:55

Government of Uzbekistan (2007) Welfare improvement strategy of Uzbekistan. Government of Uzbekistan, Tashkent, p 141

Guadagni M, Raiser M, Crole-Rees A, Khidirov D (2005) Cotton taxation in Uzbekistan: opportunities for reform. ECSSD working paper no 41, Europe and Central Asia Region, World Bank, p 23

Hu W (1997) Household land tenure reform in China: its impact on farming land use and agroenvironment. Land Use Policy 14(3):175–186

Jacoby H, Li G, Rozelle S (2002) Hazards of expropriation: tenure insecurity and investment in rural China. Am Econ Rev 92(5):1420–1447

Kan E, Lamers JPA, Eschanov R, Khamzina A (2008) Small-scale farmers' perceptions and knowledge of tree intercropping systems in the Khorezm region of Uzbekistan. For Trees Livelihoods 18:355–372

Kandiyoti D (2002) Agrarian reform, gender and land rights in Uzbekistan. Social policy and development paper no. 11, United Nations Research Institute for Social Development, Geneva, p 89

Kerkhof P (1990) Agroforestry in Africa: a survey of project experience. Panos Institute, London, p 216

Khan AR (2005) Land system, agriculture and poverty in Uzbekistan. University of California, Riverside, p 28

Lerman Z (1998) Does land reform matter? Some experiences from the former Soviet Union. Eur Rev Agric Econ 25(3):307–330

Lerman Z (1999) Land reform and farm restructuring: what has been accomplished to date? Am Econ Rev 89(2):271–275

Lerman Z (2005) The impact of land reform on rural household incomes in Transcaucasia and Central Asia. Discussion paper no. 9.05. Department of Agricultural Economics and Management, Hebrew University of Jerusalem, Los Angeles, p 14

Lerman Z (2008a) Agricultural development in Central Asia: a survey of Uzbekistan, 2007–2008. Eurasian Geogr Econ 49(4):481–505

Lerman Z (2008b) Agricultural development in Uzbekistan: the effect of ongoing reforms. Discussion paper no. 7.08. Department of Agricultural Economics and Management, Hebrew University of Jerusalem, Los Angeles, p 29

Lerman Z (2009) Land reform, farm structure, and agricultural performance in CIS countries. China Econ Rev 20:316–326

Lerman Z, Csaki C, Feder G (2002) Land policies and evolving farm structures in Transition countries. Policy research working paper 2794. World Bank, Washington, DC, p 176

Lerman Z, Csaki C, Feder G (2004) Agriculture in transition, land policies and evolving farm structures in post-soviet countries. Lexington Books, Lanham, p 254

Müller M (2006) A general equilibrium approach to modeling water and land use reforms in Uzbekistan. PhD dissertation, Rheinische Friedrich-Wilhelms-Universität Bonn, Bonn

Neef A (2001) Land tenure and soil conservation practices-evidence from West Africa and Southeast Asia. In: Stott DE, Mohtar RH, Steinhardt GC (eds) Sustaining the global farm. Purdue University and the USDA-ARS National Soil Erosion Research Laboratory, Washington, DC, pp 125–130

Niyazmetov D (2008) Efficiency of the market Infrastructure development for farmers of the Khorezm region. MSc dissertation, Urgench State University, Urgench

Niyazmetov D, Rudenko I, Lamers JPA (2011) Mapping and analyzing service provision for supporting agricultural production in Khorezm, Uzbekistan. In: Martius C, Rudenko I, Lamers JPA, Vlek PLG (eds) Cotton, water, salts and *Soums*: economic and ecological restructuring in Khorezm, Uzbekistan. Springer, Dordrecht

OblStat (2008) Agricultural producers in Khorezm, Urgench

Pomfret R (2000) Agrarian reform in Uzbekistan: why has the chinese model failed to deliver? Econ Dev Cult Chang 48(2):269–284

Pomfret R (2007) Distortions to agricultural incentives in Tajikistan, Turkmenistan and Uzbekistan. Agricultural distortions working paper 05, World Bank, Washington, DC, p 52

Rozelle S, Swinnen J (2009) Political economy of agricultural distortions in transition countries of Asia and Europe. Agricultural distortions working paper 82, World Bank, Washington, DC, p 26

Rudenko I (2008) Value chains for rural and regional development: the case of cotton, wheat, fruit, and vegetable value chains in the lower reaches of the Amu Darya River, Uzbekistan. PhD dissertation, Hannover University, Hannover

Rudenko I, Lamers J (2006) The comparative advantages of the present and future payment structure for agricultural producers in Uzbekistan. Central Asian J Manage Econ Soc 5(1–2):106–125

Sanginov S, Nurmatov S, Sattorov J, Jumabekov E (2004) Livestock production in Central Asia: constraints and research opportunities. In: Ryan J, Vlek P, Paroda R (eds) Agriculture in Central Asia: research for development. ICARDA, Halab, pp 116–125

Spoor M, Visser O (2001) The state of agrarian reform in the former Soviet Union. Europe-Asia Stud 53(6):885–901

Suleimenov M (2000) Trends in feed and livestock production during the transition period in three Central Asian countries. In: Babu S, Tashmatov A (eds) Food policy reforms in Central Asia: setting the research priorities. IFPRI, Washington, DC, pp 91–104

Swinnen J, Mathijs E (1997) Agricultural privatization, land reform and farm restructuring in Central and Eastern Europe: a comparative analysis. In: Swinnen J, Buckwell A, Mathijs E (eds) Agricultural privatization, Land Reform and Farm Restructuring in Central and Eastern Europe. Ashgate, Aldershot/Brookfield

Veldwisch GJ (2008) Cotton, Rice & Water: transformation of Agrarian relations, irrigation technology and water distribution in Khorezm, Uzbekistan. Dissertation, Bonn University, Bonn

Veldwisch GJ, Mollinga P, Zavgorodnyaya D, Yalcin R (2011) Politics of agricultural water management in Khorezm, Uzbekistan. In: Martius C, Rudenko I, Lamers JPA, Vlek PLG (eds) Cotton, water, salts and *Soums*: economic and ecological restructuring in Khorezm, Uzbekistan. Springer, Dordrecht

Wegren S (1989) The half reform: Soviet agricultural policy and private agriculture, 1976–1989. Dissertation, Columbia University, New York

Chapter 7
Mapping and Analyzing Service Provision for Supporting Agricultural Production in Khorezm, Uzbekistan

Davron Niyazmetov, Inna Rudenko, and John P.A. Lamers

Abstract Following independence in 1991, the need for a more efficient and stable agricultural production in Uzbekistan led to the institutional transformation of the main agricultural producers, the collectively owned *shirkats*, to privately owned farms. This transformation, which was completed between 2004 and 2006, *inter alia* required corresponding changes in the agricultural service-providing organizations (ASP). The efficiency and functioning of such organizations in the Khorezm region of Uzbekistan is the main research topic of this study. Data were collected through interviews with farmers and representatives of three types of ASP: machinery and tractor parks (MTP – *машино-тракторный парк* (Russian) – since Soviet times, the fleet of tractors and machines is called machine-tractor park), distribution outlets for selling fuel and mineral fertilizers, and mini-banks. Farmers were asked to give their qualitative assessment of the service provision and functioning of these service organizations. The ASP representatives provided both a qualitative and a quantitative assessment of their interaction with farmers. The financial performance of the ASP was analyzed to assess their ability to render full and timely services to farmers. Research results indicate serious flaws in the examined current service provision to farmers. Most problems can be grouped around (i) poor enforcement of property rights, because in particular local authorities frequently interfered with the internal decision-making process of both ASPs and farmers, thus affecting their activities; (ii) poor corporate management because the elected heads of the ASPs needed approval of their nomination from local authorities, which led to their direct accountability to local authorities irrespective of the business interests of the ASPs;

D. Niyazmetov (✉)
Western Illinois University, Macomb, IL, USA
e-mail: dniyazmetov@gmail.com

I. Rudenko • J.P.A. Lamers
ZEF/UNESCO Khorezm Project, Khamid Olimjan str., 14,
220100 Urgench, Uzbekistan
e-mail: inna@zef.uznet.net; j.lamers@uni-bonn.de

C. Martius et al. (eds.), *Cotton, Water, Salts and Soums: Economic and Ecological Restructuring in Khorezm, Uzbekistan*, DOI 10.1007/978-94-007-1963-7_7,
© Springer Science+Business Media B.V. 2012

(iii) weak contractual relations owing to violations of the contractual terms by both ASPs (caused by for instance the lack of machinery or other inputs of ASPs) and farmer/clients (caused by the lack of working capital). This, in turn, caused delays in service provision to farmers and untimely payments for the services rendered. The services were frequently not completed. Moreover, the results of the analyses show that commercial banks and their mini-bank branches in the villages constantly faced cash and liquidity problems, whilst farmers did not have direct access to their bank accounts. It is therefore argued that local authorities should attempt to interfere less with the activities of the ASPs and private farms.

Keywords Aral Sea Basin • Agricultural service-providing organizations • Uzbekistan • Mini-banks • Distribution outlets • Property rights • Informal rules

7.1 Introduction

The transition from a command- to a market-oriented economy, which is an ongoing process in all former USSR countries (Spoor 1999), has imputed institutional reforms at all levels and sectors including the agricultural sector. In countries with a high share of rural population, agricultural reforms took place based on the necessity to find a compromise between increasing production efficiency and increasing rural employment. Since its independence in 1991, Uzbekistan has been exploring options to establish an efficient model for its rural production.

Between 1998 and 2004, agricultural production in Uzbekistan was mandated to *shirkats*. These agricultural cooperative enterprises were established after independence as the production units succeeding the *kolkhozes* and *sovkhozes* (cf. Djanibekov et al. 2011). However, many of the *shirkats* inherited not only the corresponding infrastructure, but also the usually poor performance of their predecessors – the *kolkhozes* and *sovkhozes* – and thus continued to underperform. For example, in the Khorezm province in northwest Uzbekistan, the number of loss-making *shirkats* amounted to 62% in 2001, 96% in 2002 and even 98% in 2003 (CER 2004). This was the main motivation for further change and reforms (cf. Djanibekov et al. 2011). Consequently, between 2004 and 2006, and given the legal basis, all *shirkats* were transformed into private farms.

The replacement of *shirkats* by numerous private farms (about 50–80 farms instead of one *shirkat*) also required a fundamental change in the provision of agricultural services, since the organization and infrastructure needed for a large number of small private farm units substantially differs from that needed for a small number of large *kolkhozes* and *sovkhozes*. This refers to, for example, outlets for seed and fertilizer provision, MTPs and repair shops, soil and plant laboratories, veterinary services, food safety and phyto-sanitary controls, and credit opportunities, etc. Hence, the nationwide agricultural reforms aimed to establish the necessary production, banking, insurance, consulting, trade, veterinary and other services to farmers, and to improve farmers' access to markets, processors and customers (CER 2004). Table 7.1 gives an overview of the ASPs in Khorezm in 2006, i.e., at the end

7 Mapping and Analyzing Service Provision for Supporting Agricultural...

Table 7.1 Overview of agricultural service-providing organizations (ASP) in Khorezm in 2006

Type of ASP	Provided services to farmers
Machine-Tractor Park (MTP), Alternative MTP (AMTP)	Agricultural mechanization services. Agricultural equipment can be rented for farm and field activities such as tillage, land leveling, harrowing, mowing, harvesting, etc.
KhorezmQishloqXo'jalikKimyo, outlet for distribution of mineral fertilizers	Sale of mineral fertilizers
Urgench Neftebaza: outlet for distribution of fuel and lubricants	Sale of fuel and lubricants
Commercial banks (bank affiliations and mini-banks)	Credit provision, money transfer, bank payments
Water Users Associations (WUA)	Irrigation water provision to its members
Network for sale of seeds, chemical protectants	Provision with seeds, chemicals and pesticides
Outlet for the sale of pedigree cattle and veterinary services	Provision of cattle and veterinary services
Network for the procurement of agricultural production	Storage and procurement of agricultural production
Network for information provision and consulting	Business-training, consulting services
Commodity Exchange – *birja*	Marketing of goods such as fuel, mineral fertilizers, etc. for the cases of insufficient input provision by other networks
Insurance company	Crop, leasing insurance and other insurance services

of the *shirkat* era. This was the initial transformation stage of the ASPs, which had been designed to support the large-scale producers only, e.g., *shirkats* and their predecessors.

The favorable environment for the establishment of the private and partly state-controlled ASPs was initiated by the national authorities, whilst the process itself was closely administrated and guided by the local authorities (*Hokimiat*).

The number of ASPs decreased between 2007 and 2008 owing to the low demand of the producers for these services (networks for the procurement of agricultural production by 47%; networks for farmer training and consulting by 24%; Table 7.2). Some ASPs – such as MTPs, outlets for fuel and mineral fertilizers, and banks – served all farmers regardless of their specialization. Other ASPs, such as veterinarians, mainly supported specialized farmers, i.e., animal husbandry farms. Several types of ASP are described and analyzed in more detail in the following sections.

7.2 Machine-Tractor Parks (MTP)

The former *shirkats* had inherited the agricultural implements from the *sovkhozes and kolkhozes*, and thus were sufficiently equipped for agricultural activities in large areas. When *shirkats* were dismantled, their entire machinery assets were excluded

Table 7.2 Number of agricultural service-providing organizations (ASP) over time in Khorezm

Type	2007	2008	Growth between 2007 and 2008	Number of Farms per ASP
Farms	18,120	19,014	894	n.a.
MTP	11	11	0	1,728
AMTP	119	116	−3	164
Outlets for selling mineral fertilizers	72	85	13	224
Outlets for selling fuel and lubricants	105	109	4	174
Mini-banks	119	99	−20	192
WUAs	115	112	−3	170
Outlet for the sale of pedigree cattle and veterinary services	128	147	19	129
Network for the procurement of agricultural production	32	17	−15	1,118
Network for information provision and consulting	46	35	−11	543

Source: Regional Department of Agriculture and Water Resources Management (*OblVodkhoz*)
n.a. not applicable

from privatization and were instead handed over to MTPs with the aim of rendering fee-based services such as plowing, tillage and crop harvest operations to the newly established private farmers on a contractual basis.

Until 2009, MTPs existed under two ownership types, i.e., either as joint-stock companies (MTP/JSC) or as alternative MTPs (AMTP) in the collective property of farmers. Both types operated according to their own charter and were considered a private entity with its own bank account, official seal, and administration. The MTP/JSCs had a state share, ranging from 4% to 35%, in their charter capital. Furthermore, all MTP/JSC were exempt from the payment of value-added tax until 2007, but the AMTPs had no government support or subsidies. A MTP/JSC was headed by a chairperson, who was elected for 1 year by the board of shareholders and who could be re-elected. The AMTP heads were formally elected by their founders (farmers), but were informally "appointed" by the local *Hokimiat*.

A region-wide inventory following the establishment of MTPs in 2006 revealed the low quality of the entire fleet of agricultural machinery and equipment that the new owners had inherited. Until today, this situation has changed little, since only a few MTP/AMTP managed to purchase new, modern and efficient equipment. The low quality of the equipment of dismantled *shirkats* partly explains the 23% decrease in the stock of agricultural equipment of the AMTPs and the 8% decrease of the MTP/JSCs between 2006 and 2008 (Table 7.3).

However, the commonly practiced liquidation of obsolete equipment is not the only reason for this decline. In numerous cases, a sale out of functional machinery was needed to obtain the cash necessary for paying off the accumulated debts inherited from the former *shirkats*. This in turn explains the 63% growth over the same period

Table 7.3 Change in volume of agricultural machinery in the Khorezm region 2006–2008

Machinery owners	2006			2007			2008		
	Total	Including Tractors*	Combine harvester	Total	Including Tractors*	Combine harvester	Total	Including Tractors*	Combine harvester
MTP/JSC	338	248	90	302	227	75	312	241	71
Alternative MTP	4,771	4,589	182	3,800	3,673	127	3,680	3,558	122
Farmers	1,896	1,824	72	2,752	2,649	103	3,094	2,973	121
Others	6,987	6,732	255	8,099	7,761	338	8,355	7,986	369
Total	13,992	13,393	599	14,953	14,310	643	15,441	14,758	683

Source: Technical Supervision Department, own estimations

Tractors* – including tractors for transportation

Table 7.4 Tillage services provided by agricultural service providing organizations (ASP) Khorezm 2006–2007

| | 2006 | | 2007 | | Change in |
ASP	Area (ha)	Share (%)	Area (ha)	Share (%)	shares (%)
MTP/JSC	28,052.1	13.8	32,990.1	16.2	18
AMTP	43,897.7	21.6	17,179	8.4	−61
Others (private)	131,646.2	64.6	153,426.9	75.4	17
Total	203,596	100	203,596	100	

Source: Khorezm regional department on demonopolization, support of competition and entrepreneurship

in the privately owned agricultural equipment and machinery (Table 7.3). Overall, this underlines that various farmers had been able to generate sufficient farm capital for investments. In addition, it indicates that farmers wished to become independent from the unreliable MTPs and AMTPs and thus bypass the frequent bottlenecks in equipment availability during the labor peaks, e.g., during planting and harvesting (Müller 2006).

In the period 2006–2007, there was a drastic decrease (61%) in the total amount of services officially provided by AMTPs to farmers (Table 7.4). Yet to conclude that about 75% (ca. 204,000 ha) of the total agricultural areas in 2007 had been cultivated using private equipment is not realistic. The private farmers' fleet is clearly insufficient to cultivate more than 150,000 ha. Anecdotal evidence frequently suggests that not only private farmers but also AMTP/MTPs had been serving farmers based on private and informal arrangements that had not been included in the official accounting.

MTP finances: To analyze the financial management of AMTP/MTPs, their total revenue and net profits were compared for the period of 2005–2006. A clear reduction of ca. 75% in total (officially received) revenues had been experienced by both MTP/JSC and AMTPs (Table 7.5). This may be due to a decline in the demand for services by farmers, but a more likely explanation is incorrect accounting.

Hence, in their current condition, MTPs are neither competitive nor a promising solution. In fact, almost all MTPs are insolvent and as a result, regional tax agencies have confiscated most equipment, often through court decisions, which has increased the malfunctioning of their operations and reduced client trust. Moreover, the extremely outdated fleet further constrains the business revenues, and requires growing investments for upgrading and repairs.

Farmers' perceptions: To capture farmers' perception of the advantages and drawbacks in MTP activities, interviews were conducted with 20 farmers and with MTPs staff. The findings reveal an overwhelming dissatisfaction of the farmers with the present set up and services provided by the MTPs (Table 7.6).

Most farmers complained that there was not enough equipment available at the MTPs to meet their demands, and that most of the equipment was out-dated and did not function properly. Despite contractual arrangements with MTPs, the farmers usually did not receive the required services in due time and of the required quality.

Table 7.5 Comparative data on the economic activity of machinery and tractor parks (MTPs) of the Khorezm region 2005–2006

MTP	Number of MTPs	Number of MTPs with annual revenue of > 100 mln UZS		Number of MTPs, with an income in 2006 lower than in 2005				
		2005	2006	Units	Reduction of a max. 25%	Reduction between 25% and 50%	Reduction between 50% and 75%	Reduction of at least 75%
MTP/JSC	11	10	10	8	7	1	0	0
Total, %	100	91	91	73	64	9	0	0
AMTP	78	33	18	59	8	22	26	3
Total, %	100	42	23	75	10	28	33	4

MTP	Number of MTPs	Number of non-profitable MTPs		Number of MTPs with profitability or revenue not less than 5%	
		In 2005	In 2006	In 2005	In 2006
MTP/JSC	11	0	0	1	4
Total, %	100	0	0	9	36
AMTP	78	22	12	9	16
Total, %	100	28	15	12	21

Source: Statistics department (Adapted)

Table 7.6 Farmers' perception of the quality of services provided by MTP/JSC and AMTP

ASP	Number of farmers surveyed	Low quality and untimely delivery of services (%)	Shortage of machinery and its deterioration (%)	Informal costs (obtaining permission, repair of machinery) (%)	Prefer to possess own machinery (%)
MTP/JSC	20	70	80	25	70
AMTP	20	80	90	45	70

Source: Niyazmetov (2008)

As a consequence, they made additional, non-contractual payments to obtain the services they had officially requested.

MTP staff perceptions: The MTP staff members often complained about the delay in payments by farmers for the services rendered.

At present, the common perception prevails that both MTPs and AMTPs operate mainly under informal rules, do not comply with contractual arrangements, and usually respond to unscheduled requests from others, including local authorities. The identified critical causes for the present inefficiency of MTPs and AMTPs are:

- *Poor enforcement of property rights*, which is expressed in a general interference from outside in the day-to-day operations of MTPs, in particular by the local administration. MTP heads (chairmen) and staff frequently have little decision-making power, for example, regarding which services are to be provided first;
- *Poor management and lack of corporate management*. Although the heads of MTPs used to be elected by shareholders, in practice their appointment had to be approved by the *Hokimiat*. This has often resulted in a lack of interest of the heads in the efficient functioning of the MTPs. They are more interested in satisfying their benefactors and patrons;
- *Lack of sound contractual relations*. Contracts are not fulfilled properly because of both untimely payments by farmers and untimely and poor service provision by MTPs. The decrease in the officially reported volume of services rendered is caused partly by the high level of shadow incomes that MTPs generate by supplying informal, non-accountable services to farmers based on direct cash payments;
- *Insolvency*. Nearly all MTPs are insolvent due to a chronic lack of reserve capital to meet their liabilities in time. In 2008, 37 AMTPs were declared bankrupt by the economical court of Khorezm, and 32 had been liquidated by June 2008;
- *Low level of adequate agricultural equipment*. A general lack of enterprise capital makes it difficult for the MTPs to maintain and renew their fleet. This has generated a vicious circle; in the absence of adequate equipment, the MTPs cannot serve their client farmers, generate income or build up the capital needed for investments in new equipment. As the equipment deteriorates more and more, the service quality worsens further, driving farmers away and reducing the MTP's revenues even further;
- *Prevalence of informal over formal rules*. Farmers and MTPs in most cases prefer to settle eventual conflicts individually and informally, causing income losses to the MTPs.

7 Mapping and Analyzing Service Provision for Supporting Agricultural... 121

The MTPs are one of the most important components of the present agricultural infrastructure in the irrigated areas of Uzbekistan. The efficiency of their services in particular influences the activities of producers and their financial status. Though farmers increasingly wish to own their own machinery, and most of MTPs are likely to become bankrupt in the near future, it is unlikely that the institution MTP will disappear completely. Unless the present farms increase in size, a need will remain for specialized structures such as MTPs in the agricultural sector in Uzbekistan, since not all small-scale farmers will be able to obtain their own set of agricultural machinery, nor may it be profitable for them to own all agricultural equipment.

7.3 Distribution Outlets for Mineral Fertilizers and Oil Products

Next to agricultural equipment, the cost structure of small-scale farms is dominated by expenses for the acquisition of fuel and lubricants, mineral fertilizers and pesticides, which account for almost 40% of the total costs of raw cotton production (Rudenko 2008). Unrestricted access to such production factors represents one of the major preconditions for ensuring the autonomy of the newly established farms. In the reformation period from 1991 to 2006, numerous distribution outlets for fuel and mineral fertilizers had been established throughout Khorezm, however with different characteristics (Table 7.7).

The staff at the outlets for fuel and mineral fertilizers is authorized to conclude contracts with farmers/customers (i) only on behalf of the district departments of *Urgench NefteBaza* and *KhorezmQishloqXo'jalikKimyo*, which are scattered all over Khorezm, and (ii) partly through commodity exchange. Hence, in practice, these sale points serve mainly as warehouses and intermediates between farmers and the district departments of *Urgench NefteBaza* and *KhorezmQishloqXo'jalikKimyo*.

Direct agreements for fuel and lubricants are made between agricultural producers and *Urgench NefteBaza* and for mineral fertilizers and pesticides between agricultural producers and *KhorezmQishloqXo'jalikKimyo*. Under these agreements, fuel, lubricants and mineral fertilizers are to be sold according to the allocated and state-imposed limits for the production of the state target crops cotton and winter wheat only, and at fixed and state-determined prices. For fuel and mineral fertilizers this is against 60% and 100% prepayment, respectively. Payments of the outstanding debts are due within 60 days upon delivery.

Commodity exchange: Farmers who desire (additional) fertilizers or fuel for the production of crops other than cotton and winter wheat can gain these through auctions at the commodity exchange (*birja*). The *birja* charges market prices for agricultural inputs, which usually are higher than the state-fixed prices.

While the monetary value of transactions with fertilizers and fuel increased by 51% and 16% in 2006 and 2007, respectively, the physical amount (in tons) decreased by 10% and 17% (Table 7.8). This very likely was due to the rapid increase in sale prices for both mineral fertilizers and fuel in the same period (Table 7.9).

Table 7.7 Organizational status and functions of the outlets for mineral fertilizers and fuel

Features	Outlets for mineral fertilizers	Outlets for fuel and lubricants
Form of ownership	Not registered as a legal entity and has no chartered fund or complete balance sheet	Not registered as legal entity and has no chartered fund or complete balance sheet
Responsible legal entity	*KhorezmQishloqXo'jalikKimyo* Regional JSC, Khorezm branch	*Urgench Neftebaza* Unitary company
Main services	Store, sell and deliver mineral fertilizers (nitrogen, phosphate, potash, etc.) to designated customers	Store, sell and deliver oil products (fuel, lubricants) to designated customers
Type of financial arrangements	Cashless settlement with 100% prepayment	Cashless settlement with 60% prepayment

Table 7.8 Sale of mineral fertilizers and oil products to farmers in the Khorezm region 2006–2007

Period	Sale of mineral fertilizers			Sale of oil products		
	Contracts	(mln UZS)	(tons)	Contracts	(mln UZS)	(tons)
2006	9,027	21,927	36,951	11,399	13,852	29,957
2007	9,262	33,212	33,083	12,687	16,072	24,878
Change	3%	51%	−10%	11%	16%	−17%

Source: Regional department of agriculture and water resources management
UZS Uzbek Soum. Average exchange rate in 2007 was US$/UZS = 1,254

Table 7.9 Price dynamics for mineral fertilizers and fuel in the Khorezm region

	Average price (UZS kg^{-1})			
	2005	2006	2007	Change, 2007 *vs.* 2006 (%)
Mineral fertilizers	160.3	205.9	231.3	12
Fuel	369.2	546.6	690.1	26

Source: Regional office of *QishloqXo'jalikKimyo* and *Urgench Neftebaza*
UZS Uzbek Soum

The monopolistic status of the outlets often allowed them to provide farmers with smaller amounts than contractually agreed upon. The 'left over' fertilizers were then sold by *QishloqXo'jalikKimyo* to the (same) farmers, but at prices higher than officially declared. In 2007, *QishloqXo'jalikKimyo* was therefore accused by the demonopolization department of fraud and speculation with state-owned mineral fertilizers. The amount of illegally received income by *QishloqXo'jalikKimyo* was estimated at about 370,000,000 UZS (approximately 295,000 US$) in 2007 alone, which however was fully transferred back to the farmers according to the instructions of the demonopolization department.

Informal interviews with the staff of the district departments of *Urgench NefteBaza* and *KhorezmQishloqXo'jalikKimyo* and their distribution outlets indicate a one-sided view of the problems occurring in the process of selling agricultural inputs

Table 7.10 Farmers' view on the services rendered by outlets for fertilizers and fuel

	Number of farmers surveyed	Low quality of services, untimely delivery of inputs (%)	Supply of resources in amounts less than stated in contracts (%)	Informal costs (e.g. obtaining permission) (%)	Prefer non-state private input sources (%)
Mineral fertilizers	20	40	30	25	65
Fuel	20	65	55	45	75

such as fertilizers and fuel. Key informants mainly mentioned the delayed delivery of credits to the farms, which in turn meant that the farmers could not pay for fuel and fertilizers in time. Furthermore, interviewees pointed out that after interference of the local authorities that were concerned in the first place that farmers fulfilled their state order, commodities had to be supplied to farmers without the usual prepayments.

Farmers in turn had a number of complaints about the rendered services and mentioned various drawbacks occurring when purchasing mineral fertilizers and fuel. They expressed in particular their dissatisfaction about the distribution outlets in the districts, as staff regularly violated contractual terms (Table 7.10). Nevertheless, none of the farmers had officially submitted a case to the regional court, but rather preferred an amicable, informal settlement of conflicts directly with the outlet staff. This was mainly driven by the necessity of a future collaboration with the outlet in question and thus the wish to avoid conflicts.

The critical points in the process of obtaining fuel, lubricants and mineral fertilizers can be summarized as follows:

- *Monopolistic status of the enterprises selling mineral fertilizers and oil products.* The absence of competition in commodity provision had led to distortions in the market mechanisms, and external interference in the sales activities, distribution and payment for the resources and in particular in price setting;
- *Improper fulfillment of contractual terms.* Contracts often were improperly fulfilled due to the untimely and poor quality of the distribution services to farmers;
- *Informal rules prevail over formal regulations.* Farmers often pursued informal and non-contractual arrangements to obtain agricultural inputs, often at additional costs. Yet, none of the farmers had submitted a court claim for violation of the contractual terms; they preferred to settle eventual conflicts amicably and informally.

The present infrastructure of the outlets for selling mineral fertilizers, fuel and lubricants does not meet the demands of the farmers. The efforts to establish an efficiently operating wholesale and retail network for the trade of these commodities need to be increased. A major reason for the inefficiency of the outlets is the retention of the centralized distribution system, the main aim of which is to provide the cotton and wheat farmers with the resources they need for fulfilling the state-order production level. Hence, the present strategy of different levels of support for different crops has a negative effect on the business operations of both farmers and ASPs. It should, therefore, be considered to abandon this kind of support or to place equal importance on all crops.

7.4 Mini-banks

The rising number of newly established private farms increased the demand for banking services, and hence branches of existing banks were established throughout the Khorezm region: the so-called mini-banks. Their main function is to represent regional commercial banks, which reduces the farmers' transaction costs for timely banking services (Table 7.11).

Of the 13 commercial banks in the Khorezm region, only 8 render services to farmers through their mini-bank branches. The dominance of the *Pakhta bank* is explained by the strong interconnection (servicing and financing) between this bank and the production of cotton, which is the main source of regional export revenues (Rudenko 2008).

The mini-banks proved to be profitable mainly due to their low operational costs and the high revenues from the allocation of privileged credits. However, the results of the survey aiming at capturing the farmers' perception of the services rendered by the mini-banks are not very flattering (Table 7.12).

Most farmers stated that there was a chronic violation of the contracts by the commercial banks and their respective mini-banks; among others violations were:

- *External interference in the relations between mini-banks and farmers.* This is mostly shown in the attempts of certain local authorities to regulate the reallocation of privileged credits (for cotton and wheat) to cover certain expenses (e.g., mineral fertilizers) without permission of the farmers. Furthermore, the farmers mentioned informal obstacles when attempting to obtain credits as a major reason for their dissatisfaction with the mini-banks;
- *Management of farmers' accounts.* Farmers reported that they had limited access to their own accounts and that they could hardly withdraw cash for their everyday needs;
- *Shortage of cash funds.* Many commercial banks had a chronic and serious problem with cash and liquidity, which had led to untimely cash payments to farmers, even when payments to cover their regular expenses such as salaries were concerned;
- *Prevalence of informal rules.* No farmer had filed a court claim regarding a commercial bank's violation of contractual terms. Similar to the other cases, farmers pursued informal agreements.

7 Mapping and Analyzing Service Provision for Supporting Agricultural...   125

Table 7.11 Organizational status and functions of the mini-banks

Features	Mini-bank
Form of ownership	Not registered as legal entity and has no chartered fund or complete balance sheet
Responsible legal entity	Regional branch of commercial bank
Main services	Opening of accounts to private farms; transferring financial resources from one farmer to another; allocation of privileged and commercial credits to farmers and other types of banking services necessary for agricultural production
Type of settlement	Cashless

Table 7.12 Farmers' perception of the quality of services rendered by mini-banks

Name	Number of farmers surveyed	Can not freely manage their accounts (%)	Channel funds from farmers accounts for "undesirable" purposes (%)	Informal costs (obtaining credits, cash funds, etc.) (%)	Prefer non-state private banking services (%)
Mini-banks	20	90	70	60	65

7.5 Summary and Conclusions

After the dissolution of the *shirkats*, agricultural land was distributed to private, small-scale farms on the basis of long-term leases. This in turn increased the demand for and the changes in agricultural service provision, and made it necessary to not only increase the number of the existing ASPs, but also to establish new ASPs such as AMTPs, mini-banks and water user associations. The MTPs, outlets for sale and distribution of mineral fertilizers and oil products, and mini-banks are key players in the agricultural institutional landscape because their services encompass and influence nearly all farm types. However, MTPs – either in the form of joint stock companies or as alternative machine-tractor parks – are not able to meet the demand of farmers due to the outdated machinery, lack of investments, and insolvencies caused by poor management and regular interference by the local administration in the day-to-day operations.

Outlets for the sale and distribution of mineral fertilizers and oil products belong to the state monopolist companies – *KhorezmQishloqXo'jalikKimyo* and *Urgench NefteBaza* – that can therefore informally dictate their conditions of input supply to the farmers. Inputs, for example, are usually supplied to farmers in amounts that are on average about 10% lower than stipulated in the contracts.

The mini-banks, responsible for providing banking services to farmers and for the allocation of privileged credits, did not meet farmers' demand either. Common practices include a handling of the accounts without request and consent of the holder, i.e., the farmer.

With regard to all ASP structures examined, farmers/clients typically tried to settle conflicts by informal arrangements rather than by claiming rights through courts and justice. Although the ASPs are well regulated on paper, in practice the relationship between these and the farmers is seriously disturbed by institutional failures, caused mainly by external involvement in the activities of the ASPs. This has led to poor enforcement of property rights of the ASPs, regular violation of contractual terms by both ASPs and farmers, and prevalence of informal over formal rules. Not only has this fuelled the overall distrust of the farmers in the legal system, it has also forced the farmers to seek for services at any cost in the peak agricultural seasons, often through informal negotiations and agreements.

In summary, the current interactions of the ASPs and farmers represent at most a hybrid institutional set up, which is characterized by the fact that most ASPs and all farmers are formally private but informally still under strong control of the local administration. Obviously, this dualism undermines all certainty regarding property rights, which in turn reduces any incentive to enforce and improve the institutional environment.

One key recommendation for an 'institutional' improvement of ASPs in Khorezm, and presumably in all of Uzbekistan would be to minimize the external interference in the business of both the ASPs and farmers and to enforce the rule of law. Further benefits may be obtained from privatization of the MTP/JSCs and commercial banks, implementation of corporate management rules in the activities of the ASPs, and improved fulfillment of the contracts between ASPs and farmers. An increase and improvement in the legal support to farmers could encourage them to settle eventual conflicts through juridical procedures rather than by pursuing informal practices that are in the long run unsustainable.

References

CER (2004) The reorganization of agricultural cooperative enterprises (shirkats) into farming enterprises, Tashkent
Djanibekov N, Bobojonov I, Lamers JPA (2011) Farm reform in Uzbekistan. In: Martius C, Rudenko I, Lamers JPA, Vlek PLG (eds) Cotton, water, salts and *Soums*: economic and ecological restructuring in Khorezm, Uzbekistan. Springer, Dordrecht/Berlin/Heidelberg/New York
Müller M (2006) A general equilibrium approach to modeling water and land use reforms in Uzbekistan. PhD dissertation, Rheinische Friedrich-Wilhelms-Universität Bonn, Bonn
Niyazmetov D (2008) Efficiency of the market infrastructure development for farmers of the Khorezm region. MSc dissertation, Urgench State University, Uzbekistan
Rudenko I (2008) Value chains for rural and regional development: the case of cotton, wheat, fruit, and vegetable value chains in the lower reaches of the Amu Darya River, Uzbekistan. PhD dissertation, Hannover University, Hanover
Spoor M (1999) Agrarian transition in former soviet Central Asia: a comparative study of Kazakhstan, Kyrgyzstan and Uzbekistan. Institute of Social Studies, WP 298:25

Chapter 8
Politics of Agricultural Water Management in Khorezm, Uzbekistan

Gert Jan Veldwisch, Peter Mollinga, Darya Hirsch, and Resul Yalcin

Abstract On the basis of intensive fieldwork in the period 2002–2006, which combined interviews with direct observations, the implementation of two policies in the field of agricultural water management in Uzbekistan is analysed: the reform of the water bureaucracy along basin boundaries and the establishment of Water Users Associations. It is shown that the Uzbek government used these policies creatively for addressing some pressing issues, while the inherent decentralisation objective was pushed to the far background. Both reforms are used to strengthen the state's grip on agricultural production regulation. The latter is at the centre of day-to-day agricultural water management dynamics. It is shown that decentralisation policies originally developed in society-centric policy processes cannot be easily applied in countries with state-centric politics such as Uzbekistan.

Keywords Irrigation • Water Users Associations • Water reform • Agrarian change • State-centric politics

---

G.J. Veldwisch (✉)
Irrigation and Water Engineering (IWE) Group, Wageningen University
and Research Centre (WUR), P.O. box 47, Wageningen 6700AA, The Netherlands
e-mail: gertjan.veldwisch@wur.nl

P. Mollinga
School of Oriental and African Studies (SOAS), Thornhaugh Street
Russell Square, London WC1H 0XG, UK
e-mail: pmollinga@hotmail.com

D. Hirsch
Institute for Environment and Human Security, United Nations University (UNU),
UN-Campus Hermann-Ehlers-Str. 10, Bonn D-53113, Germany
e-mail: hirsch@ehs.unu.edu

R. Yalcin
Independent Researcher, London, UK
e-mail: resul.yalcin@virgin.net

C. Martius et al. (eds.), *Cotton, Water, Salts and Soums: Economic and Ecological
Restructuring in Khorezm, Uzbekistan*, DOI 10.1007/978-94-007-1963-7_8,
© Springer Science+Business Media B.V. 2012

8.1 Introduction

Decentralisation policies, establishment of Water Users Associations (WUAs)[1] and the reorganisation of the water bureaucracy along basin boundaries are implemented in numerous countries around the world. In Uzbekistan, the state is exceptionally strong and controls all aspects of its society. This makes it a special case when assessing the implementation process of policies that have liberal-democratic roots. The analysis of how the Uzbek state deals with the implementation of such water policies shows its creativity and dynamism with regard to seeking to reform minimally while remaining firmly in control, which may be exemplary for state-centric politics (Grindle 1999).

This chapter presents the results of field research undertaken between 2002 and 2006 in the context of two projects in the Khorezm region of Uzbekistan,[2] an irrigated region that is used largely for cotton and wheat production. Extensive field research was conducted that mainly consisted of a combination of interviews and direct observations. Due to the controlling nature of the state, it is difficult to gain access to the Uzbek society (Wall 2006; Schlüter and Herrfahrdt-Pähle 2007; Veldwisch 2008a; Schlüter et al. 2010). In spite of this difficult research environment, we were able to identify and describe a large number of relevant social and political processes.

We investigate three aspects of agricultural water management in Khorezm. In Sect. 8.2 we discuss the way in which bureaucratic reforms in the fields of agriculture and water management were introduced, and analyse the effects they have had. In Sect. 8.3, the establishment of Water Users Associations (WUAs) and their functioning is discussed. In Sect. 8.4, we describe and analyse the main processes underlying the day-to-day agricultural water distribution, which is strongly related to the state regulation of agricultural production.

The analysis of these developments provides insight on change processes in state-centric politics (Grindle 1999) as opposed to society-centric politics, which have been studied and documented in the Western policy-process literature (Yalcin and Mollinga 2007). In Sect. 8.5 we draw conclusions on the political nature of agricultural water management in Khorezm.

---

[1] The authors are aware that these have been renamed as "Water Consumers Associations" in 2009, but this analysis is based on the period before this change.

[2] The first project is 'Economic and Ecological Restructuring of Land and Water Use in Khorezm', under an agreement between the German Ministry of Education, represented by the Centre for Development Research (ZEF), the United Nations Educational, Scientific and Cultural Organisation (UNESCO), and the Government of Uzbekistan, represented by the Ministry of Agriculture and Water Resources (MAWR). The second project is NEWater, funded by the European Union in its sixth framework program. It joins 37 universities and institutes from Europe, Central Asia and Africa to study new methods for integrated water resources management in seven river basins in the world. The Amudarya river basin is one of its seven case studies.

8.2 The Creation of a New Bureaucracy[3]

After Uzbekistan gained independence in 1991, governance reform involved the creation of a joint Ministry of Agriculture and Water Resources in 1997 out of two separate ministries, and subsequently the process of transforming the region- and district-based administrative water management system, into an irrigation basin water management system based on hydrological principles. The latter involved the creation of the Irrigation Basin System Management Authorities in 2003. This section analyses these two organisational changes in the structure of water governance.[4]

8.2.1 Changes in the Structure of Water Resources Governance

Decree No. 419 of 26 November 1996 by the Cabinet of Ministers abolished the Ministry of Agriculture and the Ministry of Melioration and Water Management of Uzbekistan, which had existed since 1927–1928. In their place, a new, centralized, single organization, the Ministry of Agriculture and Water Resources (MAWR) of Uzbekistan, was established. The Uzbek President's decree No. UP-3226 of 24 March 2003, followed by decree No. 320 of 21 July 2003 of the Cabinet of Ministers, restructured the existing water resources management system from an administrative-territorial-based management to a basin-based water resources management system. The 13 region-based and 163 district-based water management departments, and more than 40 management organizations of inter-district canals and other water-sector organizations of the MAWR, established in 1997, were abolished. They were replaced by 10 main water and one main canal irrigation basin management authorities, and 52 subdivided offices of the irrigation basin management authorities. Unlike the organizations established in 1996–1997, the latter organizations are, administratively speaking, not responsible to the local governors (*hokims*) nor to the regional (*viloyat*) and district (*tuman*) offices of the MAWR, but directly to the water resources department of the MAWR in Tashkent. This department is headed by one of the deputies of the minister responsible for water resources management in Uzbekistan.

---

[3] See Yalcin and Mollinga (2007) for details.

[4] This analysis does not include the recent Law of the Republic of Uzbekistan No 240 from 25.12.2009 "On the introduction of amendments and supplements to certain legislative acts of the Republic of Uzbekistan in connection with deepening of economic reforms in agriculture and water industry."

8.2.2 The 1996/1997 Merger

Immediately after independence in 1991, the agriculture and water resources ministries that were used to receiving orders from Moscow were suddenly autonomous at the state level of the newly independent Uzbekistan republic. The constitution of a new central authority in the form of the Uzbekistan national government, parliament, President and nation-building processes was quick. However, the organizational and institutional translation of this new centrality took several years. For agriculture, the centralization process was consolidated in 1996/1997, with the merger of the two above-mentioned ministries. One of the main aims of the new Ministry of Agriculture and Water Resources is 'to implement the government's agricultural policies for achieving the state ordered crop production quotas, namely for cotton and wheat' (see Cabinet of Ministers Degree No. 376, 16 November 1989; No. 5, 11 January 1991; and No. 419, 26 November, 1996: 21–29). The creation of a single ministry solved the problem of power struggles between the agriculture and water ministries that were manifest in 1991–1996, about who should be controlling water distribution, particularly in times of water scarcity.

As Wegerich (2005) has also suggested, the merger thus sought to achieve better coordination between agricultural and water management, and to re-established the dominance of agriculture. However, we disagree with Wegerich (2005) when he suggests that achievement of equitable water distribution was an important concern in this reform. The objective to geographically spread water equitably derives from a concern to maximize cotton and wheat production (and probably to avoid social unrest and economic loss caused by regionally skewed water distribution as in the drought years 2000 and 2001) and not from a normative concern. Equitable water distribution is not a stated concern in the official decrees that were the basis of the merger of the two ministries.

Secondly, Wegerich's suggestion that the merger was part of a typical power struggle between water and agriculture ministries misinterprets the position of these two ministries in both the Soviet political economy and post-Soviet Uzbekistan. In the first years after 1991, the political elites in Uzbekistan developed a new national ideology of independence to create a conceptual scheme for internal politics. The continuation of highly centralized governance and control of societal processes was part of this conceptual scheme. However, the Moscow-based centralized governance and control system ceased to exist after independence, requiring substantial socio-economic and organizational/administrative adaptation. The merger of the agriculture and water ministries reproduces the centrality of agricultural planning of the Soviet period at the level of the new independent state of Uzbekistan, part of which is a rationalization of the bureaucracy in terms of staffing pattern and budget allocation, triggered by the financially and economically dire straits that Uzbekistan landed in after independence (Spoor 2004). The 1991–1996 periods seems to have suggested to the Uzbek political leadership that two separate ministries without firm guidance by the agricultural production interest was a problematic configuration.

8.2.3 The 2003 Separation of Tasks

The water management system based on hydrological boundaries, as introduced in 2003, is as centralized as the earlier one, with control located in Tashkent. The new water management organizations are directly responsible to the water resources department of the MAWR in Tashkent. The heads of these irrigation system basin management authorities are selected and appointed by the Minister of Agriculture and Water Resources, and thus dominance of the agricultural planning remains.

The change in our interpretation is not, as suggested by Wegerich (2005: 462), a move towards the separation of the Ministry of Water from the Ministry of Agriculture. That would be a de-merger. It is a separation of tasks within a single ministry. It has also been a move of the MAWR as a whole to reduce its dependency on *hokims* (governors). In May 2006, the regional department of the MAWR in the Khorezm region moved out of the *hokimiat's* building to a different part of Urgench city. We propose that the reform of the organizational structure of irrigation management has to be understood in the context of broader and longer-term changes in the nature of the overall governance system. It can be read as an attempt to depoliticise certain sectors to achieve more effective and efficient planning and management, i.e., as a movement in the direction of separating functional/technocratic governance and political governance at the operational level. More generally, and more speculatively, it may be interpreted as illustrative for the emergence of a development trajectory that tries to combine economic and technological growth and modernization with the maintenance of high degrees of centralized political control.

The organizational change was fed by the actual practice of agricultural and water management at the regional level. Water resources management, allocation and use before 2003 were under the operational control of MAWR province- and district-level offices, as well as under the frequent interference of the local authorities (*hokimiats*). This approach to water management led to a peculiar polycentric management system creating frictions and conflicts of interest between agricultural enterprises, water and agricultural institutions and the local authorities. All institutions at the local level considered themselves to be responsible for agriculture, water and canal management. This situation of unclear division of responsibilities reinforced the arbitrary interference of the local *hokims* in the day-to-day management of water resources at the province and district level. In some years, such as during the water-deficit years of 1999–2000 and 2000–2001, a balanced water distribution was not possible, and this resulted in a highly skewed distribution of water along the Amudarya and Syrdarya rivers to reduce the anticipated enormous economic losses (Abdullaev 2005; Hayashi 2005). The central authorities wanted to break the power of the *hokims* over the water management and to reduce the competition between the districts over water distribution.

The Deputy Minister of Agriculture and Water Resources responsible at the time for water management took advantage of the atmosphere of change and promoted the idea of an irrigation basin water management system based on *hydrological* principles in a centralized fashion. The hydrological principle as unit of organisation also was in accordance with the principles adopted by the Inter-State Committee for Water Coordination (ICWC). The discussions at that level produced legitimacy for a system of water management in Central Asia organised according to this principle.

8.3 Water Users Associations in Uzbekistan

The first WUAs in Uzbekistan were created in the year 2000. Zavgorodnyaya (2006) has shown that ideas raised by international donors landed well with the Uzbek local administration. The state administration saw a need for reform in the context of land reform experiments in which virtually all *shirkats*[5] were dismantled and divided among privatized farmers (*fermers*) (Wegerich 2000; Zavgorodnyaya 2006; Trevisani 2008). The redistribution of land into smaller units created numerous water users at the level of the former collective farm, which in turn created new challenges for water distribution (Abdullaev et al. 2006; Zavgorodnyaya 2006; Veldwisch 2007); Zavgorodnyaya (2006: 14) also notes that WUAs "were established as a 'bridge' between state irrigation management organisations and [privatised] water users".

The Government of Uzbekistan chose Khorezm as a region of early experimentation. Between 2000 and 2002, four pilot WUAs were established with the same boundaries of the *shirkats* that were dismantled at the same time. By mid 2003, all *shirkats* in Yangibazar district were dismantled, and land was redistributed among *fermers* (Trevisani 2008). At the same time, WUAs were established along hydro-graphic boundaries (Zavgorodnyaya 2006). Finally, in the period from 2005 to 2006, the last *shirkats* were dismantled. In contrast to the experiment with hydro graphic boundaries in Yangibazar, WUAs were now established with largely the same boundaries as the former *shirkats* (Veldwisch 2008b).

8.3.1 Nature of Uzbek WUAs

By a formal definition, WUAs in Uzbekistan are voluntary, non-governmental, non-profit entities, established and managed by a group of water users located along one or several watercourse canals (Zavgorodnyaya 2006; Yalcin and Mollinga 2007; Wuaconsult 2008). The WUAs are established by the state in a top-down fashion

---

[5]*Shirkat* is the name given to the cooperative large farm enterprises that were established after Uzbekistan's independence in place of the former *sovkhozes* and *kolkhozes*.

8 Politics of Agricultural Water Management in Khorezm, Uzbekistan

(Zavgorodnyaya 2006; Veldwisch 2008b). Despite being officially run by farmers, WUAs experience strong interference by the state. Thus, the "WUA becomes a place in the strongly hierarchical structure that still is controlled by the government" (Zavgorodnyaya 2006: 79).

In many cases, authorities have essentially enforced the establishment of WUAs. Evidence indicates that several WUA chairmen were appointed by regional (or district) departments of the MAWR, or that elections were influenced. In turn, not only were the democratic principles of the WUAs undermined, but these practices have also contributed to the farmers' hesitant attitude towards these new institutions.

It is important to explore how the idea of establishing non-governmental, self-financing irrigation organisations was transformed into the establishment of organisations still heavily controlled by the government for the purpose of regulating agricultural production.

The introduction of WUAs and the later countrywide implementation of these new organizations were favoured by several circumstances and incidents. First, the collapse of the Soviet Union and the emergence of a large number of individual farmers through land reforms created a demand for new water management organisation at the level below the district. Second, several international agencies such as the World Bank, the Asian Development Bank, United States Agency for International Development (USAID), and the International Water Management Institute (IWMI) lobbied for the decentralization of water management in general and the creation of WUAs in particular (Zavgorodnyaya 2006; Yalcin and Mollinga 2007; Veldwisch 2008b). They offered financial support for the introduction of WUAs. Third, apart from external pressures and international experience, there was support for the WUA concept from the MAWR. The MAWR organized field trips for its officials to Italy and Turkey to learn from international experience in water management issues, and later lobbied for the introduction of WUAs (Zavgorodnyaya 2006; Schlüter and Herrfahrdt-Pähle 2007; Yalcin and Mollinga 2007).

8.3.2 WUAs and the State

The creation of WUAs in Uzbekistan was based on (provisional) decrees by the Cabinet of Ministers. Decree No. 8 of 5 January 2002 specifies according to UNDP (2007: 66): *'(i) the procedure for establishment of WUAs on the territory of agricultural enterprises which are being reorganized; (ii) the management structure of the WUA; (iii) the standard agreement about water users integration and establishment of WUAs; (iv) the standard charter of the WUA; and (v) the standard agreement between the WUA and farmers for provision of chargeable water delivery services and works.'*

Parts of the decree are included in the charters of the WUA prescribed by the MAWR. For example, the standardized charters define that a WUA is financed by user fees, that the general assembly is the highest decision-making body of the WUA, and that the WUA chairman should be elected by this body. A WUA is established as a non-commercial organization and is considered a legal entity. A WUA has its own stamp, letterhead paper, bank account and its own property. The members of a WUA can be legal and physical persons who own or lease property in the territorial boundaries of a WUA, use its services, and pay user fees to WUA.

Despite the written regulations, the state did not transfer full authority to the WUAs. The *hokim* still interferes with the water schedule, especially in water-deficit situations, as he remains responsible for the fulfilment of the state order by the end of the growing season. Farmers know about this responsibility and have been known to use their personal connections to the *hokim*, circumventing the WUA leaders and staff, who are supposed to be responsible for water allocation and distribution (Zavgorodnyaya 2006; Veldwisch 2008b; Veldwisch and Spoor 2008).

Veldwisch (2008b) showed that state interference in WUA management is an active strategy that consists of three main elements. Firstly, *hokims* and officials from the MAWR actively interfere in (or overrule) internal WUA processes. Secondly, strict boundary conditions on inflows, outflows and what the WUA is allowed to handle on its own are set by the state. This greatly restricts the functioning of the WUAs. Thirdly, only tasks that would be costly and difficult for the state to implement, like the continuous monitoring of water delivery up to field level, are actually handed over to the WUA and even then they are monitored by the state.

This shows that there is not just 'the WUA' – a bearer of democratic principles and self-management – but that WUA structures can be created to pursue very different objectives. The WUA fulfils an important role in the implementation of (state) control over water distribution and agricultural production, as elaborated in the following section.

8.4 Water Distribution and Regulation of Agricultural Production

Agricultural production in Uzbekistan is strongly regulated by the state, and the production of cotton and wheat plays an important role in this regulatory system (Wall 2006; Trevisani 2008; Veldwisch 2008b). Parallel to this, the distribution of irrigation water is also managed through an authoritarian, top-down system. Control over water distribution is used to control agricultural production in general and state-ordered cotton and wheat production in particular. The two other main forms of production (Veldwisch 2008b; Veldwisch and Spoor 2008), i.e., commercial rice production and household production, are regulated more indirectly. However, also in these production systems water delivery plays an important regulatory role. Water demands at field level are communicated to water managers differently for each form of production (Veldwisch 2008b).

8.4.1 Peasants, Farmers and the State

Since de-collectivisation, Uzbek (and Khorezmian) rural society has had two groups of agricultural crop producers: private farmers (*fermers*) and peasants (*dehqons*). Roughly 20% of the arable land is held by *dehqons*, who make up 90–95% of the rural households. Their average land holding size is 0.21 ha,[6] which is intensively cropped with a variety of fruits and vegetables. Typically, 50% of the area is double cropped to wheat and short-duration rice. While the production on these small plots provides for a basic living, these families often also develop other economic activities like animal rearing, working on land of a *fermer*, taking paid jobs in the district or provincial capital, or engaging in labour migration outside Uzbekistan (Wall 2006; Veldwisch 2008b).

The other 5–10% of the rural households are *fermers*. They have long term-land leases on about 70–80% of the arable land.[7] *Fermers* have land of various sizes and types (Trevisani 2008; Veldwisch 2008b). *Fermers* that mainly crop large areas of cotton, wheat and rice, and in addition also crops like carrots, maize, melons and potatoes, have the strongest influence on water distribution.

8.4.2 Three Forms of Production

The Khorezmian crop production system can be understood as consisting of three interrelated forms of socio-political organisation of production, or in short "forms of production" (Veldwisch 2008b; Veldwisch and Spoor 2008):

1. State-ordered production of cotton (and wheat)
2. Commercial production of rice
3. Household production of food crops.

These forms of production are distinguished on the basis of an analysis of empirical material on socio-political processes (i.e. the organisation of production and the distribution of water). Each form of production has its own 'economic logic', which can for instance be recognised in the way in which the various input and output relations are organised. These governance categories are closely related to certain agronomic production systems, but the categories are in the first place about how this production is socio-politically organised. They thus differ from those presented by Djanibekov et al. (2011).

Cotton production, and to some extent also wheat production, is organised in the same way as a plan economy. Cotton can only be sold to the state, and payments are

---

[6]From a survey among 684 households in 4 WUAs (Veldwisch 2008b).

[7]The 0–10% of land not accounted for here has remained in the hands of various state organisations.

made to state-controlled settlement accounts, a sort of bank account from which it is difficult for the holder to withdraw cash (cf. Niyazmetov et al. 2011). With the money on these settlement accounts, state-subsidised inputs and services can be bought from state and semi-state organisations. *Fermers* are obliged to allocate about 40–60% of their land to cotton and 20% to wheat, and the way in which they manage the crop is controlled by state organisations. For a long time, this state-controlled production was the basis for a net transfer out of the cotton sector, i.e., an implicit taxation.[8] *Fermers* benefit from a stable and predictable market, while the financial margin is very small and sometimes even negative (Djanibekov 2008).

In contrast to cotton, rice can be freely sold on the market, which makes it a good source of cash for farmers. This free market is, for instance, characterised by the way in which fertilizers for this crop are acquired and the way in which workers on rice fields are paid, i.e., both are paid in cash and at fluctuating, negotiated market prices. For *fermers* to gain access to allowances for commercial rice cultivation, providing a satisfactory production of cotton to the state is essential. While cotton is used as political capital, rice is used as a proxy currency, i.e., in economic transactions it frequently replaces cash.

Dehqons produce fruits, vegetables, potatoes, rice and wheat on their small plots, mainly for home consumption.[9] Self-sufficiency and barter is revealed by their practiced nutrient management (no fertilizer but dung from domestic cattle), seed acquisition (re-use from previous harvest, exchange with friends and family), labour organisation (purely family labour), and consumption (at home and solidarity exchange).

Because the different forms of production are associated with particular crops, they are also associated with different water demands and water-use practices. Rice is cultivated in basins, which are continuously filled with water. Deep percolation losses lead to a rising groundwater table, and this can negatively affect other crops. The inflow does not need continuous supervision, and rice can thus be easily irrigated during the night. Cotton needs less water than rice, but it has specific requirements regarding the timing of irrigation. Especially in July, when cotton is flowering, farmers prefer to irrigate at a very specific moment, i.e., they want the cotton plants to experience a certain degree of water stress but not too much. Often this window of opportunity is limited to one or 2 days. The water is supplied in furrows between the beds (cf. Pulatov et al. 2011). Cotton requires good field drainage and a not too shallow groundwater table. In line with the diverse cropping patterns in the household form of production, *dehqons* irrigate irregularly and often only small parts of their fields at any given time (Veldwisch 2008b).

---

[8] See e.g. Spoor (2004) and World Bank (2005). Both Müller (2006) and Rudenko (2008) show that the cotton sector as a whole has actually been subsidised in recent years while individual producers are often still being implicitly taxed.

[9] This is one of the predominant forms of production. It certainly does not mean that *dehqons* only produce for household consumption; they market some produce on local markets and are also strongly involved in the other two forms of production, the latter mainly through labour provision.

8.4.3 Adapting to Demands

Khorezm, although it lies in the lower reaches of the Amudarya basin, has always experienced a general situation of water abundance (Conrad 2006; Veldwisch 2008b, 2010). The relatively high availability of water reduces conflicts over this resource.

Allocation and scheduling of irrigation water in Khorezm is integrated in the calculations of the so-called 'limit', an irrigation schedule that prescribes discharges in every canal and to every WUA for each 10-day period. Calculations are made on the basis of prescribed, state-regulated cropping patterns. In contrast to this top-down procedure, actual water distribution is informed by demand from below. *Fermers* and *dehqons* communicate their needs through the WUAs to the district level and higher. For each form of production, the strategy for accessing irrigation water is different. These processes become more articulate when water availability gets low.

Cotton *fermers* primarily depend on the state network through which it is relatively easy to get water when needed. There is a constant pressure on *fermers* and the supporting state agencies to irrigate cotton appropriately, both in terms of timing and quantity. The WUAs take special care of cotton irrigation, as they are made co-responsible for a good cotton yield (Veldwisch 2008b; Veldwisch and Spoor 2008). During interviews, both WUA staff and *fermers* mentioned that obtaining water for cotton is easier than for other crops. Also, WUA workers were observed to irrigate cotton fields for *fermers* when ordered to do so by the district authorities (Veldwisch 2007). When *fermers* had difficulties accessing water for cotton, they complained to the water management department or directly at the district governor's office (*hokimiat*). It was mostly enough just to threaten to do this to get water.

The *hokimiat* does not freely issue allowances to grow rice. Even when *fermers* gain access to rice production, they have to mobilise the required water through informal networks. As this water is not allocated in the official 'limit', people use their socio-political and economic power to gain access to water for rice production. These transactions are surrounded by secrecy, but it is probable that the transfer of money and other benefits play a role. The gatekeepers to water for rice are often the same people that control water for cotton, but when approached in the context of rice production the nature of the contact is different.

Dehqons are small players in the struggle over water, but we observed that they benefited from the generally accepted norm that basic food production has priority over commercial and state production. Also, it seemed technically difficult to limit their water consumption due to their large number and relatively small diversions. While our own field research during 2004–2006 did not indicate this, in years of serious water shortages such as the 2008 agricultural season, household production could suffer severely as a result of conflicting claims.

The three forms of agricultural production each have their own system of water management. This means that in each of these three forms of production, water is acquired and managed in a different way. In case water is short, each form has its own typical way of securing its access to water: State-ordered cotton *fermers* call on the state organisations, commercial rice *fermers* depend on their personal connections, and, at least in the period under research, household production water users seemed politically untouchable, as household production provides for the basic livelihood security of the majority of the rural population.

The wider structures of state regulation of agricultural production thus strongly influence the social dynamics of the distribution of water. In the case of state-ordered cotton, this is strictly organised and expressed in formal rules and regulations. In rice production, these are more informal, and for household production they are more implicit and hidden and based on locally defined ethics.

8.5   Conclusions

This chapter explored the socio-political aspects of three developments in the field of agricultural water management in Khorezm, Uzbekistan. Each of these developments reflects the nature of the Uzbek state, which continuously re-invents itself, but in essence remains authoritarian and neo-patrimonial. Reform processes in administration and policy are implemented in a top-down way and emerge from within the official bureaucracy. The main characteristics of this state-centric policy process (Grindle 1999) are as follows:

– Policy initiatives emerge within the official bureaucracy and largely reflect the actions and perceptions of elites within the government. They do not emerge in the public domain.
– Both the bureaucratic reform processes in agriculture at the province and district levels and the policies to establish WUAs had an apparent objective of decentralisation. When studied in detail, it showed that the reforms were creatively used for addressing existing bottlenecks. With the political elite firmly in the driving seat, decentralisation was in practice pushed to the far background.
– The adaptive capacity of the highly centralized Uzbek system was apparently underestimated by the international supporters of reforms, who anticipated that the restructuring would lead to a decentralized water management system.
– Seen from the perspective of the state, water management is primarily to serve crop production. Both reforms at the WUA level and bureaucratic reforms at the provincial and district levels are used to strengthen this function. When needed, also WUAs are used as instruments to ensure agricultural production targets.
– In the practice of day-to-day water management, very different processes are at play. The state's differentiated interest in agricultural production that favours cotton and wheat has led to an agrarian structure in which forms of agricultural production can be distinguished on the basis of their different socio-political control systems and economic nature. The dynamics in agricultural water distribution to a large extent take shape around the three forms of production. i.e. state-ordered production of cotton (and wheat), commercial production of rice and household production of food crops.

Decentralisation policies developed in society-centric policy processes cannot be easily applied in a state like Uzbekistan. As an alternative to directly pushing administrative and policy reforms, we suggest identifying those sections of the government apparatus that support and drive a modernisation agenda. Within the Uzbek bureaucracy, there are certainly interest groups involved in extensive consultation, negotiation, consensus building and sometimes bargaining between elites and various government departments for policy or institutional change. Further research on the political economy of land and water governance reform might fruitfully look at how different interest groups and sections of the government relate to the maintenance of centralised political control of the society at large.

References

Abdullaev I (2005) Development of water management institutions in the environmental sensitive Karakalpakistan Area of Central Asia. Internal draft document. IWMI, Tashkent

Abdullaev I, Manthrithilake H, Kazbekov J (2006) Water security in Central Asia: troubled future or pragmatic partnership? In: International conference "The last drop? Water, security and sustainable development in Central Eurasia". Institute of Social Studies, The Hague

Conrad C (2006) Fernerkundungsbasierte Modellierung und hydrologische Messungen zur Analyse und Bewertung der landwirtschaftlichen Wassernutzung in der Region Khorezm (Usbekistan). Dissertation. Universität Würzburg. http://opus.bibliothek.uni-wuerzburg.de/volltexte/2006/2079/index.html

Djanibekov N (2008) A micro-economic analysis of farm restructuring in Khorezm region, Uzbekistan. PhD dissertation, Bonn University, Bonn

Djanibekov N, Bobojonov I, Lamers JPA (2011) Farm reform in Uzbekistan. In: Martius C, Rudenko I, Lamers JPA, Vlek PLG (eds) Cotton, water, salts and *Soums*: economic and ecological restructuring in Khorezm, Uzbekistan. Springer, Dordrecht

Grindle MS (1999) In quest of the political: the political economy of development policy making. CID working paper no. 17. Center for International Development at Harvard University, Cambridge, MA

Hayashi Y (2005) Assessment report of the current development situation of the republic of Karakalpakistan with the emphasis on water and drought, (the final draft). ACA/DC/DSSP/UNDP Uzbekistan, Foundation for Advanced Studies on International Development

Müller M (2006) A general equilibrium approach to modeling water and land use reforms in Uzbekistan. PhD Dissertation, Rheinische Friedrich-Wilhelms-Universität Bonn, Bonn

Niyazmetov D, Rudenko I, Lamers JPA (2011) Mapping and analyzing service provision for supporting agricultural production in Khorezm, Uzbekistan. In: Martius C, Rudenko I, Lamers JPA, Vlek PLG (eds) Cotton, water, salts and *Soums*: economic and ecological restructuring in Khorezm, Uzbekistan. Springer, Dordrecht

Pulatov A, Egamberdiev O, Karimov A, Tursunov M, Kienzler S, Sayre K, Tursunov L, Lamers JPA, Martius C (2011) Introducing conservation agriculture on irrigated meadow alluvial soils (Arenosols) in Khorezm, Uzbekistan. In: Martius C, Rudenko I, Lamers JPA, Vlek PLG (eds) Cotton, water, salts and *Soums*: economic and ecological restructuring in Khorezm, Uzbekistan. Springer, Dordrecht/New York/Berlin/Heidelberg

Rudenko I (2008) Value chains for rural and regional development: the case of cotton, wheat, fruit, and vegetable value chains in the lower reaches of the Amu Darya River, Uzbekistan. PhD dissertation, Hannover University, Hanover

Schlüter M, Herrfahrdt-Pähle E (2007) Resilience and adaptability in the lower Amudarya river basin, Central Asia. Springer, Berlin/Heidelberg/New York

Schlüter M, Hirsch D, Pahl-Wostl C (2010) Coping with change: responses of the Uzbek water management regime to socio-economic transition and global change. Environ Sci Policy 13(7):620–636

Spoor M (2004) Inequality, poverty and conflict in transition economies. Kluwer, Dordrecht/Boston/London

Trevisani T (2008) Land and power in Khorezm. PhD Dissertation, Institut für Ethnologie, Freie Universität, Berlin

UNDP (2007) Water: critical resource for Uzbekistan's future. United Nations Development Program, Tashkent

Veldwisch GJA (2007) Changing patterns of water distribution under the influence of land reforms and simultaneous WUA establishment: two cases from Khorezm, Uzbekistan. Irrig Drain Syst 21(3–4):265–276

Veldwisch GJA (2008a) Authoritarianism, validity, and security: researching water distribution in Khorezm, Uzbekistan. In: Wall CRL, Mollinga PP (eds) Field work in difficult environments: methodology as boundary work in development research. Lit, Berlin, pp 161–181

Veldwisch GJA (2008b) Cotton, rice & water: transformation of agrarian relations, irrigation technology and water distribution in Khorezm, Uzbekistan. PhD dissertation, Bonn University, Bonn

Veldwisch GJA (2010) Adapting to demands: allocation, scheduling and delivery of irrigation water in Khorezm, Uzbekistan. In: Spoor M, Arsel M (eds) Water, environmental security and sustainable development. Routledge, London/New York

Veldwisch GJA, Spoor M (2008) Contesting rural resources: emerging 'forms' of agrarian production in Uzbekistan. J Peasant Stud 35(3):424–451

Wall CRL (2006) Knowledge management in rural Uzbekistan: peasant, project and post-socialist systems of agricultural knowledge in Khorezm. PhD dissertation, Bonn University, Bonn

Wegerich K (2000) Water user associations in Uzbekistan and Kyrgyzstan: study on conditions for sustainable development. Occasional paper no. 32. Water issues study group, School of Oriental and African Studies (SOAS). University of London, London

Wegerich K (2005) What happens in a merger? Experiences of the state department for water resources in Khorezm, Uzbekistan. Phys Chem Earth 30:455–462

World bank (2005) Cotton taxation in Uzbekistan: opportunities for reform, Washington

Wuaconsult (2008) What is a WUA? http://www.wuaconsult.uz/index.php?option=com_content&task=view&id=1&Itemid=2&lang=english

Yalcin R, Mollinga PP (2007) Institutional transformation in Uzbekistan's agricultural and water resources administration: the creation of a new bureaucracy. Working paper no. 22. ZEF, Bonn, Germany. http://www.zef.de/fileadmin/webfiles/downloads/zef_wp/WP22_Yalcin-Mollinga.pdf

Zavgorodnyaya D (2006) Water user associations in Uzbekistan: theory and practice. PhD dissertation, Bonn University, Bonn

Chapter 9
Water and Sanitation-Related Health Aspects in Khorezm, Uzbekistan

Susanne Herbst, Dilorom Fayzieva, and Thomas Kistemann

Abstract The severe man-made degradation of water and soil in the Aral Sea basin is considered to cause serious human health problems. Hence, the Center for Development (ZEF) in collaboration with the Institute for Hygiene and Public Health at the Bonn University Clinics, Germany, conducted a study on diarrhoeal diseases and its risk factors in the Khorezm Region between May 2003 and March 2004. In the study, we investigated the risk factors water, sanitation and hygiene for diarrhoeal disease by applying a combination of quantitative (exposure assessment, epidemiological data) and qualitative methods (standardized interviews, structured observations).

Diarrhoeal disease incidence in children was much higher than expected and even exceeded, with 4.6 episodes per person year for children under 5 years old, global median numbers. For those under 2 years of age, mean person-time incidence was extremely high with 6.7 episodes per child per year. Although seasonality of the age stratified person-time incidence was found, overall diarrhoeal disease incidence was high for all seasons with disease rates for the winter follow-up assessment ranging 30–59% below summer rates.

Multiple linear regression analysis revealed that visible contamination of drinking water during storage and the absence of anal cleansing materials were significantly associated with the number of diarrhoeal episodes per household, pointing to the importance of hygiene in prevention of faecal-oral disease transmission.

S. Herbst (✉) • T. Kistemann
Institute for Hygiene and Public Health, University of Bonn,
Sigmund-Freud-Strasse 25, 53105 Bonn, Germany
e-mail: suse.herbst@gmx.de; boxman@ukb.uni-bonn.de

D. Fayzieva
Institute of Water Problems, N. Khadjibayev Street 49,
100041, Tashkent, Uzbekistan
e-mail: dfayzieva@gmail.com

C. Martius et al. (eds.), *Cotton, Water, Salts and Soums: Economic and Ecological
Restructuring in Khorezm, Uzbekistan*, DOI 10.1007/978-94-007-1963-7_9,
© Springer Science+Business Media B.V. 2012

Exposure estimations according to drinking water monitoring assessed that roughly one quarter of the population was exposed to faecally contaminated drinking water at the source and an estimated 12% were at risk from excess nitrate in drinking water. A large part of the population was expected to ingest about 5–7 $g\,d^{-1}$ of salt from drinking water in 2003.

Overall, the findings show that at the time of the study the domestic domain played a major role with regard to faecal-oral disease transmission in Khorezm, Uzbekistan. Unhealthy excreta handling and disposal habits as well as unsafe drinking water treatment and handling practices have to be urgently tackled in order to break the faecal-oral transmission route.

Keywords Diarrhoea • Drinking water quality • Hygiene • Risk factor analysis • Sanitation

9.1 Introduction

Although the water needed globally to cover drinking water demands and domestic hygiene purposes amounts only to 8% of the annual global water consumption, still about 1,000,000,000 people had no access to a safe drinking water supply in 2002. At the same time about 2,600,000,000 people worldwide lacked improved sanitation and 1,600,000 deaths were directly attributed to unsafe water supply, sanitation and hygiene. Since safe drinking water supply and sanitation are two sides of the same coin, only improvements in both facilitate substantial health benefit (Anonymous 2004). Thus, the Millennium Development Goal (MDG) 7 includes halving the proportion of people without access to improved drinking water supply and sanitation by 2015. An improved drinking water source adequately protects – by the nature of its construction – the source from outside contamination, in particular with faecal matter, and an improved sanitation facility hygienically separates human excreta from human contact (WHO/UNICEF 2010).

According to the MDG monitoring report by the 'Joint Monitoring Program on Water Supply and Sanitation' for 2002, a total of 89% (97% urban, 84% rural) of the Uzbek population had access to improved drinking water supply and 53% (85% urban, 33% rural) had a household connection to the drinking water supply network (WHO/UNICEF 2004). In Khorezm, 51% of the population – 18% more than the national average – had no access to piped water and utilized drinking water from tube wells or dug wells. The coverage with improved sanitation was 57% (73% urban, 48% rural).

Recently, the transmission of typhoid fever by ingestion of non-boiled surface water has been described in the Samarkand region (Srikantiah et al. 2007). A randomized intervention study by (Semenza et al. 1998) also revealed the high incidence of diarrhoeal diseases in Karakalpakstan, the downstream district that neighbors Khorezm, where this was associated with contaminated water. For the period since

9 Water and Sanitation-Related Health Aspects in Khorezm, Uzbekistan 143

the independence in 1991, official data show a strong decline for all acute intestinal infections for Khorezm. In contradiction to that, the number of drinking water samples above the critical limit for faecal pathogens constantly increased. Hence, the present study aimed to (i) create active monitoring data on the incidence of diarrhoeal disease, (ii) identify risk factors for diarrhoeal disease, (iii) identify water and health-related behavioral habits, and (iv) monitor other water- and sanitation-related health aspects such as nitrate and salinity levels of water.

9.2 Methods and Material

To meet the needs of the study dealing with the interrelated risk factors water, sanitation and hygiene, a combination of quantitative and qualitative methods was applied, to determine indicator variables for the underlying risk factors. Epidemiological data collection was carried out using a self-reported monitoring system of diarrhoeal diseases, which was conducted over a 12-week period in the summer of 2003 and a 4-week winter follow-up in February 2004. Each of the 186 randomly selected households entered all diarrhea episodes on a daily basis into a diarrhea diary form, which was checked and exchanged by interviewers weekly.

Households were interviewed using standardized questionnaires on water source, water collection, household drinking water treatment and storage as well as on sanitation and health-related water use. Hygienic conditions of the drinking water storage vessels and the sanitation facility were checked using structured observations (spot checks). Between May and August 2003, in total 463 samples from water sources and storage vessels of the surveyed households were monitored for faecal indicator bacteria. Additionally, the chemical parameters nitrate and salinity were tested. Nitrate levels were determined with the help of the Quantofix® test stick, which facilitates detection of a nitrate level ranging from 10 to 500 mg L^{-1}. Salinity was determined using the Multi measuring probe S/N 70941 produced by Consort Eijkelkamp.

9.3 Results

9.3.1 Epidemiological Data and Risk Factor Analysis

Comparing official numbers using passive surveillance data with active surveillance data from this study reveals that active surveillance can tremendously increase reported diarrhea incidence. According to passive surveillance data from Regional Centre of Sanitation and Epidemiology (OblSES) in Urgench between 1991 and 2002, diarrhoeal diseases showed a strong seasonality peaking each year in July.

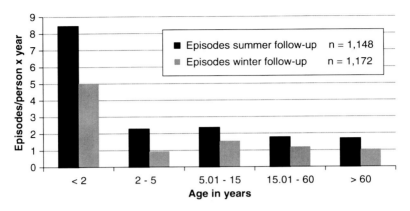

Fig. 9.1 Person-time incidence according to age strata

Over the same period diarrhea incidence decreased about 86% in Khorezm, but still peaked in July 2002 with 37 cases per 100,000 population. In contrast, active surveillance found an incidence of 16,494 case per 100,000 population per month in July 2003, while a peak of diarrhoeal disease in July could not be confirmed. A seasonality effect with higher incidence rates of 30%, but varying for the age strata, could be shown for the summer time.

The person-time incidence for diarrhea peaked in summer with 8.4 episodes per person year for those aged under two, falling to 2.3 for children aged between 2 and 5 and further decreasing to 1.7 for the age group of the over 60 year-olds. In the winter follow-up survey, the same trend for age distribution was determined with 5.0 episodes per person per year for the youngest age group and within a range from 0.9 to 1.6 episodes per person per year for the other age groups. In winter, the overall incidence for children up to 5 years was 2.5 episodes per person year. In both seasons, diarrhea incidence peaked among those aged under two, with a 41% decline in winter. Children between 2 and 5 years of age experienced 59% less diarrhea in winter. For the other age groups, the seasonal difference amounted to about 30%. Despite seasonality in occurrence of diarrhoeal disease, the person-time incidence shows high to very high figures for all studied seasons (Fig. 9.1).

Some 43 variables concerning the most studied environmental risk factors for diarrhoeal disease – water, sanitation and selected aspects of their hygiene – were included to the stepwise multiple linear regression analysis with the number of diarrhoeal disease episodes per household (n = 171) as the dependent variable. The findings of the multivariate linear model demonstrated the association between the number of diarrhea episodes per household and the exposures to (a) availability of toilet accessories such as toilet paper, newspaper and water in the latrine or toilet and (b) visible contamination of stored drinking water. This indicates that the promotion of simple interventions to break the faecal-oral disease transmission route by establishing a primary barrier is expected to have great impact on diarrhoeal disease reduction.

Table 9.1 Classification of drinking water quality according to magnitude of contamination (WHO 1997)

CFU^a/L^{-1}	Category	health risk	Proportion (%) of samples							
			Piped water				Tube well		Dug well	
			Inside		Outside					
			FC^b	E^c	FC	E	FC	E	FC	E
0	I	No	87	69	54	52	82	75	41	23
1–10	II	Intermediate to high	13	26	35	38	11	20	39	41
11–100	III	High to very high	0	5	10	10	6	5	16	24
>100	IV	High to very high	0	0	0	0	1	1	4	11

^{a}*CFU* Colony forming units
^{b}*FC* Faecal coliforms
^{c}*E Enterococci*

9.3.2 Quality of Water Sources and at Point-of-Use Water Quality

At the time of the study in Khorezm about 50% of the population – mostly in urban regions – had access to piped drinking water. In rural areas groundwater or shallow groundwater from tube wells or dug wells were the predominant drinking water sources. A tube well consists of an perforated iron pipe, which is driven into the earth until a water-bearing stratum is reached and equipped with an electrical or manual suction pump at the upper end. Dug wells are dug out by hand or by an auguring machine, they are usually old, large in diameter and often cased by concrete, a piece of pipe or hand-laid bricks. The microbiological drinking water quality differed for each of the sources. Drinking water taken from piped water taps outside of the house contained substantially more faecal indicators than from taps inside the house. Samples from tube wells were less often and also to a lesser extend contaminated than samples from dug wells (Table 9.1).

Concerning the microbiological quality of drinking water and household water needs, it is important to distinguish source quality and point-of-use quality. Even excellent source water can deteriorate during collection, transport, treatment and storage (Clasen and Bastable 2003; Wright et al. 2004; Hoque et al. 2006) before ingestion or use for household needs. Recent studies and systematic reviews have shown that point-of-use drinking water treatment can be very effective (Clasen et al. 2007) and might have even a higher impact on diarrhea outcome than source quality (Quick et al. 1999; Crump et al. 2005; Fewtrell et al. 2005; Clasen et al. 2006; Arnold and Colford 2007). Results from this study show that the percentage of tube well samples contaminated with faecal coliforms and *enterococci* was double that of those wells with hand pumps that needed priming (Table 9.2).

Household drinking water was treated by 56% of the households in summer and by 48% in winter. Measures applied to improve drinking water quality comprised boiling, settling and filtering or a combination of these activities. Due to higher turbidity households using treated piped water applied drinking water treatment more often than those utilizing water from dug wells or tube wells.

Predominant storage vessels were buckets (76%), 75% of which were covered. Usually, short handled dippers were used for ladling. An optimal storage container

Table 9.2 Microbiological quality of tube well water

Priming needed	Yes		No	
	No of samples	%	No of samples	%
	37		85	
Faecal coliforms > 0 CFU[a]/L^{-1}	9	24	7	8
Enterococci > 0 CFU/L^{-1}	14	38	16	19
Heterotrophic plate count 100 CFU/L^{-1}	36	97	65	76

[a]Colony forming units

with a narrow opening and an outlet could not be found in the households enrolled in the study.

9.3.3 Sanitation and Hygiene Issues

The vast majority of the population (88%) in Khorezm had access to improved sanitation by means of pit latrines. But what does 'potentially improved sanitation' mean in terms of environmental faecal-oral pathogen load under the given excreta management practices?

Pit latrines were emptied by 78% of the survey households at least once per year. Most commonly, the faecal sludge was dug out by own labor. Flushing out latrines by diverting surface waters into the pit was reported for 6% of the households. The most common ways of faecal sludge deposition included a burial it in the garden (50%), an application on the vegetables in the garden (22%) or a deposition onto agricultural land (12%).

Taking into account a minimal safety distance recommendation of 15 m (Cave and Kolsky 1999), in 9% of the households this safety distance between latrines and drinking water sources was violated. Since on-site sanitation facilities can create hazards to personal and public health via groundwater contamination and unsanitary conditions, application of this basic rule is essential.

Overall, open deposition of children's faeces was a common practice. Only, half of the survey households deposited children's faeces via latrine or flush toilet, a 25% buried it in the garden and the other 25% practiced open deposition in the garden. In the remaining households small children were free to defecate. There is observational evidence that washing hands at most critical times was not regarded as essential in personal hygiene.

9.3.4 Nitrate

Nitrate concentration in piped water (median 0, mean 2.14, range 0–10 mg L^{-1}) was always below the critical limit of 50 mg L^{-1} (European Union 1998; WHO 2004). Drinking water from dug wells contained more nitrate than piped water but was on the average also below the critical limit (median 10, mean 16.9, range 0–50 mg L^{-1}). The median for nitrate in drinking water samples from tube wells was 25 mg L^{-1} (mean 45, range 0–250 mg L^{-1}), but 11 samples showed concentrations substantially above the 50 mgL^{-1} threshold. A separate analysis of dug well (n = 7) and tube well (n = 13) samples showed that four out of the seven dug wells contained less than 10 mg L^{-1} nitrate (median). For two dug wells the median concentration varied around the critical limit. The nitrate concentration in drinking water from tube wells varied substantially between different sampling points. Three out of the 13 tube wells permanently contained excess nitrate above the critical limit, which poses a risk to infant health.

9.3.5 Salinity of Water Sources

The salinity was measured twice for all drinking water sampling points except for the storage vessels. All drinking water sources located in the Urgench, Khiva and Kushkupir districts contained increased levels of salt. The taste threshold of about 200 mg L^{-1} (WHO 2004) was exceeded roughly around tenfold in all well water samples. The Uzbek State Standard sets limits at 1 g L^{-1} for piped water and 1.5 g L^{-1} for decentralized sources such as wells. Piped water contained on average 1.3 g L^{-1} salt (median) ranging from 0.7 to 1.7 g L^{-1}. For dug wells the salt level ranged from 1.9 to 3.4 g L^{-1} with the median of 2.4 g L^{-1}. The median salt concentration of tube well water was the same as for the water in dug wells, but the range in the tube well sampling points was, with 1.9–4.6 g L^{-1}, much broader.

9.4 Discussion

9.4.1 Diarrhoeal Disease and Related Water, Sanitation and Hygiene Aspects

Overall population-based estimates assess the incidence of diarrhea to 0.2–1.4 episodes per person per year for industrialized countries (Wheeler et al. 1999; Herikstad et al. 2002). A study conducted in Cherepovets, Russia, including all age groups found an incidence of 1.7 episodes per person per year (Egorov et al. 2003), which coincides well with the 1.8 episodes per person per year for all ages found in Khorezm. Thus, the situation in Khorezm may not be much different from other rural areas in Central Asia and Russia.

The children's global median incidence for the period between 1990 and 2000 was highest among those under 1 year of age with 4.8 episodes per child per year, falling progressively to 1.4 episodes per child per year for the 4-year-olds. The global median number of episodes for all children under 5 is 3.2 per child per year (Kosek et al. 2003) ranging between 6 and 7 episodes per child in low-income countries and one or two episodes per child per year in high-income countries (Thapar and Sanderson 2004). The person-time incidence rates for children in Khorezm exceeded median global estimates for non-industrialized countries. Estimates for mean person-time incidences have been carried out assuming 6-month summer and 6-month winter conditions. Thus, in Khorezm each individual under the age of five suffers on average from 4.6 episodes per child per year. Children aged under two face 6.7 episodes per child per year, which decreases to 1.6 episodes per child per year for those aged between 2 and 5.

When comparing the magnitude of diarrhea for children under the age of five in Khorezm and elsewhere according to the estimated yearly person-time incidence rates, it became evident that the incidence of diarrhea is tremendously high in the study region. The current personal and economic burden that the high incidence of childhood diarrhea imposes on the children's families is expected to be enormous, but may even be exceeded by the unknown long-term effects and costs of regular early childhood diarrhea. Recent studies on the long-term effects of regular early childhood diarrhea have proven consequences such as growth faltering (Moore et al. 2001) and impaired physical and cognitive development in later life (Guerrant et al. 2002; Niehaus et al. 2002). The results therefore give reason for increased alarm and actions with the aim to interrupt the vicious cycle of diarrhea, malnutrition and impaired child development.

Assuming that conditions of drinking water supply similar to those found in this study are typical for the Khorezm region, an estimated 85,000 households (24% of population or about 340,000 persons) are being exposed to faecally contaminated drinking water, which demands for immediate action to reduce faecal-oral disease transmission. Home chlorination of drinking water with high salinity – as proven for wells in Khorezm – is thought to create a 'taste composition', which will not be accepted by consumers in the long run. However, the high number of sunlight hours (3,000–3,100 h year^{-1}) and the high total solar radiation in Uzbekistan (in flat areas 130–160 ccal cm^{-1} year^{-1}) suggest that simple solar disinfection systems could be installed. Producing low budget systems locally could thus offer microbiologically safe drinking water and opportunities for local enterprises. Another low budget option would be point-of-use water treatment by means of locally produced simple ceramic filters.

The population in Khorezm has access to basic sanitation, but a huge share is unaware of health hazards and environmental pollution due to hygienically unsafe sewage discharge, excreta and solid waste management. For example widely practiced fertilization of vegetable gardens with untreated human excreta – before being safely deposited in improved sanitation facilities – leads to a contamination with faecal-oral pathogens. In fact, this means that a lot of potentially safely disposed human excreta are brought back to the domestic and public environment posing in turn a hazard to public and personal health. Another unacceptable practice is the deposition of toddlers faeces in the garden. Because due to the very high incidence of diarrhoeal disease in children aged under two, toddlers faeces contaminate neighborhoods. Hence, being a major source for faecal-oral disease transmission.

However, three facts promise to be a good starting point for the implementation of sustainable sanitation. Firstly, the already practiced hygienically unsafe use of human excreta as fertilizer shows that the social acceptance of nutrient recycling is high. Secondly, the socio-economic situation demands for affordable fertilizers, and thirdly, the environmental situation requires for sustainable solutions.

Overall, a fundamental lack of knowledge about interrelationships between sanitation, drinking water quality, hygiene and health and their implications triggers unhealthy habits in the Khorezm region. Because of the very low awareness of associated health risks accompanying health hygiene education deems essential.

9.4.2 Nitrate

In some surveyed households the heavy fertilization of orchards in spring led to high nitrate levels in wells. The primary health concern regarding excess nitrate and nitrite in drinking water is methaemoglobinaemia, also known as 'blue baby syndrome'. The reduced oxygen level in the infant's blood causes exertional dyspnoea, cyanosis and respiratory depression. In serious cases, stupor and asphyxia will develop which may lead also to death of the individual (Hunter 1997).

Gastro-intestinal infections exacerbate the conversion from nitrate to nitrite, especially in children (Knobeloch et al. 2000). Methaemoglobinaemia affects hundreds of infants in the United States each year (Knobeloch et al. 2000) and is regarded as a common public health problem in Eastern Europe (Ayebo 1997). Case studies prove this serious medical condition in infants under 6-months that are bottle-fed with formulas prepared with nitrate-contaminated water from wells. Some sources report that the water used for preparation of formulas contained even less than 50 mg L^{-1} of nitrate (Knobeloch et al. 1993, 2000; Knobeloch and Proctor 2001).

In the present study, 25% of all wells at least occasionally had nitrate concentrations higher than 50 mg L^{-1} (60% > 25 mg L^{-1}). On average, about 50% of the population utilized shallow groundwater for drinking water purposes. If the rate of contamination is similar in the non-studies districts of Khorezm, water supplies in an estimated number of 44,500 households failed to meet the standard. Based on current yearly birth data, 6,290 infants (under 6-months of age) are expected to live in homes that have, at least from time to time, nitrate-contaminated water supplies. Assuming a threshold concentration of 25 mg L^{-1}, about 15,100 children are at risk to be exposed.

9.4.3 Salinity

Another water quality problem of health concern in the Khorezm region is the high salinity in all drinking water sources. The (American Heart Association 2005) recommends a total daily sodium intake of 2.3 g, which corresponds to 5.8 g salt (sodium chloride). Assuming a medium to low drinking water intake of 2 L per day under the prevailing arid conditions, a larger part of the population of Khorezm, may take in about 5–7 g d^{-1} of salt via drinking water alone.

Recent epidemiological studies focused mainly on the links between diets induced sodium consumption and hypertension, which has been subject of considerable scientific controversy over decades. Though, recent studies confirmed a clear link between sodium intake, hypertension and cardiovascular diseases (Cook et al. 2007; Walker et al. 2007; Cook 2008). The impact of sodium on human body fluid and salt homeostasis, bone density, renal disease and loss of muscle mass was also only recently studied (Jones et al. 1997, 2007; Frings-Meuthen et al. 2008). There is even evidence for a so far unknown osmotically inactive sodium storage mechanism in the human body, which removes sodium from the circulation if the daily sodium intake is above 5.7 g (250 mmol) (Heer et al. 2000; Gerzer and Heer 2005). The consequences of these hitherto unknown sodium storage mechanisms for human health are completely unknown.

While a significant and positively correlated risk between elevated sodium levels in the diet and coronary heart disease has been established (Tuomilehto et al. 2001; He and MacGregor 2002), there is insufficient evidence of the impact of high sodium levels on cardiovascular illnesses when the source of the sodium is from long-term intake of water. An epidemiological study – in the village Gandimiyon close to Khiva (Khorezm) – showed that the prevalence of hypertension among the examined population group was 29 per 1,000 population on average. It was significantly higher ($P < 0.001$) among respondents using well water (45/1,000), than those respondents using tap water (13/1,000). At the same time, the increased prevalence of hypertension among examined subjects was associated with an increased volume of consumed drinking water (Fayzieva et al. 2002).

9.5 Recommendations

Domestic hygiene, sanitation and health-related behavior play a major role in faecal-oral disease transmission in Khorezm. Hence, the recommendations mainly aim at establishing primary barriers in faecal-oral disease transmission. These include:

- Upgrading of sanitation facilities and hygiene education in public domains such as kindergartens, schools, and universities;
- Hygienically safe pit emptying techniques, nutrient recycling and domestic wastewater disposal in the private domain;
- Health education on causes, prevention and proper management of diarrhea with regard to sanitation, personal hygiene and strategies on safe drinking water storage in institutions such as kindergartens, schools, universities;
- Elimination of open defecation by children and promotion of safe disposal of children's excreta;
- Conduct of a representative, structured observation on hand washing behavior – to verify the results of this study – and subsequent promotion of proper hand washing after contact with faecal matter;

- Improvement of protection of water sources by keeping safety distances between latrines and wells, providing dug wells with covers and avoiding priming in hand pumps.

Because most people have financial constraints, any intervention should set incentives to invest energy and money into hygiene and sanitation and not simply spread regulations and lectures. Therefore, only a participatory or social marketing approach seems to be appropriate to meet those requirements (Water and Sanitation Program 2004; Tearfund 2010).

Taking into account local drinking water taste problems, one recommendation concerning secondary barriers of faecal-oral disease transmission should target the development of sustainable and affordable point-of-use treatment measures of drinking water. Here, the local production of simple ceramic filters and/or small-scale solar systems facilitating desalinization and disinfection would be an optimal solution for Khorezm with its high amount of solar irradiation.

Based on the observations during the course of the study and the results, the following further research needs could be identified: (i) long-term ingestion of saline drinking water and its effect on human health, (ii) incidence of methaemoglobinaemia in rural areas due to high nitrate contamination of decentralized drinking water sources (>50 mg L^{-1}), (iii) impact of irrigation with sewage-contaminated surface water, (iv) helminthic diseases and child malnutrition.

References

American Heart Association (2005) Sodium recommendations. www.americanheart.org/presenter.jhtml?identifier=4708. Accessed 27 June 2008

Anonymous (2004) Clean water alone cannot prevent disease. Lancet 364:816

Arnold BF, Colford JM (2007) Treating water with chlorine at point-of-use to improve water quality and reduce child diarrhea in developing countries: a systematic review and meta-analysis. Am J Trop Med Hyg 76:354–364

Ayebo A (1997) Infant methemoglobinemia in the Transylvania region of Romania. Int J Occup Environ Health 3:20–29

Cave B, Kolsky P (1999) WELL study – groundwater, latrines and health. London school of hygiene & tropical medicine, UK WEDC. Loughborough University, London

Clasen TF, Bastable A (2003) Faecal contamination of drinking water during collection and household storage: the need to extend protection to the point of use. J Water Health 1:109–115

Clasen TF, Brown J, Collin SM (2006) Preventing diarrhoea with household ceramic water filters: assessment of a pilot project in Bolivia. Int J Environ Health Res 16:231–239

Clasen T, Schmidt WP, Rabie T, Roberts I, Cairncross S (2007) Interventions to improve water quality for preventing diarrhoea: systematic review and meta-analysis. Br Med J 334:782

Cook NR (2008) Salt intake, blood pressure and clinical outcomes. Curr Opin Nephrol Hypertens 17:310–314

Cook NR, Cutler JA, Obarzanek E, Buring JE, Rexrode KM, Kumanyika SK, Appel LJ, Whelton PK (2007) Long term effects of dietary sodium reduction on cardiovascular disease outcomes: observational follow-up of the trials of hypertension prevention (TOHP). Br Med J 334:885

Crump JA, Otieno PO, Slutsker L, Keswick BH, Rosen DH, Hoekstra RM, Vulule JM, Luby SP (2005) Household based treatment of drinking water with flocculant-disinfectant for preventing diarrhoea in areas with turbid source water in rural western Kenya: cluster randomised controlled trial. Br Med J 331:478

Egorov AI, Naumova EN, Tereschenko AA, Kislitsin VA, Ford TE (2003) Daily variations in effluent water turbidity and diarrhoeal illness in a Russian city. Int J Environ Health Res 13:81–94

European Union (1998) Council directive (98/83/EC) of 3 November 1998 on the quality of water intended for human consumption. Official Journal of the European Commission L330/32

Fayzieva DK, Arustamov DL, Nurullaev RB, Klyopov YY, Kudyakov R (2002) Study on the quality of potable water, rate of the urolithiasis and hypertension in the Aral Sea Area. In: Proceedings of the 5th international congress "ECWATECH-2002, Water: ecology and technology", Moscow, 4–7 June 2002, p 493 (in English), pp 687–688 (in Russian)

Fewtrell L, Kaufmann RB, Kay D, Enanoria W, Haller L, Colford JM (2005) Water, sanitation, and hygiene interventions to reduce diarrhoea in less developed countries: a systematic review and meta-analysis. Lancet Infect Dis 5:42–52

Frings-Meuthen P, Baecker N, Heer M (2008) Low-grade metabolic acidosis may be the cause of sodium chloride-induced exaggerated bone resorption. J Bone Miner Res 23:517–524

Gerzer R, Heer M (2005) Regulation of body fluid and salt homeostasis-from observations in space to new concepts on Earth. Curr Pharm Biotechnol 6(4):299–304

Guerrant RL, Kosek M, Moore S, Lornitz B, Brantley R, Lima AA (2002) Magnitude and impact of diarrheal diseases. Arch Med Res 33:351–355

He FJ, MacGregor GA (2002) Effect of modest salt reduction on blood pressure: a meta-analysis of randomized trials. Implications for public health. J Hum Hypertens 16:761–770

Heer M, Baisch F, Kroop J, Gerzer R, Drummer C (2000) High dietary sodium chloride consumption may not induce body fluid retention in humans. Am J Physiol Renal Physiol 278:F585–F595

Herikstad H, Yang S, Van Gilder TJ, Vugia D, Hadler J, Blake P, Deneen V, Shiferaw B, Angulo FJ (2002) A population-based estimate of the burden of diarrhoeal illness in the United States: foodNet, 1996–7. Epidemiol Infect 129:9–17

Hoque BA, Hallman K, Levy J, Bouis H, Ali N, Khan F, Khanam S, Kabir M, Hossain S, Shah Alam M (2006) Rural drinking water at supply and household levels: quality and management. Int J Hyg Environ Health 209:451–460

Hunter P (1997) Waterborne disease: epidemiology and ecology. Wiley, Chichester/New York/Weinheim

Jones G, Beard T, Parameswaran V, Greenaway T, von Witt R (1997) A population-based study of the relationship between salt intake, bone resorption and bone mass. Eur J Clin Nutr 51:561–565

Jones G, Dwyer T, Hynes KL, Parameswaran V, Udayan R, Greenaway TM (2007) A prospective study of urinary electrolytes and bone turnover in adolescent males. Clin Nutr 26:619–623

Knobeloch L, Proctor ME (2001) Eight blue babies. Off Publ State Med Soc Wis 100:43–47

Knobeloch L, Krenz K, Andderson H, Howell C (1993) Methemoglobinemia in an infant. Morb Mortal Wkly Rep 42:217–219

Knobeloch L, Salna B, Hogan A, Postle J, Anderson H (2000) Blue babies and nitrate-contaminated well water. Environ Health Perspect 108:675–678

Kosek M, Bern C, Guerrant RL (2003) The global burden of diarrhoeal disease, as estimated from studies published between 1992 and 2000. Bull World Health Organ 81:197–204

Moore SR, Lima AA, Conaway MR, Schorling JB, Soares AM, Guerrant RL (2001) Early childhood diarrhoea and helminthiases associate with long-term linear growth faltering. Int J Epidemiol 30:1457–1464

Niehaus MD, Moore MD, Patrick PD, Derr LL, Lorntz B, Lima AA, Guerrant RL (2002) Early childhood diarrhea is associated with diminished cognitive function 4 to 7 years later in children in a northeast Brazilian shantytown. Am J Trop Med Hyg 66:590–593

Quick RE, Venczel LV, Mintz ED, Soleto L, Aparicio J, Gironaz M, Hutwagner L, Greene K, Bopp C, Maloney K, Chavez D, Sobsey M, Tauxe RV (1999) Diarrhoea prevention in Bolivia through point-of-use water treatment and safe storage: a promising new strategy. Epidemiol Infect 122:83–90

9 Water and Sanitation-Related Health Aspects in Khorezm, Uzbekistan

Semenza JC, Roberts L, Henderson A, Bogan J, Rubin CH (1998) Water distribution system and diarrheal disease transmission: a case study in Uzbekistan. Am J Trop Med Hyg 59:941–946

Srikantiah P, Vafokulov S, Luby S, Ishmail T, Earhart K, Khodjaev N, Jennings G, Crump J, Mahoney F (2007) Epidemiology and risk factors for endemic typhoid fever in Uzbekistan. Trop Med Int Health 12(7):838–847

Tearfund (2010) Adoption of community-led total sanitation guidance for programming of CLTS in tearfund-supported projects. http://www.communityledtotalsanitation.org/sites/communityledtotalsanitation.org/files/Tearfund_CLTS_guidelines.pdf. Accessed 11 Jan 2011

Thapar N, Sanderson IR (2004) Diarrhoea in children: an interface between developing and developed countries. Lancet 363:641–653

Tuomilehto J, Jousilahti P, Rastenyte D, Moltchanov V, Tanskanen A, Pietinen P, Nissinen A (2001) Urinary sodium excretion and cardiovascular mortality in Finland: a prospective study. Lancet 357:848–851

Walker J, Mackenzie AD, Dunning J (2007) Does reducing your salt intake make you live longer? Interact Cardiovasc Thorac Surg 6:793–798

Water and Sanitation Program (2004) Field note – the case for marketing sanitation. http://siteresources. worldbank.org/INTWSS/Resources/case_marketing_sanitation.pdf. Accessed 11 Jan 2011

Wheeler JG, Sethi D, Cowden JM, Wall PG, Rodrigues LC, Tompkins DS, Hudson MJ, Roderick PJ (1999) Study of infectious intestinal disease in England: rates in the community, presenting to general practice, and reported to national surveillance. The infectious intestinal disease study executive. Br Med J 318:1046–1050

WHO (1997) Guidelines for drinking water quality – surveillance and control of community supplies, vol 3, 2nd edn. World Health Organisation, Geneva

WHO (2004) Guidelines for drinking water quality – recommendations. World Health Organization, Geneva

WHO/UNICEF (2004) Meeting the MDG drinking water and sanitation target: a mid-term assessment of progress. Joint monitoring programme for water supply and sanitation, Geneva

WHO/UNICEF (2010) Joint monitoring programme for water supply and sanitation. Progress on sanitation and drinking-water: 2010 update. http://www.wssinfo.org/resources/documents. html. Accessed 27 Sep 2010

Wright J, Gundry S, Conroy R (2004) Household drinking water in developing countries: a systematic review of microbiological contamination between source and point-of-use. Trop Med Int Health 9:106–117

Chapter 10
Price and Income Elasticity of Residential Electricity Consumption in Khorezm

Bahtiyor Eshchanov, Mona Grinwis, and Sanaatbek Salaev

Abstract Price or income elasticity of demand is the percentage change in electricity demand resulting from a one-percent change in its price or income. Information on price and income elasticity of demand for electricity is crucial for formulating appropriate reform policies. In Central Asian economies with a less-developed industrial sector, the share of residential electricity consumption is relatively high. Residential electricity consumers are known to be more flexible with respect to price- and income-related changes and to adapt to these changes relatively fast. Knowledge about price and income elasticity of electricity demand is crucial for implementing, for instance, tariff reforms. The aim of this study is to estimate the short-term price and income elasticity of residential electricity demand using panel data based on time series of 6 years for the 12 districts of the Khorezm region, Uzbekistan, under the condition of data limitations. Using a unique dataset, the study attempts to identify the best proxy for the unobservable income variable. As for the price of electricity, nominal and real electricity prices tended to be growing between 2002–2007 from 7.0 Uzbek *Soums* (UZS) to 42.3 UZS (nominal), and 7.0 to 33.5 UZS (real), respectively, while the US$ equivalent of the nominal price decreased from 6.23 US cents in 2002 to 3.35 US cents in 2007. Residential electricity demand is relatively price inelastic (between 0 and −1) in the short run. The study further tests the validity of indicators such as industrialization and growth in residential area as alternative measures of economic development that

---

B. Eshchanov (✉) • M. Grinwis
Faculty of Economic, Political and Social Sciences and Solvay Business School,
Vrije Universiteit Brussel, Pleinlaan 2, Brussels 1050, Belgium
e-mail: bahtiyor.eshchanov@vub.ac.be; mona.grinwis@vub.ac.be

S. Salaev
Urgench State University, Khamid Olimjan str., 14,
Urgench 220100 Khorezm, Uzbekistan
e-mail: s_sanat@list.ru

C. Martius et al. (eds.), *Cotton, Water, Salts and Soums: Economic and Ecological Restructuring in Khorezm, Uzbekistan*, DOI 10.1007/978-94-007-1963-7_10,
© Springer Science+Business Media B.V. 2012

cannot be captured by income. The results suggest that the income elasticities have a low value. The industrialization rate variable has a high value of demand determination. The growth in residential area used in this study as explanatory variable requires more precise data to be conclusive.

Keywords Electricity demand • Short-run elasticity • Panel data • Transition economies • Central Asia

10.1 Introduction

The availability of affordable, secure and environmentally friendly energy is a key to welfare, economic growth and sustainable development (Karimov 1998). Because of the scarcity of this kind of energy source, it is undoubtedly important to use the present sources effectively (Figueres and Philips 2007). This is particularly true in developing and transitional countries that usually have an enormous potential for increasing energy efficiency, which in turn could be exploited through adequate and appropriate reform policies.

Among the many differences between developed and developing countries, specific factors such as the use of traditional types of energy,[1] process of transition to modern energies, and structural deficiencies in various sectors of the economy, inadequate investment decisions and misleading subsidies (Urban et al. 2007) are to be underlined in line with the commonly accepted inefficiency in electricity generation, distribution and consumption. Moreover, in developing countries, residential electricity demand is characterized by rapid growth over the last two decades of the last century (Filippini and Pachauri 2002), and the price and income elasticity of electricity consumption has been shown to be unstable over time (Holtedahl and Joutz 2004). This is illustrated in particular in countries of the former Soviet Union, which were previously part of a centralized system and subjected to objectives dictated by Moscow. Since their independence in 1991, these countries have had to reform their energy sector (IMF 2002). For example, the rich oil and gas reserves in Central Asian countries (CAC)[2] had been exploited during the Soviet era for serving overall objectives but had not aimed at supporting the economic development and growth of the individual countries.

In this regard, CAC transition economies can be grouped separately among the developing countries. They share unique similarities such as (i) slow growth in electricity consumption despite their recent rapid economic development and population growth, (ii) identical energy generation systems built during the Soviet era (e.g., in terms of technological performance and efficiency levels), (iii) introduction of energy sector reform and pricing policies with a high level of subsidization,

---

[1] Such as firewood, cotton stalks, reeds, etc.

[2] Kazakhstan, Kyrgyzstan, Tajikistan, Turkmenistan and Uzbekistan.

10 Price and Income Elasticity of Residential Electricity Consumption in Khorezm

(iv) pace of macro-economic stability (e.g., GDP growth rate, inflation), and (v) structure of national economies with a high share of the agriculture sector (Dowling and Wignaraja 2006) except for Kazakhstan. Although the newly independent CAC announced energy sector reform policies in the aftermath of independence in 1991, these have been introduced with different intensity and pace. Also, until today, energy saving opportunities have not been tackled, although they could be achieved through relatively simple policy measures. A correct pricing that leads to an efficient consumption of electricity is among these measures.

Given that residential electricity consumption constitutes on average more than one fourth (27.5%) of the total global energy demand (EarthTrends 2005) with an even higher share in most non-industrialized developing countries, it is particularly important to consider this component in overall electricity consumption. In Uzbekistan, for instance, the share of total primary energy consumed by the residential sector was as high as 39% in 2005 (EarthTrends 2005) and therefore much higher than the global average. Therefore, developing and applying price related measures for the residential electricity consumers increases efficiency and consequently enhances the welfare of the people (Dodonov et al. 2004). As increasing the efficiency of electricity consumption is considered to be part of the solutions to the scarcity of energy sources it is important to identify the determinants of residential electricity consumption such as price and income. Knowing price and income elasticities is crucial for the development of a new energy policy.

10.2 Residential Electricity Demand Modeling

Knowledge about residential electricity demand is vital for formulating energy pricing policies, given the significant share of residential electricity consumption in the overall electricity demand. However, electricity producers in CAC are known for oversupplying (Atakhanova and Howie 2007) or undersupplying (Komilov 2002) residential consumers. Consequently, the current residential consumption does not reflect the real demand. In a free market, such excess supply or demand would lead to price changes and in the end could clear such market disturbances.

Once the determinants of demand and magnitude of price and income elasticity of residential electricity consumption are known, the authorities can make better informed decisions to introduce necessary reforms in electricity generation, distribution and consumption efficiency. However, data on most variables are only available for short periods of a maximum of 10 years, while demand functions need longer time series of a period of 30 years and more (Dilaver and Hunt 2011), which are unavailable for most of the post Soviet countries. Consequently, to obtain plausible estimates for the short-run price and income elasticity of electricity consumption, the use of more sophisticated econometric methods presently seems a sensible approach to fill data gaps. Although residential energy demand modeling received a strong impetus after the early 1970s, mainly triggered by the global energy crisis, to date there is no all-encompassing modeling approach (Bhattacharyya and Timilsina 2010). Most of the models developed refer to pioneering work on residential electricity demand in the USA by Fisher and Kaysen (1962) and consider developed economies, whereas empirical approaches specifically designed for developing countries are still in their early stages of development and require a more spatially explicit approach of each specific geographic location.

Residential electricity demand models are classified into two groups: (i) models based on an end-users approach (input-output analysis) and (ii) macro-economic models based on an econometric approach (Bhattacharyya and Timilsina 2010). While end-user models are known for their higher data requirements and need of human resources to handle the overall process of modeling, they are recommended for developed countries due to their higher precision (World Bank 2004). The econometric approach is presently most widely applied due to its ability to cope with data limitations. Furthermore, due to their simplicity, macro-economic models are therefore presently preferred. Fisher and Kaysen (1962) used growth rate in electricity utilizing appliance stock as a proxy to income, together with price and income of households for 1946–1957 in the USA, and determined income and price elasticity of the short- and long-run demand for electricity. Halvorsen (1975) included natural gas price as a substitute good price in an electricity demand model to capture the interrelation between demand for electricity and demand for natural gas. Therefore, he introduced area and temperature-related data to capture the fundamental demand components. In contrast, Holtedahl and Joutz (2004) used urbanization as a proxy for a stock of electric appliances to identify economic development characteristics and changes in the stock of electric appliances, thus avoiding the shortcomings of applying Fisher's and Kaysen's approach to developing countries. Halicioglu (2007) confirmed the validity of the urbanization variable as a proxy for the stock of electric appliances in Turkey. In a study funded by the Bank World (2004), the aggregated electricity demand in CAC was estimated through the total sales in GWh for the electricity sector as a whole, not distinguishing the different consumer categories such as the industry, agriculture, services and residential consumers. Atakhanova and Howie (2007) used the industrialization rate of the provinces of Kazakhstan to capture economic development factors that are not captured by income. The industrialization rate was measured by the contribution of the regional industry sector in the gross regional product (GRP). Aside from the study by Atakhanova and Howie (2007), who modeled electricity demand for Kazakhstan and estimated the price and income elasticity of demand for electricity for the industrial, service and residential consumers, we know of no other study that has addressed the price and income elasticity of electricity in CAC.

The objective of the present study is to estimate the price and income elasticity in the Khorezm region in northwest Uzbekistan through econometric modeling under conditions of data limitations, especially with respect to the unobservable income variable. Apart from fundamental determinants of demand (price and income proxies), this study also uses proxies to capture economic development that possibly cannot be captured with income proxies. Data on price of electricity, income, industrialization rate and increase in residential areas were used as key determinants to estimate electricity consumption in this region. Thus, this study attempts to address the special characteristics of electricity demand such as the potential relation with the size of new dwellings.

10.3 Methodology

Elasticity of energy demand is a measure that denotes the responsiveness of the quantity demanded of energy (electricity in this case) to a change in its price or consumers' incomes. It gives the percentage change in quantity demanded in response to a 1% change in price or income while holding everything else constant. A large number of empirical studies have estimated the price and income elasticity of energy demand. The interest in price and income elasticity of demand is high because these factors are vital when forecasting energy demand (Medlock 2009). The price and income elasticity of demand are unit-free measures. In general the price elasticity of demand is negative meaning that an increase in the price of a good leads to a decrease in the quantity demand. With the income elasticity of demand this is different: an increase in income tends to increase the quantity demanded, depending on the kind of goods. So usually the income elasticity of demand will be positive.

10.3.1 Description of the Study Region

Twelve administrative districts (ten district and two city administrations) in the Khorezm region in northwest Uzbekistan form the basis of our study. This region is representative of both Uzbekistan and the neighboring countries. From the approximate 1,500,000 people (as of 2008), about 75% are rural and 25% are considered urban. The income of the rural population depends to a large extent on income from agricultural activities, notably from cotton production (Rudenko et al. 2008). Cotton is the dominant crop in the region. Khorezm shares with other regions in CAC a common culture and history that directly impacts energy consumption patterns. The region also has similar climatic conditions with short, cold winters and long, hot summers that directly influence household demand for energy for heating, lighting and cooking (Conrad 2006; Conrad et al. 2011). The average size of a household in Khorezm is 6–10 persons.

10.3.2 Measuring Determinants of Electricity Demand

In previous studies, the panel-data approach was successfully applied to estimate either the short-term elasticity of demand or to capture the cross-sectional variation of electricity consumption behavior within sub-regions (e.g. Garcia-Cerrutti 2000;

Narayan et al. 2007; Cebula 2010; Eskeland and Mideksa 2010; Nakajima 2010). For key indicators such as income and urbanization rate, panel data of the 12 districts for 2002–2007 were used to cope with the incompleteness of the long-term dataset. This panel data consists of time series of 6 years for all 12 districts and thus of 72 observations for each variable.

The analysis used data from three sources. Residential electricity supply data were provided by the only regional electricity distribution company,[3] and included information on a district level of annual residential electricity consumption (in kWh) and annual average price (in local currency, Uzbek *Soums*, UZS[4]) of electricity. Since these data were collected for accounting purposes, they are very likely to be accurate. This dataset was matched with data from the regional statistical authority (UZISTIQBOLSTAT) on yearly observations of population monetary income, yearly increase in the housing space at district levels for 2002–2007, district level breakdown of the annual Gross Regional Product (GRP), cotton yields for the same period, and per capita monetary income per district. Next, per capita residential electricity consumption, per capita monetary income, per capita gross district income, and per capita increase in housing area were calculated. Since monetary income and price variables are nominal, the annual GDP deflator (World Bank 2010) was used to estimate real monetary income and price. Official average annual exchange rates of UZS-US$ (Wikipedia 2011) were used as an alternative for the GDP deflator method. Hence, a balanced database was developed for 2002–2007.

To overcome the main empirical challenge in modeling residential electricity in CAC transition economies, the incomplete information on income was taken care of by estimating a standard energy demand function, with alternative imperfect proxies for income.

The following functional form of residential electricity demand was adopted from Halicioglu (2007) and Atakhanova and Howie (2007):

$$\ln C_{t,i} = a_0 + a_1 \ln Y_{t,i} + a_2 \ln P_{t,i} + a_3 \ln Ind_{t,i} + a_4 \ln NewDwl_{t,i} + \varepsilon_{t,i} \quad (10.1)$$

where:

$C_{t,i}$ – residential electricity consumption per capita for the tth year in the ith district;
$Y_{t,i}$ – income per capita for the tth year in the ith district;
$P_{t,i}$ – real residential electricity price for the tth year in the ith district;
$Ind_{t,i}$ – industrialization rate for the tth year in the ith district;
$NewDwl_{t,i}$ – increase in residential dwelling area per capita for the tth year in the ith district.

---

[3]"KhorezmElektrTarmoqlari" is Khorezm's regional department of the state-owned monopole electricity distribution network company.
[4]Uzbek *Soum* (UZS) exchange rate rose from 111.90 UZS in 2002 to 1 263.80 UZS in 2007 for 1 US$.

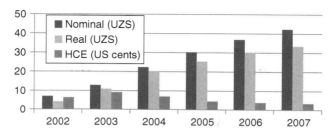

Fig. 10.1 Dynamics of the residential electricity price in Uzbekistan, 2002–2007 (in UZS and US cents kWh^{-1}). HCE denotes hard currency equivalent of the nominal price (Source: Based on data from KhorezmElektrTarmoqlari and IMF)

The dependent variable is residential electricity consumption per capita (in kWh). Since alternative sources of energy, such as solar and wind energy, are not developed, it is assumed that electricity consumption per capita alone accurately represents the volume of electricity consumed by the residential consumers.

Two alternative proxies were used to measure the independent variable income. The real price (GDP deflator method) and the US$ equivalent of real price were used for price estimations. Although the price of residential electricity increased steadily during 2002–2007, the hard currency equivalent (HCE) price in US cents decreased from 6.23 US cents kWh^{-1} in 2002 to 3.35 US cents kWh^{-1} in 2007 (Fig. 10.1)

The level of industrialization was estimated as the contribution of the district's industry sector to the GRP as previously suggested by Atakhanova and Howie (2007) for the regions of Kazakhstan. Given the low rate of industralisation in Khorezm (Rudenko et al. 2008), this seems an adequate proxy.

The model mainly served to test the fixed against random effects, since the impacts of district- or time-specific effects are unknown. The model with the best fit was selected.

In the following, the real monetary income of the population is estimated in 1,000 UZS per capita whilst the income proxy cotton yield is in tons per capita. The real price of electricity is in UZS kWh^{-1}, and the US$ equivalent of the nominal price in US cent kWh^{-1} according to the corresponding year's average exchange rate. The share of each district's industry branch in the GRP is considered as industrialization rate and is given in 1,000 UZS per capita. "New dwellings" is a measure of per capita annual increase in residential area in square meters. This last variable is introduced as it is assumed that an increase in the quantity of housing will have an impact on electricity demand.

10.4 Results and Discussion

The estimations of the model are presented in Table 10.1. As the findings represent the short-run elasticity for 2002–2007, the magnitude of the estimated price elasticity was higher than that in previous studies in developing countries (e.g. Atakhanova and Howie 2007; Halicioglu 2007) and varied from −0.5161 to −0.8991. When the

Table 10.1 Regression findings of demand for electricity under alternative estimation techniques

	Fixed effects model				Random effects model			
	Real monetary income of the population (1,000 UZS per capita)		Cotton yield (t per capita)		Real monetary income of the population (1,000 UZS per capita)		Cotton yield (t per capita)	
	Real price	HCE	Real price	HCE	Real price	HCE	Real price	HCE
Constant	−0.6330	5.8246	−0.9421	3.4927	−1.2565	5.7875	−0.9558	5.2609
	(0.56)	(0.01)*	(0.40)	(0.01)*	(0.12)	(0.00)*	(0.25)	(0.00)*
Income	0.0456	−0.2682	−0.1708	−0.0990	−0.0437	−0.0938	0.0958	0.2058
	(0.70)	(0.15)	(0.25)	(0.52)	(0.01)*	(0.00)*	(0.08)**	(0.00)*
Price	−0.5436	−0.8991	−0.6087	−0.6446	−0.5473	−0.6898	−0.5381	−0.5161
	(0.00)*	(0.00)*	(0.00)*	(0.00)*	(0.00)*	(0.00)*	(0.00)*	(0.00)*
Industrialization	1.3745	0.0090	1.6465	0.2127	1.4435	0.0431	1.3201	−0.0590
	(0.00)*	(0.95)	(0.00)*	(0.21)	(0.00)*	(0.70)	(0.00)*	(0.64)
New Dwellings	−0.1726	0.3648	−0.1797	0.3555	−0.0597	0.1311	−0.1098	0.0190
	(0.56)	(0.01)*	(0.23)	(0.02)*	(0.43)	(0.09)**	(0.25)	(0.81)
N	72	72	72	72	72	72	72	72
F-statistics	27.51	25.14	28.40	24.00	141.34	120.49	131.50	93.70
R2 within	0.6628	0.6423	0.6698	0.6315	0.6529	0.6209	0.6483	0.5824
R2 between	0.0002	0.7582	0.0171	0.4109	0.8071	0.7574	0.7500	0.5960
R2 overall	0.5350	0.4952	0.5382	0.2516	0.6784	0.6427	0.6625	0.5831

Note: p values are in parenthesis; * and ** indicate 5% and 10% levels of significance respectively

10 Price and Income Elasticity of Residential Electricity Consumption in Khorezm

hard currency equivalent (HCE) of the nominal price was used, demand was more price elastic compared to when the real price[5] was used, suggesting that even the real price may undervalue the true price of electricity. All coefficients of the real price and the HCE price were statistically significant at $p < 0.001$. Although all models tested had statistically significant price coefficients and the expected signs, other estimates did not fit some criteria. All the models had at least one of the following flaws: (a) statistically insignificant coefficients, (b) unexpected signs, or (c) a coefficient with a magnitude that cannot be explained from an economic point of view.

Uzbekistan's population of 28,000,000 is the largest in Central Asia and about 45% of the entire population in the region. Currently, the price of electricity in Khorezm and in the whole of Uzbekistan does not include the true cost of the natural resources used to produce this electricity. About 80% of the total electricity in the country is produced using natural gas, which is expected to be depleted within the next 25–30 years.

The contribution of renewable energy sources in generating electricity is very small (less than 1%). To achieve long-term sustainability, the authorities in CAC must have a strong interest in achieving a correct pricing of electricity, unless they accept further subsidization of this sector. As long as energy is significantly undervalued, as is presently the case, price mechanisms to encourage energy savings will have only a limited impact, if at all. Furthermore, under priced electricity will dis-incentivize the introduction of renewable electricity sources, which currently would mean relatively high capital investment costs for the population in CAC. These countries currently have gross national per capita incomes that vary from 700 US$ in Tajikistan to 870 US$ in Kyrgyzstan, 1,100 US$ in Uzbekistan, 3,420 US$ in Turkmenistan, and 6,920 US$ in Kazakhstan (World Bank 2009).

Furthermore, the present efficiency level of residential electricity consumption is rather low, which is due to outdated equipment (spiral-based light bulbs, valve-based TV sets, and old refrigerators). Consequently, it can be expected that an increase in income could slightly reduce the per capita electricity consumption. This is evidenced by the income coefficients of the random effects models with real monetary income as an income proxy, which were statistically significant with $p = 0.010$ under real price and $p < 0.001$ under HCE of the nominal price. As for the case of cotton yield as an income proxy, coefficients were statistically significant with a positive sign and relatively higher than in case of real monetary income. Therefore the signs in these statistically significant models were controversially under the two income proxies used (real monetary income and cotton yield). Only real monetary income supported the initial hypothesis of a negative relation between income and electricity consumption. Thus, due to the sign variation of the coefficients in the alternative models, the results for the income variable are inconclusive. Given the wide variety of non-monitored income-generating activities pursued by the rural population (Djanibekov 2008), and the unknown share of remittances from family members abroad to rural households (Conliffe 2009) for securing incomes and

---

[5]Real value is an adjusted price for the effect of inflation on the nominal price.

livelihoods, the true income of the population is very likely underestimated when based solely on cotton yields. This explains in part the low fit of the income findings based on cotton yields. Despite this discrepancy, all other criteria are well supported by the random effects model with the real price. Moreoever, Hausman's specification test results imply that fixed effects estimates should be abandoned in favor of random effects estimates. This indicates that an increase in income may also lead to an increase in per capita residential electricity consumption. The positive effect of the income growth on decreasing per capita residential electricity consumption is also proven by the findings of the current short-run elasticity estimation.

The industrialization rate was positively elastic and statistically significant, but only under the conditions of a real price, and ranged from 1.3201 to 1.6465. This indicates that every 1,000 UZS per capita increase in industrial production will lead to more than 1 kWh per capita increase in residential electricity consumption.

It was hypothesized that per capita increase in dwelling areas leads to an increase in per capita residential electricity consumption. For example, if a family in Khorezm with an average of six members formerly living together separates into two families living in two houses, the electricity consumption level (for one TV set, one refrigerator and ten light bulbs for six people) will rise with the split, as will the per capita dwelling area. In Uzbekistan, with its relatively high birth rate and rapidly increasing dwelling areas, this variable may be one of the important explanatory variables of residential electricity consumption. However, only three models out of the eight (moreover using HCE values of the price) supported the assumption of a high explanatory value of the dwelling area, as evidenced by the range of coefficients from 0.1311 to 0.3648 (Table 10.1). In the models with real price, the coefficients of the New Dwelling variable were statistically insignificant and had the opposite sign (an increase in dwelling area decreases the demand for electricity). Given the limitations of the present dataset and analyses, a more in-depth study of the current variable could clarify the present evidence that an increase in residential dwelling area leads to a significant increase in electricity consumption.

The overall goodness-of-fit of the models varied from 0.4952 to 0.6784, with real monetary income and real price under random effects providing the best fit (Table 10.1). In general, the random effects variations were captured better than those of the fixed effects, implying that variation over time is captured more precisely compared to the variation in consumption behavior within districts. This suggests that (a) price and income elasticity of electricity demand over time is not stable, and (b) the distribution of income proxies within the districts is uneven.

Central Asian economies may be able to meet the domestic demand until about 2020 through the implementation of measures aiming at reducing electricity loss, rehabilitation of the existing generative capacity and regional trade. Given the expected rapid depletion of natural gas – at present the main source of energy in Uzbekistan – the implications of price-related measures (increase in price) could be fourfold:

1. The currently low efficiency level would be significantly improved, facilitating the process of the transition to electricity-efficient high-tech equipment;

2. Further expansion of the fossil-based electricity generation sector would be dampened, leading to a more sustainable fossil-fuel use together with a lower increase in emissions;
3. The revenues to be expected from a higher price for electricity would generate more funds, which could be spent on diversifying the electricity supply sector, for instance by increasing the share of solar, wind and other renewable sources in places where these sources could generate a technically feasible capacity;
4. The increase in price, in line with facilitating an efficient consumption by the residential consumers, could prepare the ground for introducing renewable energy, making it more competitive with centrally supplied electricity.

A possible side effect of price-related changes may be an increased non-payment. A study by the Bank World (2004) estimated only 70–85% billing of the total electricity in CAC in 2002. Some non-residential consumers (i.e., mining industry) are not billed but use an exchange method of payment for the consumed electricity (Bank World 2004). Residential consumers are 100% billed. Avoiding this gap in revenue generation may require an increase in metering and accounting efforts.

10.5 Conclusions

An electricity price increase in Uzbekistan will definitely result in increased consumption efficiency and hence lead to a significant decline in per capita residential electricity consumption. But, even in that case, UzbekEnergo[6] will need to increase the residential electricity supply due to the steady population growth. However, as the price of electricity for consumers other than residential ones increased simultaneously, it is necessary to determine the effect of price increase on the industrial, agricultural and other consumers to achieve a comprehensive energy reform policy. This requires, becoming acquainted with the characteristics of the energy consumption of these consumer groups.

The income elasticity of residential electricity consumption in the short run is assumed to be negatively elastic, which is due to the transition to high-tech electricity consumption by households. This hypothesis was not sufficiently well supported by the findings. On the contrary, empirical evidence demonstrates that income has a positive elasticity with small coefficients. As the standard of living of the population increases, its demand for electricity will also be growing. In addition, increases in incomes can change the price elasticity of the electricity demand by decreasing the effect of future price changes. Therefore, in line with price-related demand-regulation mechanism, UzbekEnergo needs to seek alternative control tools as well as measures for meeting the future residential energy demand. Introduction of renewable energy resources in the remote areas can be one of these measures. In some of the eight estimated models both industrialization and increase in dwelling space revealed

---

[6]State-owned monopole electricity supplier for all customers.

(through their statistical significant coefficients) to be interesting new explanatory variables. Identifying such determinants of the electricity consumption is crucial for policy makers and demand forecasting, and could also be another important contribution for residential electricity demand modeling. Increased knowledge about the reform policy of UzbekEnergo, its current strategy, and short- and long-term planning would allow better appraising the present policy better and supporting the formulation of recommendations than presently possible.

References

Atakhanova Z, Howie P (2007) Electricity demand in Kazakhstan. Energy Policy 35:3729–3743
Bhattacharyya SC, Timilsina GR (2010) Modelling energy demand of developing countries: are the specific features adequately captured? Energy Policy 38:1979–1990
Cebula RJ (2010) Recent evidence on determinants of per residential customer electricity consumption in the U.S.: 2001–2005. J Econ Finance. doi:10.1007/s12197-12010-19159-12192
Conliffe A (2009) The combined impacts of political and environmental change on rural livelihoods in the Aral Sea region of Uzbekistan. PhD dissertation, Oxford University, Oxford
Conrad C (2006) Remote sensing based modeling and hydrological measurements to assess the agricultural water use in the Khorezm region, Uzbekistan. PhD dissertation, University of Würzburg, Würzburg
Conrad C, Schorcht G, Tischbein B, Sultonov M, Davletov S, Eshchanov R, Lamers JPA (2011) Agro-meteorological implications and trends of recent climate development in Khorezm. In: Martius C, Rudenko I, Lamers JPA, Vlek PLG (eds) Cotton, water, salts and *Soums*: economic and ecological restructuring in Khorezm, Uzbekistan. Springer, Berlin/Heidelberg/New York/Dordrecht
Dilaver Z, Hunt L (2011) Industrial electricity demand for Turkey: a structural time series analysis. Energy Econ 33: p 426
Djanibekov N (2008) A micro-economic analysis of farm restructuring in Khorezm region, Uzbekistan. PhD dissertation, Bonn University, Bonn
Dodonov B, Opitz P, Pfaffenberger W (2004) How much do electricity tariff increases in Ukraine hurt the poor? Energy Policy 32:855–863
Dowling M, Wignaraja G (2006) Central Asia after fifteen years of transition: growth, regional cooperation, and policy changes. ADB working paper series on regional economic integration no. 3, Asian Development Bank, July 2006
EarthTrends (2005) Energy consumption by sector for Uzbekistan in 2005. Retrieved on 22 Jan 2011, from Earth trends environmental information of the World Resources Institute. http://earthtrends.wri.org/pdf_library/data_tables/ene3_2005.pdf
Eskeland GS, Mideksa TK (2010) Electricity demand in a changing climate. Mitig Adapt Strateg Glob Chang 15(8):877–897
Figueres C, Philips M (2007) Scaling up demand-side energy efficiency improvements through programmatic CDM. ESMAP technical paper 120/107
Filippini M, Pachauri S (2002) Elasticity of electricity demand in urban Indian households. CEPE working paper, 16. Centre for Energy Policy and Economics Swiss Federal Institutes of Technology
Fisher FM, Kaysen C (1962) A study in econometircs: the demand for electricity in the United States. North-Holland Publishing House, Amsterdam
Garcia-Cerrutti LM (2000) Estimating elasticity of residential energy demand from panel county data using dynamic random variables models with heteroskedastic and correlated error terms. Resour Energy Econ 22:355–366

10 Price and Income Elasticity of Residential Electricity Consumption in Khorezm

Halicioglu F (2007) Residential electricity demand dynamics in Turkey. Energy Econ 29:199–210

Halvorsen R (1975) Residential demand for electric energy. Rev Econ Stat 57(1):12–18

Holtedahl P, Joutz FL (2004) Residential electricity demand in Taiwan. Energy Econ 26:201–224

IMF (2002) Energy sector quasi-fiscal activities in the countries of the former Soviet Union. International Monetary Fund working paper. Collective authors, March 2002

Karimov I (1998) Uzbekistan on the threshold of the twenty-first century: challenges to stability and progress. Palgrave Macmillan, New York p 163

Komilov A (2002) Renewable energy in Uzbekistan, potential and perspectives. Renewable Energy for Sustainable Development, DAAD and Carl Von Ossietzky Universitat Press, Oldenburg, pp 246–256

Medlock KB III (2009) Energy demand theory. In: Evans J, Hunt L (eds) International handbook on the economics of energy. Edward Elgar, Cheltenham/Northampton

Nakajima T (2010) The residential demand for electricity in Japan: an examination using empirical panel analysis techniques. J Asian Econ 21:412–420

Narayan PK, Smyth R, Prasad A (2007) Electricity consumption in G7 countries: a panel cointegration analysis of residential demand elasticity. Energy Policy 35:4485–4494

Rudenko I, Grote U, Lamers JPA (2008) Using a value chain approach for economic and environmental impact assessment of cotton production in Uzbekistan. In: Qi J, Evered KT (eds) Environmental problems of Central Asia and their economic, social, and security impacts, NATO science for peace and security series – C: environmental stability. Springer, Dordrecht, pp 361–380

Urban F, Benders RJ, Moll HC (2007) Modelling energy systems for developing countries. Energy Policy 35:3473–3482

Wikipedia (2011) Tables of historical exchange rates to the USD. Wikipedia – the free online encyclopedia, Retrieved on 02 Feb 2011 from The World Bank website: http://en.wikipedia.org/wiki/Tables_of_historical_exchange_rates_to_the_USD

World Bank (2004) Central Asia regional electricity export potential study. World Bank, Washington, DC

World Bank (2009) GNI (Atlas method). Country statistics, The World Bank database, Retrieved on 20 Mar 2011, from The World Bank website: www.data.worldbank.org

World Bank (2010) World development indicators. The World Bank, Retrieved on 02 Feb 2011, from The World Bank website: http://data.worldbank.org/data-catalog/world-development-indicators?cid=GPD_WDI

Part IV
Land Management
Improvement Options

Chapter 11
Optimal Irrigation and N-fertilizer Management for Sustainable Winter Wheat Production in Khorezm, Uzbekistan

Nazirbay Ibragimov, Yulduz Djumaniyazova, Jumanazar Ruzimov, Ruzumbay Eshchanov, Clemens Scheer, Kirsten Kienzler, John P.A. Lamers, and Maksud Bekchanov

Abstract The efficiency of the nitrogen (N) application rates 0, 120, 180 and 240 kg N ha^{-1} in combination with low or medium water levels in the cultivation of winter wheat (*Triticum aestivum* L.) cv. *Kupava* was studied for the 2005–2006 and 2006–2007 growing seasons in the Khorezm region of Uzbekistan. The results show an impact of the initial soil N_{min} (NO_3-N + NH_4-N) levels measured at wheat seeding on the N fertilizer rates applied. When the N_{min} content in the 0–50 cm soil layer was lower than 10 mg kg^{-1} during wheat seeding in 2005, the N rate of 180 kg ha^{-1} was found to be the most effective for achieving high grain yields of high quality. With

---

N. Ibragimov (✉) • Y. Djumaniyazova • J.P.A. Lamers
ZEF/UNESCO Khorezm Project, Khamid Olimjan Str., 14, 220100 Urgench, Uzbekistan
e-mail: nazar@zef.uznet.net; yulduz@zef.uznet.net; j.lamers@uni-bonn.de

J. Ruzimov
ZEF/UNESCO Khorezm Project/Urgench State University, Khamid Olimjan Str., 14, 220100 Urgench, Uzbekistan
e-mail: jumanazar@zef.uznet.net

R. Eshchanov
Urgench State University, Khamid Olimjan Str., 14, 220100 Urgench, Uzbekistan
e-mail: ruzimboy@mail.ru

C. Scheer
Institute for Sustainable Resources, Queensland University of Technology, 2 George Street, QLD 4000 Brisbane, Australia
e-mail: clemens.scheer@qut.edu.au

K. Kienzler
International Bureau of the Federal Ministry of Education and Research at the Project Management Agency c/o German Aerospace Center (DLR), Heinrich-Konen-Str.1, 53227 Bonn, Germany
e-mail: kirsten.kienzler@dlr.de

M. Bekchanov
Center for Development Research (ZEF), Walter-Flex-Str. 3, 53113 Bonn, Germany
e-mail: maksud@uni-bonn.de

C. Martius et al. (eds.), *Cotton, Water, Salts and Soums: Economic and Ecological Restructuring in Khorezm, Uzbekistan*, DOI 10.1007/978-94-007-1963-7_11, © Springer Science+Business Media B.V. 2012

a higher N_{min} content of about 30 mg kg^{-1} as was the case in the 2006 season, 120 kg N ha^{-1} was determined as being the technical and economical optimum. The temporal course of N_2O emissions of winter wheat cultivation for the two water-level studies shows that emissions were strongly influenced by irrigation and N-fertilization. Extremely high emissions were measured immediately after fertilizer application events that were combined with irrigation events. Given the high impact of N-fertilizer and irrigation-water management on N_2O emissions, it can be concluded that present N-management practices should be modified to mitigate emissions of N_2O and to achieve higher fertilizer use efficiency.

Keywords Nitrogen rate • Aral Sea Basin • N_2O emissions • Productivity

11.1 Introduction

Winter wheat is a major staple crop in Uzbekistan. It is grown under irrigation throughout the country in rotation with cotton, except for some rainfed areas in the mountainous part of the country. As irrigation water always has been a scarce resource and is gradually becoming scarcer, also throughout Central Asia, water-efficient irrigation practices are vital to sustain agriculture in this region. Generally, the number of irrigation events in wheat production depends on the daily temperatures, and can be as few as four and as many as six or more times during the growing season (Mansurov et al. 2008).

One of the most crucial development stages for irrigating winter wheat is the flowering stage. Drought or water stress during this critical stage will immediately reduce quantity and quality of the grain (Abayomi and Wright 1999; Moussavi-Nik et al. 2007). Moreover, surplus irrigation at a later growth stage cannot compensate the impact of water deficits at earlier growth stages (Rawson and Gomez-Macpherson 2000). In addition, nitrogen (N) fertilizer applications will have a limited effect when the available water supply during flowering is insufficient. Over- and deficit irrigation can also lead to losses of N fertilizer due to volatilization, nitrification/denitrification and leaching, and thus to insufficient N uptake by the wheat crop. Thus, the amount, intensity and frequency of irrigation as well as the optimal allocation, timing and composition of N fertilizer applications should be carefully managed to minimize N-fertilizer loss, and consequently to increase the economic benefits and reduce environmental hazards such as groundwater pollution and global warming.

An analysis of benefits and costs associated with the present irrigation and fertilizer management practices should be economically profitable before one can expect it to be adopted by farmers. This is particularly the case for irrigated agriculture, since most experience with N fertilization and recommendations for its use predominantly originate from rainfed wheat production, which takes place on higher altitudes in the country. Optimal N-fertilization schemes for winter wheat production in Khorezm have been studied extensively by local research institutes.

11 Optimal Irrigation and N-fertilizer Management for Sustainable Winter Wheat...

However, the effect of irrigation amounts in combination with different N-fertilizer levels on N losses, yield production and quality of wheat grain and stover has received little attention. During a long-term study, a model was elaborated by ZEF/UNESCO in the Khorezm region to study the response of winter wheat to different levels of N fertilization in combination with varying irrigation water levels, while concurrently assessing the gaseous losses from winter wheat fields to the environment.

11.2 Materials and Methods

Experimental design: The winter wheat variety *Kupava* R2 was seeded at a rate of 250 kg ha^{-1}. Nitrogen (N) fertilizer was applied as NH_4NO_3 at 0, 120, 180 and 240 kg N ha^{-1} in a randomized complete block design with four replications. In addition, a basal dressing of single super phosphate (SSP) at a rate of 100 kg P_2O_5 ha^{-1} and of potassium chloride (KCl) at a rate of 70 kg K_2O ha^{-1} was applied manually before seeding to prevent P and K deficiency during the experiments. Nitrogen fertilizers were hand-applied in three splits, i.e., before seeding ($N_{20\%}$, $P_{100\%}$, $K_{100\%}$), at tillering ($N_{40\%}$) and at booting ($N_{40\%}$) directly followed by irrigation according to the recommendations provided by the Uzbekistan Scientific Production Center of Agriculture (2000). Irrigation water was applied to maintain soil moisture levels either at 65% (WL1) or 75% (WL2) of field capacity. Water amounts were measured at every irrigation event with a Chipoletti weir (Table 11.1). Each of the 32 plots sized 120 m^2.

Observations and recordings: Grain maturity was visually estimated at the point of complete loss of the green color of the grumes as previously suggested (El-Hendawy et al. 2005). At this stage, three sub-plots of 1 m^2 each were harvested in each of the 32 plots. Following harvest, the main spike was separated from the other spikes of each harvested plant. Next to the number of plants per m^2, also spike length and spikelet number were recorded for each subplot. Ears were threshed and dried at 70°C for 48 h to determine the dry weight (DW) of all plant fractions. Grain yield, 1,000-grain weight (TGW) and stover weight were determined.

Economic analyses: To assess the economic feasibility of each treatment combination, yield, grain and straw market prices, and fertilizer costs were collected for the observed years. This allowed the comparison of the experimental results based on the total revenue (TR), total costs (TC), net benefit (NB) and rate of return (RR). Real prices for N fertilizers and wheat grain were obtained from market surveys. The TR was calculated as the harvest multiplied by the output price as determined at the Urgench local market in 2005–2007. The TC included the variable and fix costs. The NB was calculated as the difference between TR and TC. The economic efficiency of the different treatments was estimated via the associated rate of return (RR), which shows the profit received per each invested monetary unit (RR = NB/TC). All estimates were computed in the local currency (UZS) and converted

Table 11.1 Amounts of irrigation water and frequency of application during two growing seasons

	2005/2006		2006/2007	
Irrigation level	Amount (mm)	Frequency	Amount (mm)	Frequency
WL 1 (low water level)	521	6	473	6
WL 2 (medium water level)	469	78	448	7

into US\$ based on the average exchange rate of 1 US\$ = 1,114.5 UZS in 2005 and 1 US\$ = 1,200 UZS in 2006.

Statistical analysis: ANOVA and t-tests were conducted with the software packet SAS version 9.1 (SAS Institute 2005) for yields and yield determining components, i.e., grain yield, TGW, spike length, shoot stand and kernel number per spike. The least significant difference test (LSD) was used to separate for significant treatment effects.

11.3 Results and Discussion

Grain yield and yield components: The lower amounts but the higher frequencies of water supplied under WL2 than under WL1 increased yield components under no N application (N0; Table 11.2). The combined application of lower amounts of irrigation water with higher frequencies and different N fertilizer rates tended to increase shoot stands, as was also found in other studies (Loveras et al. 2004). The same trend was observed for spike length and number of kernels per spike, albeit the differences among treatments within each and between water levels were insignificant.

Significant effects of N fertilization on various yield-determining components such as productive tillers per m^2, spike population, number of grains per spike, and TGW have been recurrently reported (Jan and Khan 2000; Ali et al. 2003; Maidl et al. 2009). In 2006, the TGW decreased significantly under WL2 as compared to WL1, which is in line with the expectations based on previous studies (Ellen 1989) and that was caused by the higher shoot density in WL2.

In both study years, grain yield under WL2 was higher than under WL1, although the differences between these treatments were not statistically confirmed. The absence of clear differences could be due to the different initial soil conditions. Native soil N_{min} (NO_3-N + NH_4-N) level at wheat seeding in autumn 2006 was about 30 mg kg^{-1} in the 0–50 cm soil layer, which decreased the efficacy of the N applications (120, 180 and 240 kg N ha^{-1}). Therefore, significant yield differences were obtained only with 120 kg N ha^{-1} in comparison with no N application (N0), while yield differences between rates of 120 and 180 kg N ha^{-1}, and between 180 and 240 kg N ha^{-1} were insignificant irrespective of the irrigation water amount. Irrespective of water level, when the N_{min} content in the same soil layer was lower than 10 mg

11 Optimal Irrigation and N-fertilizer Management for Sustainable Winter Wheat... 175

Table 11.2 Impact of Water Level (WL) and Nitrogen (N) rate on wheat grain yield and yield components during two growing seasons

Irrigation	N fertilizer (kg ha^{-1})	Shoot stand (piece m^{-2})	Spike length (cm)	Kernel number/ spike (piece)	1,000-grain weight (gram)	Yield (t ha^{-1})
2005/2006						
WL1	0	381 (84)[c]	6.7 (0.13)[b]	35 (2.6)[b]	35 (1.3)[e]	2.00 (0.2)[d]
WL1	120	483 (97)[ab]	8.5 (0.5)[a]	52 (5.9)[a]	41 (0.8)[a]	3.89 (0.5)[c]
WL1	180	429 (55)[bc]	8.7 (0.29)[a]	53 (1.9)[a]	40 (2.8)[ba]	4.45 (1.2)[bc]
WL1	240	452 (26)[abc]	8.7 (0.62)[a]	54 (4.1)[a]	41 (1.5)[a]	5.37 (1.6)[ab]
WL2	0	379 (113)[c]	6.9 (0.56)[b]	36 (6.1)[b]	36 (1.3)[de]	1.99 (1.1)[d]
WL2	120	510 (121)[ab]	8.9 (0.88)[a]	52 (2.4)[a]	36 (2.2)[de]	4.34 (0.8)[bc]
WL2	180	500 (50)[ab]	8.9 (0.5)[a]	53 (2.4)[a]	37 (0.5)[dc]	5.61 (0.7)[a]
WL2	240	531 (78)[a]	8.9 (0.22)[a]	52 (3.7)[a]	38 (1.0)[bc]	5.82 (0.2)[a]
2006/2007						
WL1	0	386 (40)[f]	8.5 (0.57)[c]	42 (3.8)[c]	36 (1.0)[b]	2.61 (0.8)[d]
WL1	120	472 (75)[ed]	9.4 (0.47)[b]	50 (3.0)[a]	37 (0.8)[ba]	5.11 (0.9)[c]
WL1	180	535 (87)[dc]	9.8 (0.57)[ba]	48 (2.2)[ba]	37 (1.0)[a]	5.83 (0.8)[abc]
WL1	240	641 (34)[a]	10.1 (0.46)[a]	51 (7.8)[a]	37 (1.0)[ba]	6.31 (0.6)[a]
WL2	0	425 (89)[ef]	8.7 (0.29)[c]	42 (4.6)[bc]	36 (1.2)[ba]	3.07 (0.8)[d]
WL2	120	558 (20)[bc]	9.4 (0.53)[b]	51 (3.8)[a]	37 (1.8)[ba]	6.05 (0.7)[ab]
WL2	180	622 (69)[ba]	9.6 (0.35)[ba]	47 (5.6)[bac]	37 (1.0)[ba]	6.21 (0.7)[a]
WL2	240	611 (63)[bac]	9.6 (0.59)[ba]	48 (3.4)[a]	37 (1.9)[ba]	6.53 (0.8)[a]

Data in parentheses are standard deviation. Means with the same letter in each column are not significantly different at $P < 0.1$ according to year using the t-test (LSD Alpha 0.1)

during wheat seeding in 2005, yield differences between N0, 120 and 180 kg N ha^{-1} were significant, while this was not observed for 180 and 240 kg N ha^{-1}. This apparent impact of N fertilizer on winter wheat yields has often been recorded (Fowler 2003). In both study years, grain yield differences were significant for 120 kg N ha^{-1}. The WL*N interaction at $P < 0.1$ was not significant.

The economic analyses show that changes in both N rate and irrigation scheduling had a substantial impact on the economic efficiency (Fig. 11.1). In 2005/2006, the highest (0.62) rate of return (RR) was obtained with WL2N180. The RR values were negative (-0.19 and -0.22, respectively) in the absence of N fertilization (N0) for both water levels.

The low grain yield of 2 t ha^{-1} was the main cause of these low returns and not so much the low production costs. In general, a slight increase in net benefits was observed when N and irrigation water amounts were increased. In the first season the highest benefit per ha of 375,000 UZS (336 US$) was achieved with WL2N240. With the total costs not varying significantly among the treatments, TR changed considerably and, hence, grain yields played a major role in determining the economic efficiency of the different treatments. In the second season 2006/2007, the RR (as an indicator of economic efficiency) increased substantially (Fig. 11.2), particularly owing to the increased domestic prices for wheat, which partly followed

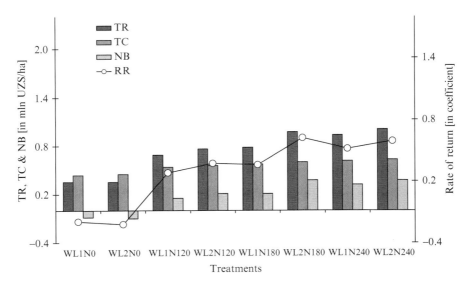

Fig. 11.1 Economic indicators of treatments as a response to changes in irrigation-water and nitrogen-fertilizer application levels in 2005/2006. Connecting line is included for better visualization only (*TR* total revenue, *TC* total costs, *NB* net benefit and *RR* rate of return)

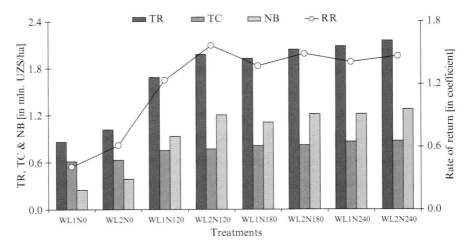

Fig. 11.2 Economic indicators of treatments as a response to changes in irrigation-water and nitrogen-fertilizer application levels in 2006/2007. Connecting line is included for better visualization only (*TR* total revenue, *TC* total costs, *NB* net benefit and *RR* rate of return)

11 Optimal Irrigation and N-fertilizer Management for Sustainable Winter Wheat... 177

the steep increase in wheat prices on the world markets.[1] The net benefits of all treatments were positive and ranged from 150,000 to 1,280,000 UZS ha^{-1} (125–1,067 US\$). The RR even of the least profitable option (WL1N0) was 0.45, while the maximum RR reached nearly 1.60 (WL2N120) in 2006/2007. Total costs incurred in 2005/2006 ranged between 440,000 and 640,000 UZS ha^{-1} (395–574 US\$) and did not increase substantially in 2006/2007 where they were in the range of 560,000–830,000 UZS ha^{-1} (467–692 US\$). But since the wheat prices increased almost two-fold compared to 2005/2006, the RR for each treatment in 2006/2007 was substantially higher than in 2005/2006.

Considerably different net benefits compared to those of N0 were achieved in the second year of research. Experiments with nitrogen application earned much higher net benefits, especially if compared to the first-year results, attributed to higher yields (above 5 t ha^{-1}), and even more so with the medium water (WL2) application level (above 6 t ha^{-1}). Treatments with a medium water level and the highest fertilizer rate (WL2N240 in Fig. 11.2) showed the highest net benefit of about 1,100,000 UZS ha^{-1} (about 917 US\$). An addition of 60 kg N ha^{-1} has the potential of only very slight increases in net benefit and hence economic efficiency. With regard to the irrigation-water level, the highest profitability was observed for WL2 (about 5,682 m^3 ha^{-1}) where yields varied between 3.07 to 6.53 t ha^{-1} for all N rates. While the profitability in 2005/2006 was mainly determined by the grain yields, in 2006/2007 the main determinants were the combination of wheat price and yield. Rapid price increases had a greater impact on wheat profitability, since average yield increases in the second year reached 30%, whereas domestic prices for wheat for sale to the state increased by about 50% and for sale at the local market by 100%.

N_2O emissions: To assess the sustainability of wheat production, it is very important to assess not only technical and economic indicators but also ecological factors. To achieve this, emissions of the greenhouse gas N_2O were monitored over the cropping season using an analytical set up (Scheer et al. 2008a).

High N_2O emissions indicate a low efficiency of fertilizer use and, at the same time, a detrimental impact on the environment. Because N_2O is a very potent greenhouse gas – about 300 times as potent as carbon dioxide – even small quantities over time and across very large areas can contribute significantly to global warming (Scheer et al. 2008b). The temporal course of N_2O emissions of winter wheat is displayed in Fig. 11.3.

The emissions were strongly influenced by irrigation and fertilization, and extremely high emissions were measured immediately after fertilizer application events that were combined with irrigation events. These "emission peaks" accounted for approximately 90% of the total N_2O emissions over the whole cycle and amounted to seasonal values of 0.6 kg N_2O-N ha^{-1} for WL1 and 0.9 kg N_2O-N ha^{-1} for WL2.

---

[1] According to International Grain Council, USDA and FAO. http://www.fao.org/es/esc/prices/ CIWPQueryServlet

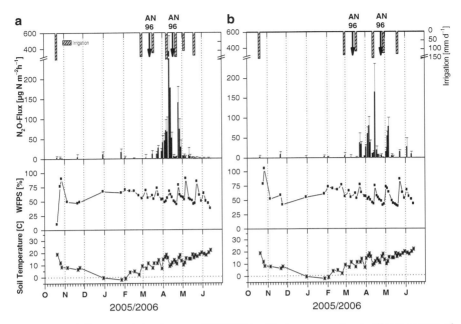

Fig. 11.3 N$_2$O flux rates, irrigation rates (*WFPS* Water filled pore space) and soil temperature of the medium (**a**) and low (**b**) irrigated winter wheat field in 2005/2006. *Arrows* indicate the events of N (kg N ha^{-1}) application to the plots (*AN* ammonium nitrate), whereas bars provide information about the amount of irrigation. *Error bars* indicate the standard error. *Connecting lines* are inserted for showing the data points more clearly

11.4 Conclusions

The official N fertilizer recommendations for irrigated winter wheat of 180 kg N ha^{-1} were found to correspond well with the potential N uptake of winter wheat measured in this study. However, as the results show, the economically and ecologically effective N fertilizer dose for winter wheat is site specific. In both winter wheat seasons, yield differences were always significant between 0 and 120 kg N ha^{-1}, but the differences were insignificant between 180 and 240 kg N ha^{-1} with both WL1 and WL2. The significance in yield differences between 120 and 180 kg N ha^{-1} for each irrigation water level apparently is due to the imposed soil moisture levels and varied with the initial mineral N content in the soil (Djumaniyazova et al. 2010). Optimal N management in winter wheat, therefore, could be achieved through further investigations, e.g., measurement of plant N status directly in the field using non-destructive or indirect methods such as leaf color chart or various optical sensors (e.g., SPAD 502, GreenSeeker) (Ibragimov et al. 2006; Djumaniyazova et al. 2007, 2008).

Given the high impact of N-fertilizer and irrigation-water management on N_2O emissions, it can be concluded that management practices should be modified to mitigate N_2O emissions and to sustain higher fertilizer use efficiency. Modifications in the amount and timing of the fertilizer application and irrigation events could reduce N losses, and concomitant N fertilization and irrigation should be avoided where possible. In general, management practices that have been shown to increase the N-fertilizer use efficiency in irrigated systems, such as sub-surface fertilizer application, fertigation and drip irrigation, will most likely also reduce the N_2O emissions and thus are expected to lead to more sustainable agriculture.

References

Abayomi YA, Wright D (1999) Effects of water stress on growth and yield of spring wheat (*Triticum aestivum L.*) cultivars. Trop Agric 76(2):120–125
Ali L, Mohy–ud–Din Q, Ali M (2003) Effect of different doses of nitrogen fertilizer on the yield of wheat. Int J Agric Biol. doi:1560-8530/2003/05-4-438-439, http://www.ijab.org
Djumaniyazova Y, Ibragimov N, Ruzimov J, Kuryazov I (2007) Fast prediction of the wheat yield. Agric Uzbekistan 12:22
Djumaniyazova Y, Ibragimov N, Ruzimov J, Xaitboyeva J (2008) Leaf diagnosis. Agric Uzbekistan 1:26
Djumaniyazova Y, Sommer R, Ibragimov N, Ruzimov J, Lamers J, Vlek P (2010) Simulating water use and N response of winter wheat in the irrigated floodplains of northwest Uzbekistan. Field Crop Res J 116(3):239–251
El-Hendawy SE, Hu YC, Yakout GM, Awad AM, Hafiz SE, Schmidhalter U (2005) Evaluating salt tolerance of wheat genotypes using multiple parameters. Eur J Agron 22:243–253
Ellen J (1989) Effects of nitrogen and plant density on growth, yield and chemical composition of two winter wheat (Triticum aestivum L.) cultivars. Department of Field Crops & Grassland Science, Wageningen Agricultural University, the Netherlands
Fowler DB (2003) Crop nitrogen demand and grain protein concentration of spring and winter wheat. Agron J 95:260–265
Ibragimov N, Ruzimov J, Ruecker G, Djumaniyazova Y (2006) Nitrogen fertilizer efficiency. Agric Uzbekistan 9:30–31
Jan MT, Khan S (2000) Response wheat yield components to type of N fertilizer, their levels and application time. Pak J Biol Sci 3(8):1227–1230
Loveras J, Manent J, Viudas J, López A, Santiveri P (2004) Seeding rate influence on yield and yield components of irrigated winter wheat in a mediterranean climate. Agron J 96:1258–1265
Maidl FX, Sticksel E, Retzer F, Fischbeck G (2009) Effect of varied N-fertilization on yield formation of winter wheat under particular consideration of mainstems and tillers. J Agron Crop Sci 1:15–22
Mansurov A, Ibragimov I, Nazarov K (2008) Characteristics and production agrotechnology of winter wheat varieties in Khorezm, Uzbekistan. Khorezm branch of Uzbekistan Cereal Research Institute for Irrigated Areas, Urgench, Uzbekistan
Moussavi-Nik M, Mobasser HR, Mheraban A (2007) Effect of water stress and potassium chloride on biological and grain yield of different wheat cultivars. In: Buck HT, Nisi JE, Salomon N (eds) Wheat production in stressed environments, Developments in plant breeding 12. Springer, New York, pp 655–658
Rawson HM, Gomez-Macpherson H (2000) Irrigated wheat. Rome, FAO, p 96
Uzbekistan Scientific Production Center of Agriculture (2000): Recommendation for cereal production technologies in irrigated and rainfed areas of Uzbekistan. Andijan, "Hayot", p 45

SAS Institute (2005) SAS for Windows [computer software]. 9.1 TS Level 2

Scheer C, Wassmann R, Butterbach-Bahl K, Lamers JPA, Martius C (2008a) The relationship between N_2O, NO, and N_2 fluxes from fertilized and irrigated dryland soils of the Aral Sea Basin, Uzbekistan. Plant Soil. doi:10.1007/s11104-11008-19728-11108

Scheer C, Wassmann R, Lamers JPA, Martius C (2008b) Does irrigation of arid lands contribute to climate change? ICARDA caravan no. 25, Dec 2008, pp 2054–2057

Chapter 12
Groundwater Contribution to N fertilization in Irrigated Cotton and Winter Wheat in the Khorezm Region, Uzbekistan

Kirsten Kienzler, Nazirbay Ibragimov, John P.A. Lamers, Rolf Sommer, and Paul L.G. Vlek

Abstract In the irrigated areas of Uzbekistan the nitrogen (N) fertilizer efficiency in crop production is low, as N is partially leached to the groundwater. The N-fertilizer use is still based on recommendations from Soviet times when fertilizer supply was subsidized to maximize production at all costs. Also irrigation water is applied sub-optimally, and groundwater levels have been reported to be of less than 1 m during the vegetation period. As substantial upward movement of salts from the groundwater is frequently observed due to high evapotranspiration rates, it can be expected that nitrate from leached N fertilizer may also move in the soil profile thus influencing the N balance of the soil. In this study we therefore estimated the groundwater contribution to N-fertilization to improve the N management while sustaining yields and quality and reducing negative environmental effects of groundwater nitrate. Nitrate in irrigation and groundwater was measured during spring and summer. Data were complemented with field measurements of groundwater levels, irrigation and

---

K. Kienzler (✉)
International Bureau of the Federal Ministry of Education and Research
at the Project Management Agency c/o German Aerospace Center (DLR),
Heinrich-Konen-Str.1, 53227 Bonn, Germany
e-mail: kirsten.kienzler@dlr.de

N. Ibragimov • J.P.A. Lamers
ZEF/UNESCO Khorezm Project, Khamid Olimjan Str., 14, 220100 Urgench,
Uzbekistan
e-mail: nazar@zef.uznet.net; j.lamers@uni-bonn.de

R. Sommer
International Center for Agricultural Research in the Dry Areas (ICARDA),
P.O. Box 5466, Aleppo, Syria
e-mail: r.sommer@cgiar.org

P.L.G. Vlek
Center for Development Research (ZEF), Walter-Flex-Str. 3, 53113 Bonn, Germany
e-mail: p.vlek@uni-bonn.de

C. Martius et al. (eds.), *Cotton, Water, Salts and Soums: Economic and Ecological Restructuring in Khorezm, Uzbekistan*, DOI 10.1007/978-94-007-1963-7_12,
© Springer Science+Business Media B.V. 2012

N-fertilizer amounts. With the CropSyst model, upward fluxes of groundwater and evapotranspiration rates were derived, as we could not measure these in the field. We calculated the contribution from the upward flux of nitrate-containing groundwater to the N content in the rooting zone. The difference between the simulated actual evapotranspiration and the irrigated water amount was 335 mm. The average nitrate content in the groundwater was low under summer crops (2 mg nitrate L^{-1}) and higher under the spring crop (24 mg nitrate L^{-1}). However, the temporal dynamics were very much linked to the irrigation and fertilization practices, and corresponded to the changes in groundwater table depth: Almost immediately after fertilization, the nitrate content increased to up to 75 mg nitrate L^{-1} in spring. At the end of the growing period, the nitrate amounts had reached levels similar to those prior to fertilization. A groundwater contribution of 355 mm and an average nitrate concentration of up to 75 mg nitrate L^{-1} would enhance the N stocks in the soil by up to 5–61 kg N ha^{-1}. This is equivalent to one single fertilizer application event. However, in case farmers would rely on the input of N through the groundwater to satisfy crop demand and consequently reduce N fertilizer application levels, the N concentrations in the groundwater would reduce and become an unreliable source.

Keywords Groundwater contribution • Supplemental N-fertilizer • Nitrate • Irrigated crop production • Evapotranspiration

12.1 Introduction

Following Uzbekistan's independence and after a history of subsidized inputs and crop production on the state collective farms, the newly established Uzbek farmers are challenged by on-going changes in land-tenure regulations (cf. Djanibekov et al. 2011), and rising costs for fertilizers, pesticides, machinery and fuel for tractors and electricity for irrigation pumps (cf. Rudenko et al. 2011). Caught between the obligation to fulfill the state's production targets for cotton and winter wheat and the burden of ensuring their livelihood, farmers decide to apply agronomically suboptimal amounts of N fertilizer and water to their crops, i.e. irrigated cotton and winter wheat. Furthermore, fertilizer recommendations date back to the time before independence when fertilizer supply was subsidized to maximize production at all costs.

Already in 1940, due to the extension of the irrigated area, a systematic increase in the groundwater level was reported for the lower reaches of the Amudarya river, e.g. the Khorezm region, Uzbekistan. Despite the step-wise construction of the drainage network that was more or less completed in 1975, the area with groundwater tables of less than 1.0 m increased (Jabborov 2005; cf. Tischbein et al. 2011). This can be attributed to malfunctioning drains, but also to over-applications of irrigation water amounts (Jabborov 2005).

Frequent irrigation events with large amounts of water increase the potential for N losses below the rooting zone or even to the groundwater via leaching of the

12 Groundwater Contribution to N fertilization in Irrigated Cotton and Winter... 183

mobile fractions (Smika and Watts 1978; Young and Aldag 1982; Riley et al. 2001; Burkart and Stoner 2002; Liu et al. 2003; Ju et al. 2006). As it is not attracted by the negatively charged clay surfaces, nitrate (NO_3) is not retained by the soil particles and is, therefore, more easily leachable than ammonium (NH_4), which is positively charged and usually bound by electrostatic attraction to the exchange sites of the clay particles (Scheffer and Schachtschabel 1998). The rate of nitrate leaching is governed by the water regime, soil textural properties, crop uptake and rooting distribution, as well as by fertilizer inputs and mineralization rates (Burns 1980; Young and Aldag 1982; Hadas et al. 1999; Riley et al. 2001; Dinnes et al. 2002; Dunbabin et al. 2003; Liu et al. 2003).

The fate of N in irrigated systems has been intensively studied. Unaccounted N losses of 8–51% observed in the Tashkent region of Uzbekistan (Ibragimov 2007) correspond with N-fertilizer inefficiencies found in other regions, e.g., 42–43% (Freney et al. 1993; Mahmood et al. 2000), or 25–50% (Chua et al. 2003), where N is assumed to be lost from the soil-crop system through denitrification and leaching. Yet, for the region of Khorezm, there has been little research conducted on the N losses particularly to the groundwater from fertilizer applications to agricultural crops.

Sustainable N-fertilizer management and irrigation practices thus are challenging, especially in the irrigated regions of Uzbekistan, where poor N and water management can inevitably influence the soil-N balance and plant-N uptake and may lead to nutrient losses to the environment via volatilization or leaching, while sacrificing crop yields and quality. This research, therefore, attempted to identifying the groundwater contribution to N-fertilization in irrigated cotton and winter wheat production in the Khorezm region to improve the N fertilization strategies for cotton and wheat while reducing threats to the environment.

12.2 Methods

Nitrate content in the irrigation water and groundwater was measured after the cotton and wheat harvest in 2007 using four observation wells (piezometers), which were installed in a transect towards the drainage canal. The piezometers consisted of 2.2-m long poly-ethylene pipes of 4 cm in diameter. The pipes were blocked at the bottom, and the lower half of the pipe was perforated. To protect the perforated holes from clogging, the pipes were wrapped in fine synthetic fiber. The groundwater data were used to approximate groundwater table dynamics throughout the season; the data were later used in the modeling (see below). Water samples, water depth and nitrate measurements were taken from the irrigation and groundwater after fertilization and irrigation events. Nitrate content in the water was determined using nitrate test sticks (Merkoquant®) and photometrically with a calibration solution (Spectroquant®).

Using the model CropSyst (version 4.09.05) and its newly developed generic routine for cotton (Sommer et al. 2007), upward fluxes of groundwater were derived, which could not have been measured in the field experiments. For the Khorezm region with changing groundwater tables and soil salinity hazards, CropSyst seemed an appropriate software for it can simulate crop yield and detailed N and soil organic matter (SOM) dynamics applying algorithms used in the CENTURY model (Parton et al. 1987; Parton and Rasmussen 1994) while considering fluctuating shallow groundwater tables and salinity stress (Ferrer-Alegre and Stockle 1999; Stockle et al. 2003).

The CropSyst water budget accounts for precipitation, irrigation, soil evaporation, canopy and residue interception and transpiration, runoff, surface storage and ponding, infiltration and deep percolation (Stockle et al. 1994, 2003). Model details for the water budget and crop growth and development are described elsewhere (e.g. Stockle et al. 1994, 1997, 2003; Jara and Stockle 1999; Stockle and Nelson 2000; Sadras 2002; Fuentes et al. 2003; Bechini et al. 2006). Given that the model calculates water transport for each soil node using a finite difference solution of the Richards' equation (Stockle et al. 1997), results from Forkutsa (2006) and Forkutsa et al. (2009a) could be integrated in the parameterization.

In the model, nitrogen leaching is related to water movement in the soil (concentration of N in the water), which is determined by the amount of soil water in each soil layer and free movable mineral N in the profile and the soil cation exchange capacity (Stockle et al. 1994; Donatelli and Stockle 1999). Simulations of infiltration and water redistribution in the profile are done via the cascade approach (Sadras 2002). Nitrate is not retained by the soil matrix, and ammonium movement is dependent on the absorption capacity of the solid soil matrix as described by Langmuir (Eq. 12.1) (Stockle et al. 1994):

$$X - NH_4 = \frac{k \times q \times [NH_4]}{1 + k \times [NH_4]} \tag{12.1}$$

where X-NH_4 is the amount of ammonium absorbed by the exchange sites (kg kg^{-1}), $[NH_4]$ is the concentration of ammonium (g L^{-1}) in the soil solution, and k and q are constants (kg kg^{-1}). Effects of diffusion and hydrodynamic dispersion are not considered (Stockle et al. 1994).

Model parameters needed for CropSyst were either estimated from field measurements or adjusted for cultivar characteristics based on literature data. Most of the components necessary for the water balance were measured in the field (i.e., irrigation water, precipitation, soil water fluxes, as described in Kienzler 2010). Those parameters not measured in the field were estimated using the model HYDRUS 1-D (Forkutsa et al. 2009a, b). Runoff was negligible as the soils were fairly levelled.

The upward flux of nitrate-containing groundwater was assumed to not be adsorbed in the soil but to contribute to the nitrate content in the rooting zone of cotton (Burns 1980). The upward movement of nitrate was thus estimated according to Eq. 12.2

$$\text{Contribution} = 3551 \text{ m}^{-2} \times 8 \text{ mgl}^{-1} = 2.84 \text{ g m}^{-2} = 28.4 \text{ kg nitrate ha}^{-1} \tag{12.2}$$

12 Groundwater Contribution to N fertilization in Irrigated Cotton and Winter... 185

using the simulated groundwater contribution of 355 mm and an average nitrate concentration in the groundwater of 8 mg L^{-1} (see results shown below). The equivalent nitrate-nitrogen amount (in kg ha^{-1}) was obtained by multiplying the respective amount of nitrate with 0.2259 (i.e. atomic mass of N divided by atomic mass of nitrate). A more detailed contribution to the subsoil nitrate content was computed using the daily water balance simulations from the model CropSyst (Kienzler 2010). The bottom flux (Vbot) was calculated using Eqs. 12.3 and 12.4:

$$V bot = F - ETa - \Delta W \qquad (12.3)$$

$$F = P + I - R - Pi - Pm \qquad (12.4)$$

where F is the infiltration (mm), ETa the actual evapotranspiration (mm), ΔW the storage change, P the precipitation (mm), I the irrigation amount (mm), R the surface runoff (mm), Pi the crop interception (mm) and Pm the mulch interception. The storage change was calculated as the daily water fluctuation in the soil between the rooting zone and the groundwater table.

12.3 Results and Discussion

12.3.1 Groundwater Nitrate Content in 2007 and 2008

The concentration of nitrate in the irrigation water during cotton and wheat growth was rather low (<0.5 mg L^{-1}). Groundwater nitrate content was monitored for the summer crops (carrot, cabbage and maize) in summer 2007 and during the whole winter wheat growth period in 2007/2008.

The average nitrate content in the groundwater under the summer crops was 1.8 mg nitrate L^{-1} (Fig. 12.1). However, the temporal dynamics were very much linked to the irrigation and fertilization practices. Almost immediately after fertilization, however, the contamination of the groundwater with nitrate increased to a maximum of 7.8 mg nitrate L^{-1} in the piezometer Pz4. At the beginning of September, the levels had reached values similar to those prior to fertilization.

The average nitrate content of the groundwater under winter wheat in 2008 was high (23.9 mg nitrate L^{-1}) (Fig. 12.2). The minimum nitrate amount of 13.8 mg nitrate L^{-1} on 29.03.08 was measured in the groundwater one week after the last irrigation event (22.03.08), while the maximum content of 44.4 mg nitrate L^{-1} on 02.04.08 was found one day after fertilization and irrigation had occurred (01.04.08).

All measurements in 2008 represent the means of the nitrate test-sticks color step, as photometric measurements were not available. Therefore, individual obser-vation wells showing 75 mg nitrate L^{-1} directly following irrigation could have had an actual nitrate content ranging from 50 to 100 mg nitrate L^{-1}. Still, the overall trend was that nitrate levels in the groundwater changed with every management activity in the field (fertilization, irrigation). Furthermore, the dynamics of the

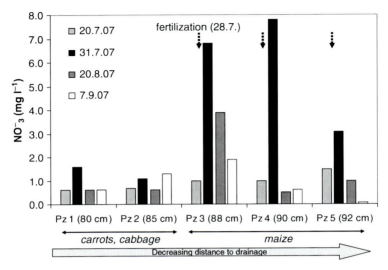

Fig. 12.1 Nitrate measurements (mg L^{-1}) in five piezometers (Pz) for four irrigation events in 2007. Average groundwater depth is indicated in brackets. Pz 1 and 2 were installed in carrot and cabbage fields, Pz 3–5 in maize fields

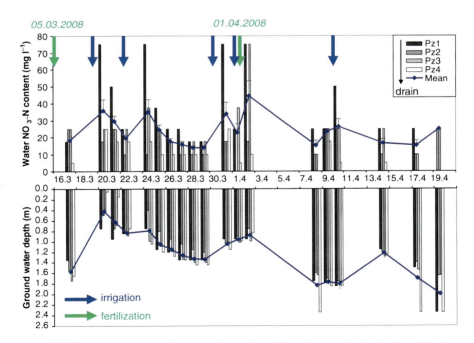

Fig. 12.2 Nitrate measurements in groundwater (mg L^{-1}) and groundwater depth (m) in four piezometers (Pz) under winter wheat in 2008. The mean of the four piezometers is indicated as *blue line*. Fertilization occurred on March 5th and April 1st

nitrate content in the water correspond to the changes in groundwater table depth: At times of shallow groundwater (due to irrigation inputs) the nitrate level in the water was also enhanced.

Both nitrate content and water table in general decreased in the direction of the drainage, i.e., from piezometer 1 (Pz1) to piezometer 4 (Pz4). Especially piezometer Pz1 reacted rapidly to fertilization and irrigation events.

Under the assumption that nitrate is not adsorbed in the soil (Burns 1980), the upward flux of groundwater could be assumed to lead to the nitrate accumulation in the rooting zone. A groundwater contribution of 355 mm (see results given below) and an average nitrate concentration in the groundwater of 8 mg L^{-1} would therefore give an approximated upward movement of 28 kg nitrate-N ha^{-1} or 6 kg N ha^{-1}. Higher peak concentrations (up to 75 mg nitrate L^{-1} as measured directly after irrigation) would consequently increase also nitrate-N in the soil. The nitrate input from the irrigation water (280 mm) was with 3 kg N ha^{-1} (at a concentration of 1 mg nitrate L^{-1}) noticeably lower. The calculated upward flux of water using the daily water balance simulations (see results given below) was 250 mm. With a groundwater flux between 250 and 355 mm and nitrate concentrations of 8–75 mg nitrate L^{-1} the nitrate contribution from the groundwater to the subsoil during the growing season was approximated to be around 23–269 kg nitrate-N ha^{-1} or 5–61 kg N ha^{-1}.

12.3.2   Crop Modeling – Water Balance

For the period of cotton growth, the FAO-56 potential evapotranspiration was 714 mm (Table 12.1). Simulated actual evapotranspiration fluctuated around 633 mm. The amount of water actually transpired by cotton was 230 mm lower. Furthermore, the simulated actual evapotranspiration (and crop transpiration) during the vegetation season was higher than the total irrigation water applied. The difference between the simulated evapotranspiration and the irrigated water amount ranged from 283 mm for non-fertilized treatments to 335 mm for treatments receiving 250 kg N ha^{-1}.

The high evaporative demand and comparatively low irrigation amounts are according to Forkutsa (2006) responsible for a strong upward movement of groundwater (capillary rise); this was confirmed by the CropSyst simulations (Figs. 12.3 and 12.4) where the predicted N leaching losses from below the rooting zone (90 cm) were low (10–12 kg ha^{-1}). This is in line with the generally low irrigation amount and a dominating upward movement of water flow during the vegetation period (cf. Awan et al. 2011; Tischbein et al. 2011). However, as the quantification of actual water volumes draining below the rooting zone was not in the scope of this study, a follow-up study should be conducted to confirm these data in the field.

Overall, the crop modeling results allow the assumption that N contributions from irrigation and groundwater influence the N balance and enhance N uptake. However, no full nitrate-N routine for irrigation and groundwater had been incorporated into the model at the time of the study that would allow detailed simulations of the

Table 12.1 Simulated potential and actual evapotranspiration (mm) and soil water drainage amount for two N fertilizer treatments (kg ha^{-1}) for the cropping season 2005 in the Khorezm region

N rate	Fertilizer	Average groundwater depth (season)	Seasonal precipitation	Total irrigation	Seasonal potential evapotrans-piration	Seasonal actual evapotrans-piration
kg ha^{-1}	–	m	mm			
0	NPK-0	1.3	27	280	714	587
250	DAP-Urea-Urea					640

DAP-Urea-Urea = 3 splits at the officially recommended cotton growth stages, using DAP, urea, and urea fertilizer

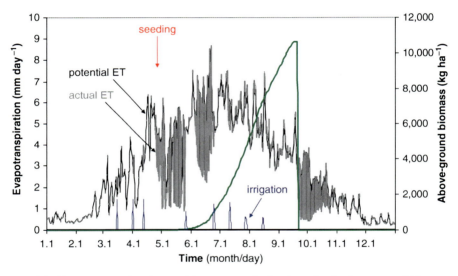

Fig. 12.3 Simulated actual and potential evapotranspiration (mm day^{-1}), above-ground biomass (*green line*, kg ha^{-1}) and irrigation (*blue line*) in 2005

N contribution via (sub-surface) water supply. Efforts should therefore be made to improve the simulations of the N balance and N uptake to improve the reliability of the model predictions.

12.3.3 Crop Water Demand and Subirrigation

While the winter wheat field was sufficiently supplied with irrigation water, water supplied to the cotton experiment turned out too low to meet the crop water demand. However, the groundwater table under the cotton experiment throughout the growing

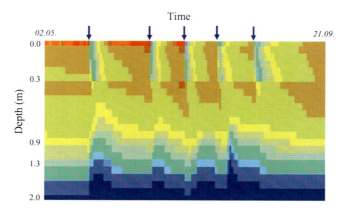

Fig. 12.4 Simulated water content in 0–2 m depth during cotton vegetation period in 2005. *Blue colors* indicate high water content (0.43–0.45 m³ m⁻³), *red colors* low (0.14–0.19 m³ m⁻³), and *yellow colors* intermediate water content (0.30–0.34 m³ m⁻³). *Blue arrows* indicate irrigation times

season was generally shallow (<1.2 m). Similar levels have also been measured by Zakharov (1957), Khaitbayev (1963), Kadirhodjayev and Rahimov (1972) and Ibragimov et al. (2007) in previous cotton experiments in Khorezm. The groundwater level during the vegetation period in these experiments ranged 0.5–1.2 m below surface. After harvest, the levels decreased to 2.0 m and deeper (Zakharov 1957; Khaitbayev 1963; Kadirhodjayev and Rahimov 1972; Ibrakhimov 2004; Ibragimov et al. 2007).

For regions with particularly shallow groundwater during the vegetation season (see Ibrakhimov 2004; Forkutsa 2006), a significant amount of crop water is supplied via upward flow of groundwater, which supplement irrigation inputs (Ayars et al. 2006), also confirmed for Khorezm by Awan et al. (2011). For loamy soils with a similar groundwater table to that in this study, Ayars et al. (2006) calculated a potential evapotranspiration of around 45%. Other studies also evidenced a substantial contribution of shallow groundwater tables to satisfy crop water demand (e.g., Chaudhary et al. 1974; Rhoades et al. 1989; Ayars and Hutmacher 1994; Ayars 1996; Ayars et al. 2006). For conditions in Khorezm, Forkutsa (2006) calculated the contribution of this so-called supplemental irrigation or subirrigation to cotton to be as high as 17–89% of the actual evapotranspiration. The capillary rise of water ranged between 92 and 277 mm depending on the irrigation management (Forkutsa 2006). Forkutsa et al. (2009b) show that none of the six irrigation events on the cotton field leached water amounts below 80 cm depth. The calculated upward flux of 250 mm from the mass balance equation (see calculated groundwater contribution results above) is, therefore, in line with these findings, whereas the assumed groundwater contribution, calculated, as actual evapotranspiration minus irrigation amounts, was much higher (355 mm). Furthermore, Conrad (2006) simulated higher actual evapotranspiration amounts over cotton fields in 2005 of 853 mm as compared to the 633 mm in this study. Although his evapotranspiration values were based on a 2-week longer growth period and included crop coefficients and input data based on information of the Central Asian Scientific Irrigation Research Institute (SANIIRI), which may have led to this high value, all results substantiate the plausibility of a large contribution of shallow groundwater to crop evapotranspiration, making even a contribution of 355 mm feasible. Thus, despite the low irrigation amounts applied to cotton during the experiment, the crop water demand was likely met, as the irrigation water amounts were complemented by shallow groundwater.

12.3.4 Supplemental Nitrogen Fertilization from Groundwater

Crops have shown enhanced N-utilization, reduced water stress, and stabilized or increased yields when subirrigated via shallow groundwater tables (Drury et al. 1997; Fisher et al. 1999; Patel et al. 2001; Elmi et al. 2002). The accumulated nitrate-N in the groundwater may thus function as supplemental N fertilizer if taken up by the crop (Steenvoorden 1989). During the vegetation period of the cotton experiment, the groundwater table was shallow enough to contribute with approximately 5–61 kg nitrate-N ha^{-1} to crop N uptake. Together with an input of 3 kg N ha^{-1} via the irrigation water, this amount is equivalent to a single N-fertilizer application. This N supply may cause the observed weak responses of cotton to the N amounts applied (Kienzler 2010). The data set collected in this study however was not aimed at quantifying the N contribution from groundwater. However, the observations do suggest that the nitrate-N dynamics in the Khorezmian system deserve further study.

12.4 Conclusions

Modern N management needs to enable farmers to obtain stable crop yields of good quality and preserve the environment. The potential and actual evapotranspiration and crop transpiration showed a potential contribution of the groundwater to crop water demand of between 283 (for non-fertilized treatments) and 355 mm (for treatments receiving 250 kg N ha^{-1}). At high nitrate levels in the groundwater, the concomitant supplemental N contribution from the groundwater influences the soil-N balance and plant-N uptake. The calculated range of 5–61 kg N ha^{-1} is substantial as it equals the amount usually applied during one single N-fertilizer application event. Particularly cotton with its long tap root may profit from those N supplies from the groundwater, which, however, depend on many factors such as groundwater depth, its nitrate content, and the field's proximity to the next drain. However, farmers should not collectively rely on this N input and reduce N applications in cotton and wheat production, as under continuously low applications of N-fertilizer, those crops would slowly mine the N resources. Clearly, the (nitrate-) N dynamics in the Khorezmian system deserve further study.

References

Awan UK, Tischbein B, Kamalov P, Martius C, Hafeez M (2011) Modeling irrigation scheduling under shallow groundwater conditions as a tool for an integrated management of surface and groundwater resources. In: Martius C, Rudenko I, Lamers JPA, Vlek PLG (eds) Cotton, water, salts and *Soums*: economic and ecological restructuring in Khorezm, Uzbekistan. Springer, Dordrecht

Ayars JE (1996) Managing irrigation and drainage systems in arid areas in the presence of shallow groundwater: case studies. Irrig Drain Syst 10:227–244

Ayars JE, Hutmacher RB (1994) Crop coefficients for irrigating cotton in the presence of groundwater. Irrig Sci 15:45–52

Ayars JE, Christen EW, Soppe RW, Meyer WS (2006) The resource potential of in-situ shallow ground water use in irrigated agriculture: a review. Irrig Sci 24:147–160

Bechini L, Bocchi S, Maggiore T, Confalonieri R (2006) Parameterization of a crop growth and development simulation model at sub-model components level. An example for winter wheat (*Triticum aestivum* L.). Environ Model Softw 21:1042–1054

Burkart MR, Stoner JD (2002) Nitrate in aquifers beneath agricultural systems. Water Sci Technol 45:19–29

Burns IG (1980) Influence of the spatial distribution of nitrate on the uptake of N by plants: a review and a model for rooting depth. Eur J Soil Sci 31:155–173

Chaudhary TN, Bhatnagar VK, Prihar SS (1974) Growth response of crops to depth and salinity of ground water, and soil submergence. I. wheat (Triticum aestivum L.). Agron J 66:32–35

Chua TT, Bronson KF, Booker JD, Keeling JW, Mosier AR, Bordovsky JP, Lascano RJ, Green CJ, Segarra E (2003) In-Season nitrogen status sensing in irrigated cotton: I. yields and nitrogen-15 recovery. Soil Sci Soc Am J 67:1428–1438

Conrad C (2006) Fernerkundungsbasierte Modellierung und hydrologische Messungen zur Analyse und Bewertung der landwirtschaftlichen Wassernutzung in der Region Khorezm (Usbekistan). Julius-Maximilians-Universität Würzburg

Dinnes DL, Karlen DL, Jaynes DB, Kaspar TC, Hatfield JL, Colvin TS, Cambardella CA (2002) Nitrogen management strategies to reduce nitrate leaching in tile-drained Midwestern soils. Agron J 94:153–171

Djanibekov N, Bobojonov I, Lamers JPA (2011) Farm reform in Uzbekistan. In: Martius C, Rudenko I, Lamers JPA, Vlek PLG (eds) Cotton, water, salts and *Soums*: economic and ecological restructuring in Khorezm, Uzbekistan. Springer, Dordrecht

Donatelli M, Stockle CO (1999) The suite CropSyst to simulate cropping systems. Short course at the Middle East Technical University – Dep. Of Economics, Ankara, Turkey, 18–22 November 1999. http://www.sipeaa.it/mdon/software/cropsyst/lectures/Lect1_Introduction.PDF

Drury CF, Tan CS, Gaynor JD, Oloya TO, van Wesenbeeck IJ, McKenney DJ (1997) Optimizing corn production and reducing nitrate losses with water table control-subirrigation. Soil Sci Soc Am J 61(3):889–895

Dunbabin V, Diggle A, Rengel Z (2003) Is there an optimal root architecture for nitrate capture in leaching environments? Plant Cell Environ 26:835–844

Elmi AA, Madramootoo C, Egeh M, Liu A, Hamel C (2002) Environmental and agronomic implications of water table and nitrogen fertilization management. J Environ Qual 31(6):1858–1867

Ferrer-Alegre F, Stockle CO (1999) A model for assessing crop response to salinity. Irrig Sci 19:15–23

Fisher MJ, Fausey NR, Subler SE, Brown LC, Bierman PM (1999) Water table management, nitrogen dynamics, and yields of corn and soybean. Soil Sci Soc Am J 63(6):1786–1795

Forkutsa I (2006) Modeling water and salt dynamics under irrigated cotton with shallow groundwater in the Khorezm region of Uzbekistan. Rheinische Friedrich-Wilhelms-Universität Bonn, Bonn

Forkutsa I, Sommer R, Shirokova YI, Lamers JPA, Kienzler K, Tischbein B, Martius C, Vlek PLG (2009a) Modeling irrigated cotton with shallow groundwater in the Aral Sea Basin of Uzbekistan: I. Water dynamics. Irrig Sci 27(4):331–346

Forkutsa I, Sommer R, Shirokova YI, Lamers JPA, Kienzler K, Tischbein B, Martius C, Vlek PLG (2009b) Modeling irrigated cotton with shallow groundwater in the Aral Sea Basin of Uzbekistan: II. Soil salinity dynamics. Irrig Sci 27(4):319–330

Freney JR, Chen DL, Mosier AR, Rochester IJ, Constable GA, Chalk PM (1993) Use of nitrification inhibitors to increase fertilizer nitrogen recovery and lint yield in irrigated cotton. Nutr Cycl Agroecosyst 34(1):37–44

Fuentes JP, Flury M, Huggins DR, Bezdicek DF (2003) Soil water and nitrogen dynamics in dryland cropping systems of Washington State, USA. Soil Tillage Res 71(1):33–47

Hadas A, Hadas A, Sagiv B, Haruvy N (1999) Agricultural practices, soil fertility management modes and resultant nitrogen leaching rates under semi-arid conditions. Agric Water Manag 42(1):81–95

Ibragimov NM (2007) Ways of increasing the nitrogen fertilizer efficiency on cotton in the belt with irrigated sierozem soils. Uzbekistan Cotton Research Institute, Habilitation (in Russian)

Ibragimov M, Khamzina A, Forkutsa I, Paluasheva G, Lamers JPA, Tischbein B, Vlek PLG, Martius C (2007) Groundwater table and salinity: spatial and temporal distribution and influence on soil salinization in Khorezm region (Uzbekistan, Aral Sea Basin). Irrig Sci 21:219–236

Ibrakhimov M (2004) Spatial and temporal dynamics of groundwater table and salinity in Khorezm (Aral Sea Basin), Uzbekistan. Rheinische Friedrich-Wilhelms-Universität Bonn, Bonn

Jabborov H (2005) Irrigation canals and collector-drainage systems in Khorezm. Center for Development Research (ZEF), Urgench, p 44

Jara J, Stockle CO (1999) Simulation of water uptake in maize, using different levels of process detail. Agron J 91(2):256–265

Ju XT, Kou CL, Zhang FS, Christie P (2006) Nitrogen balance and groundwater nitrate contamination: comparison among three intensive cropping systems on the North China Plain. Environ Pollut 143(1):117–125

Kadirhodjayev F, Rahimov M (1972) Efficiency and periods of using phosphorous fertilizers under cotton on newly developing lands of Khorezm (in Russian). In: Cotton Research Institute (ed) Chemicalization in cotton growing. XXIII edn. MCX USSR, Tashkent

Khaitbayev KK (1963) The effect of application timing, dozes and types of potassium fertilizers on cotton yield in Khorezm conditions. Samarkand Agriculture Institute, Tashkent, p 23 (in Russian)

Kienzler KM (2010) Improving the nitrogen use efficiency and crop quality in the Khorezm region, Uzbekistan. Ecology and Development Series 72, Rheinische Friedrich-Wilhelms-Universität Bonn, Bonn

Liu X, Ju X, Zhang F, Pan J, Christie P (2003) Nitrogen dynamics and budgets in a winter wheat-maize cropping system in the North China Plain. Field Crops Res 83(2):111–124

Mahmood T, Ali R, Sajjad MI, Chaudhri MB, Tahir GR, Azam F (2000) Denitrification and total fertilizer-N losses from an irrigated cotton field. Biol Fertil Soils 31(3):270–278

Parton WJ, Rasmussen PE (1994) Long-term effects of crop management in wheat-fallow: II. CENTURY model simulations. Soil Sci Soc Am J 58(2):530–536

Parton WJ, Schimel DS, Cole CV, Ojima DS (1987) Analysis of factors controlling soil organic matter levels in great plains grasslands. Soil Sci Soc Am J 51:1173–1179

Patel RM, Prasher SO, Broughton RS (2001) Upward movement of leached nitrate with subirrigation. Trans ASAE 44(6):1521–1526

Rhoades JD, Bingham FT, Letey J, Hoffman GJ, Dedrick AR, Pinter PJ, Replogle JA (1989) Use of saline drainage water for irrigation: imperial valley study. Agric Water Manag 16(1–2):25–36

Riley WJ, Ortiz-Monasterio I, Matson PA (2001) Nitrogen leaching and soil nitrate, nitrite, and ammonium levels under irrigated wheat in Northern Mexico. Nutr Cycl Agroecosyst 61(3):223–236

Rudenko I, Nurmetov K, Lamers JPA (2011) State order and policy strategies in the cotton and wheat value chains. In: Martius C, Rudenko I, Lamers JPA, Vlek PLG (eds) Cotton, water, salts and *Soums*: economic and ecological restructuring in Khorezm, Uzbekistan. Springer, Dordrecht

Sadras V (2002) Interaction between rainfall and nitrogen fertilization of wheat in environments prone to terminal drought: economic and environmental risk analysis. Field Crops Res 77(2–3):201–215

Scheffer F, Schachtschabel P (1998) Lehrbuch der Bodenkunde. Enke Verlag, Stuttgart

Smika DE, Watts DG (1978) Residual nitrate-N in fine sand as influenced by N fertilizer and water management practices. Soil Sci Soc Am J 42(6):923–926

Sommer R, Wall PC, Govaerts B (2007) Model-based assessment of maize cropping under conventional and conservation agriculture in highland Mexico. Soil Tillage Res 94(1):83–100

Steenvoorden JHAM (1989) Agricultural practices to reduce nitrogen losses via leaching and surface runoff. In: Germon JC (ed) Management systems to reduce impact of nitrates. Commission of the European Communities/Elsevier, London

Stockle CO, Nelson R (2000) CropSyst User's Manual (Version 3.0). Department of Biological Systems Engineering, Washington State University, Pullman

Stockle CO, Martin SA, Campbell GS (1994) CropSyst, a cropping systems simulation model: water/nitrogen budgets and crop yield. Agric Syst 46:335–359

Stockle CO, Cabelguenne M, Debaeke P (1997) Comparison of CropSyst performance for water management in southwestern France using submodels of different levels of complexity. Eur J Agron 7(1–3):89–98

Stockle CO, Donatelli M, Nelson R (2003) CropSyst, a cropping systems simulation model. Eur J Agron 18(3–4):289–307

Tischbein B, Awan UK, Abdullaev I, Bobojonov I, Conrad C, Forkutsa I, Ibrakhimov M, Poluasheva G (2011) Water management in Khorezm: current situation and options for improvement (hydrological perspective). In: Martius C, Rudenko I, Lamers JPA, Vlek PLG (eds) Cotton, water, salts and *Soums*: economic and ecological restructuring in Khorezm, Uzbekistan. Springer, Dordrecht

Young JL, Aldag RW (1982) Inorganic forms of nitrogen in soil. In: Stevenson FJ (ed) Nitrogen in agricultural soils. Agronomy no. 22. ACS, Madison

Zakharov PI (1957) Terms and methods of application of phosphoric fertilizers on saline lands (in Russian). In: Cotton Research Institute (ed) Proceedings of the experimental agricultural station of the Khorezm region. UnionCRI, Tashkent

Chapter 13
Introducing Conservation Agriculture on Irrigated Meadow Alluvial Soils (Arenosols) in Khorezm, Uzbekistan

Alim Pulatov, Oybek Egamberdiev, Abdullah Karimov, Mehriddin Tursunov, Sarah Kienzler, Ken Sayre, Latif Tursunov, John P.A. Lamers, and Christopher Martius

Abstract Uzbekistan's economy depends to a great extent on agriculture, particularly on revenues from irrigated cotton (*Gossypium hirsutum* L.) production. Since poor soil fertility and high soil salinity are the major obstacles in crop production, conservation agriculture (CA) may offer the potential to increase soil fertility and crop yields, reduce soil salinity and, thus, save water used for leaching salts out of the soil. Furthermore, compared to the wealth of data on CA in rain fed areas worldwide, scarce information exists on CA under irrigated conditions. This study

---

A. Pulatov (✉) • A. Karimov • M. Tursunov
EcoGIS Center, Tashkent Institute of Irrigation and Melioration,
Qori-Niyoziy Str., 39, 100100 Tashkent, Uzbekistan
e-mail: alimpulatov@mail.ru; abdullakarimov@yahoo.org; mehriddin@gmail.com

O. Egamberdiev • J.P.A. Lamers
ZEF/UNESCO Khorezm Project, Khamid Olimjan Str., 14, 220100 Urgench, Uzbekistan
e-mail: oybek72@zef.uznet.net; j.lamers@uni-bonn.de

S. Kienzler • C. Martius
Center for Development Research (ZEF), Walter-Flex-Str. 3, 53113 Bonn, Germany
e-mail: skienzler@gmx.de; gcmartius@gmail.com

K. Sayre
International Maize and Wheat Improvement Center,
CIMMYT Apdo. #370, P.O.Box 60326, Houston, TX 77205, USA
e-mail: k.sayre@cgiar.org

L. Tursunov
Biological and Soil-Science Faculty, National University of Uzbekistan,
Universitetskaya 174, 100174 Tashkent, Uzbekistan
e-mail: Tursunov_L@mail.ru

C. Martius et al. (eds.), *Cotton, Water, Salts and Soums: Economic and Ecological Restructuring in Khorezm, Uzbekistan*, DOI 10.1007/978-94-007-1963-7_13,
© Springer Science+Business Media B.V. 2012

aims at identifying the potential of reduced tillage (selected CA practices) on soil parameters and crop yields under irrigated agriculture in a cotton-wheat rotation on salinity affected areas in Khorezm. A complete randomized field experiment with four replications was conducted 2002–2005 in Khiva, a district of the Khorezm province of Uzbekistan. For the first time in Khorezm, four tillage treatments were tested and compared: conventional tillage (CT; control), intermediate tillage (IT), permanent bed planting (PB), and zero tillage (ZT). Treatments were with and without retention of crop residues (+CR and −CR), and all were furrow irrigated except for ZT that was flood irrigated. The crucial soil parameters, i.e., soil organic matter (SOM), salinity, and total nitrogen (N) were monitored, together with crop yields. Data analysis included statistical appraisals with ANOVA and multiple regression as well as mapping with ArcGIS. The results of the combined analyses show important tendencies such as an overall SOM increase with time and a reduced soil salinity increase under CA practices, and the yield-reducing effects of salinity. The ArcGIS maps reveal a certain variation in SOM over the entire experimental area, but all values remained within the "moderate" category, although wheat (*Triticum aestivum* L.) yields were reduced in the more saline areas of the large-scale experimental field. The ANOVA results show that CR retention had a slightly positive effect (yet not significant) on the SOM and N contents, and that it did not affect yields. CR retention slowed down the salinity increase over time. The SOM was significantly higher under ZT and PB, and soil salinity was significantly lower, but these differences remained below 13% compared to CT. Due to a high variability in yields, the effects on yields were insignificant, but cotton yields were very low under ZT and IT. Cotton yields were high under CT and PB (+CR), and high wheat yields were observed under PB. Wheat yields under IT were high, but in combination with the cotton yields, this system cannot be recommended. The PB practices are a good alternative to CT that may lead to yield declines in the long run due to the build-up of soil salinity. These first results from 3 years of cropping immediately after introducing CA practices hold sufficient promise for CA and residue retention (mulch) in irrigated drylands. But further studies are needed to understand long-term dynamics and to elaborate detailed land management procedures to increase the sustainability of dryland agriculture.

Keywords Crop residue • Soil tillage • Dryland • Conservation agriculture • Irrigated agriculture

13.1 Introduction

Farmers in more than 100 countries, mainly in tropical and subtropical regions, cultivate cotton (*Gossypium hirsutum* L.) on a total of about 31,000,000 ha of land. Almost 80% of all cotton is produced in China, USA, India, Pakistan, Brazil and Uzbekistan. About 75% of all cotton worldwide is produced under flood-and-furrow irrigation. Uzbekistan produces 1,100,000 tons of cotton fiber annually exclusively under irrigation, which is about 6% of the total global cotton output (Kooistra and Termorshuizen 2006). But since the mid 1990s, Uzbekistan has placed much emphasis also on wheat production to satisfy domestic demands and to become independent from food imports. Since then, 3,700,000 ha of the 4,300,000 ha of arable land in Uzbekistan have been allotted to cotton-wheat rotation. Both cotton and wheat are strategic state crops, contributing significantly to employment, food security and foreign exchange generation (cf. Rudenko et al. 2011).

While worldwide, irrigated cropland expanded globally in about one century from about 50,000,000 ha in 1900 to about 280,000,000 ha in 2003, the increase in irrigated croplands of the five Central Asian countries (Kazakhstan, Kyrgyzstan, Tajikistan, Turkmenistan and Uzbekistan) from 2,000,000 to 7,900,000 ha occurred in less than four decades (FAO/WFP 2000). This expansion, predominantly achieved during the Soviet Union era, was driven by the need for increasing irrigated cotton production. About 60% of the total land area in Uzbekistan is desert or semi-desert, and not even 10% of the territory is suitable for crop cultivation. Of this small area, about 80% depends solely on irrigation (FAO/WFP 2000).

Both cotton and wheat cultivation in Uzbekistan is based on recommendations developed mainly in Soviet times. These recommendations include intensive tillage to increase regular aeration and thus enhance microorganism activity. Due to an accelerated decomposition of soil organic matter (SOM) and a subsequent improved plant nutrient availability, such tillage operations increase soil productivity, but only temporarily. On a long-term basis, intensive tillage operations lead to increasing degradation of the physical, chemical and biological parameters of the soil (Craswell and Lefroy 2001), which translates into water logging of fields due to compaction and to declining soil fertility due to lower SOM content. Also, many fields in Uzbekistan suffer from high soil salinity due to the shallow groundwater tables (cf. Tischbein et al. 2011) and the high potential evaporation in this arid region (cf. Conrad et al. 2011). Furthermore, after independence in 1991, producers have experienced price increases for fuel, lubricants and agricultural implements, which have more than tripled production costs in the past decade under virtually stable farm-gate prices for raw cotton and wheat (Rudenko and Lamers 2006).

Since the Government of Uzbekistan wants to revert the present trends of soil degradation and water insecurity and to reclaim degraded lands and improve livelihoods, research has been directed to assessing options for introducing conservation agricultural (CA) practices. These practices have become common in many farming systems worldwide under a range of agro-ecological conditions (FAO 2007) on almost 100,000,000 ha (FAO 2006a), however mainly in rainfed areas in northern and southern America and in Australia (Sommer et al. 2007). CA consists of a combination and variation of three basic practices, namely (i) reduction in tillage and controlled field traffic, (ii) permanent soil cover with plants or plant residues, and (iii) application of economically feasible and diversified crop rotations (Ekboir 2002). In a vast number of CA studies it is argued that applying these three practices, in the short run, reduces production costs associated with tillage operations, increases SOM and crop yield, and improves soil structure which in turn leads to greater soil-moisture holding capacities. Changes in soil-chemical, biological and physical properties are long-term benefits (Knowler et al. 2001). It was therefore to be expected that such short- and long-term benefits would also contribute to improving the agricultural productivity and sustainability of land use in the irrigated drylands of Uzbekistan and the Aral Sea Basin in general. Given that agricultural practices in Uzbekistan were developed during Soviet times but continue to be followed until today, and given the declared intention of the Uzbek government to modernize land management policies, convincing science-based arguments are needed before introducing CA practices. In Uzbekistan, research on CA technologies was first conducted in Tashkent Province, leading to improved soil quality and yields of winter wheat and cotton (Pulatov 1999, 2002; Pulatov et al. 1997, 2001a, b; Choudhari et al. 2000; FAO 2007, 2009). All experiments were done comparatively on non degraded and salinity-affected soils under irrigation. This study aims at identifying the potential effect of CA practices on soil parameters and on crop yields under an irrigated cotton-wheat rotation on salinity-affected areas in Khorezm.

13.2   Materials and Methods

Data from an experiment over five vegetation cycles addressing a cotton-wheat rotation during the period 2002–2005 were analyzed and summarized with regard to selected soil and crop yield parameters under different management practices and CA technologies.

In 2002, the CA experiment was initiated in a cotton-wheat rotation on the experimental farm of the Urgench State University, located 45 km from Urgench (latitude 41.35°N, 60.310°W, and 92 m above sea level) in southwestern Khorezm. The soil texture at the 2.85 ha experimental area, which had been kept fallow for the five preceding years, was classified as coarse sandy loam to medium fine loam, or an irrigated meadow alluvial soil (ASSRI 1975) according to the Soviet classification system used in Uzbekistan, and as a gleyic calcaric (sodic) Arenosol according to the FAO classification (FAO 2006b). Physical and chemical soil properties at the onset of the experiment (Table 13.1) indicated that the soils were low in SOM, poor in total nitrogen (N), phosphorus (P) and potassium (K). Based on the total soluble salt in the topsoil and the Cl^-/SO_4^{2-} ratio, the soil was classified as a chloride-sulphate salinity type (ASSRI 2003).

Experimental design: The experimental layout was a completely randomized block design (Fig. 13.1). The treatments included four soil tillage methods, with and without crop residue retention. Soil tillage included (i) conventional tillage (CT) representing farmer practices as the control, (ii) zero tillage (ZT), (iii) permanent raised beds (PB), and (iv) intermediate tillage (IT) (see below for more details). The four tillage treatments were subjected to total retention (+CR) or total removal of residues (−CR) from the previous crops. The setup thus involved eight treatment combinations replicated four times (total 32 plots). The plots for each treatment were 11.3 m × 75 m (847.5 m²) covering a total area of 2.8 ha, including the border rows. Prior to the start of the experiment, the area was leached with 1,000 m³ water

13 Introducing Conservation Agriculture on Irrigated Meadow Alluvial Soils...

Table 13.1 Soil physical and chemical characteristics at the experimental site in 2002

Soil depth, cm	Percent (%) on weight basis										
	Sand	Silt	Clay	SOM[a]	N	P	K	TSS[a]	SO_4^{2-}	Cl^-	HCO_3^-
0–20	34.3	47.6	17.9	0.57	0.070	0.088	1.52	0.33	0.15	0.05	0.03
20–40	35.4	47.0	17.5	0.41	0.058	0.068	1.32	0.33	0.14	0.05	0.04
40–60	44.2	42.6	12.9	0.40	–	–	–	0.45	0.28	0.04	0.04

[a]*TSS* total soluble salts, *SOM* soil organic matter content

Fig. 13.1 Schematic field layout of treatments and the irrigation/drainage network at the experimental site. CT, RB, ZT and IT refer to conventional farmer practice, permanent raised bed, zero tillage and intermediate tillage practice, respectively. The symbols (+/–) refer to total retention or removal of residues of the previous crop

ha^{-1}, a common practice to remove excess salinity (cf. Tischbein et al. 2011). Following leaching, the field was tilled by shallow harrowing to prepare a fine seed bed.

Description of the tillage treatments: The Ministry of Agriculture and Water Resources of Uzbekistan (MAWR 2004) recommends to seed cotton only after several intensive soil tillage operations, including moldboard plowing to 30–40 cm soil depth, followed by chiseling to 20–25 cm soil depth, and leveling to breakdown soil clogs and to soften the seedbed. After cotton germination, a system of raised bed and furrows is installed through field operations, mainly conducted mechanically with the intensive use of tractors, but sometimes also done manually.

The cropping cycle from 2002 to 2005 was a cotton-winter wheat rotation, which started with cotton. Under conventional tillage (CT), cotton was cultivated in a bed-and-furrow system in which first one row of cotton was seeded on the flat soil at 7–8 cm depth. Next, irrigation furrows were drawn between the seeded cotton, thus creating a bed of 90 cm width. During cotton growth, fertilizers were applied three times, each time followed by irrigation. The irrigation events coincided with an additional reshaping of the irrigation furrows and the removal of weeds. After harvest, the raised beds were destroyed and the soil was prepared anew for sowing winter wheat, which is conventionally also seeded on a flat soil. Following soil preparation, seedbeds of winter wheat were shaped in the form of basins. No further tillage occurred during the following cultivation of wheat. Prior to seeding cotton again in spring, the soil was ploughed (30 cm deep) after the harvest of the winter wheat. Prior to leaching in March/April, the plots were leveled to avoid patchy irrigation and fertilizer distribution. Following chiseling, the seedbed was prepared. For winter wheat, all soil preparations were performed in autumn.

For the implementation of the intermediate tillage (IT) treatment, the soil was prepared in the same way as for the CT for cotton, except that first the beds and irrigation furrows were established and then cotton was seeded directly on the 90 cm wide beds. After cotton harvest, these beds were reused for seeding winter wheat, without any tillage of the previously established seedbed. The wheat was sown in four rows on the existing 90 cm beds with a row spacing of 15 cm. Prior to the cultivation of the subsequent crop cotton, the soil was ploughed, leveled and chiseled. The IT treatment was implemented in 2004 on plots that had been under a PB treatment with a 30 cm bed height between 2002 and 2003 (Karimov 2003). This contrasted with the standard PB treatment of 15 cm bed height in the other plots, but did not yield significantly different results. Hence, the 30 cm PB was given up and IT implemented instead.

The permanent raised bed (PB) treatments were similar to the IT treatments, but in contrast to IT where the soil was ploughed once every two cultivation cycles, the PB plots were never tilled throughout this experiment. Cotton was cultivated on the created raised beds. The beds were reshaped before planting the next crop only if necessary.

In the zero soil tillage (ZT) treatments, in general also referred to as "no tillage", no previous soil cultivation was conducted prior to seeding of cotton and winter wheat. During seeding with a specially developed zero-till planter (Tursunov 2009), the soil was slightly disturbed by thinly slicing the soil for seeding, which was subsequently closed with a press wheel. Hence, all crops were seeded each year on a flat soil. With this method, not more than about 10% of the soil surface was disturbed.

Management of the experiment: Throughout the experiment, the cotton variety "Khorezm-127" (60 kg ha^{-1}) and the wheat variety "Mars" (200 kg ha^{-1}) were seeded. Mineral fertilizers (N:P:K) were applied according to the local recommendations for both crops, i.e., 180:140:100 kg ha^{-1}. N was applied in equal splits including a basal application before irrigation. Phosphorus and K fertilizers were applied as a basal dose at seeding.

Data collection, processing and analyses: Throughout the entire experimental period, the soil was sampled six times, i.e., just before seeding and after the harvest of each crop. Soils were sampled at three locations in each of the 32 plots at 0–30, 30–50, 50–80 and 80–100 cm depths (total of 2304 samples over study period). Samples were air dried and analyzed for soil organic matter (SOM [%]), total nitrogen (N [%]), total phosphorous (P [%]), total potassium (K [%]), chloride (Cl$^-$ [%]), hydrogen carbonate (HCO$_3^-$ [%]), sulphate (SO$_4^{2-}$ [%]) and total dissolved solids (salinity [%]). SOM content was determined by the Turin method, and soil P and K by the Machigin-Protasov method, which is a modified Olsen extraction. Salinity was measured in a soil-water extract of 1:5. More details are presented elsewhere (Egamberdiev 2007).

Groundwater table fluctuations and groundwater salinity were monitored only during the vegetation periods. For this, piezometers were installed in each of the 32

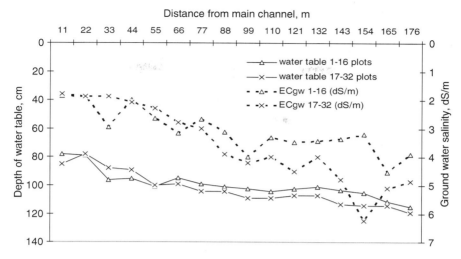

Fig. 13.2 Groundwater depth and salinity at the experimental site during vegetation period. Plots with the same distance to the major collector drain were pooled (Fig. 13.1)

plots to a depth of 250 cm. Surface to groundwater-level data were corrected to a reference level to obtain actual groundwater table values. The data from plots 1–16 and 17–32 were pooled, since these plots were located at the same distance to the major collector drain (Fig. 13.2). The groundwater table was close to the surface (80 cm) near the main canal (Fig. 13.2), but increased away from the main canal. At a distance of 176 m from the main irrigation canal, the depth was 115 cm, falling at the rate of 2 mm m^{-1} away from the main canal.

The drainage water collector influenced the level of the groundwater table. For example, it was 1–4 cm lower in the plots 17–32 that bordered the collector. Groundwater salinity varied from 1.9 to 4.5 dS m^{-1} in distant plots (1–16) and from 1.8 to 6.2 dS m^{-1} in plots near the drain (17–32). Salinity increased with the depth of the groundwater table (Fig. 13.2). A high groundwater table (near the canal) had lower salinity and was likely to contribute to crop water demands, and also resulted in lower soil salinization levels in the root zone.

Statistical analyses: Classification and regression trees (Cart Salford Systems 2000–2006) were used for first data mining (Breiman et al. 1984). The considered variables were year, month, plot number, treatment, mulch, SOM, depth [cm], N content, P, K, SO_4^{2-}, HCO_3^-, Cl$^-$, salinity, and crop yield. From these preliminary results (data not shown), five parameters were selected for subsequent analyses: SOM, salinity, N, and cotton and winter wheat yield. For each particular CART run, all remaining variables were chosen as predictor variables. Given the objective to analyze the temporal dynamics of the selected soil parameters under the different CR and soil tillage management practices, the experimental period March 2002–October 2005 was split into seven periods delineated by the time of soil sampling. These included March 2002 (before planting cotton) and October 2002 (after cotton harvest), July 2003 (after winter wheat harvest), March 2004 (before planting cotton) and October 2004 (after cotton harvest), July 2005 (after winter wheat harvest) and October 2005 (end of observation period). An ANOVA could not be performed for the data of March 2002 due to the lack of sampling replications.

Since the dependent variables were subject to repeated measurements, further analysis also included the ANOVA option 'repeated measures' to determine the existence of a significant number of changes over the experimental period. ArcGIS tools were used to visualize the obtained results in thematic maps, which depict the distribution of the measured parameters in space. This was followed by a linear regression analysis for the parameters SOM and salinity to determine whether relationships existed between plot location (expressed as the distance to the main channel) and whether plot location had an effect on these two soil parameters. Of all parameters monitored, only cotton and winter wheat yield, soil salinity, SOM and soil N content were chosen as dependent variables to determine how they were affected by the two treatment factors, namely two CR levels and four soil tillage methods. All variables were analyzed over the entire experimental period 2002–2005. The CR effects were first confounded over all treatments and subsequently considered for each soil tillage treatment. Soil salinity was determined by chemical analyses at 0–20, 20–40, 40–60 cm in 2002–2003, and at 0–30, 30–50, 50–80 and 80–100 cm depths in 2004–2005, and SOM and N content at 0–30 and 30–50 cm depths. Although plants can take up nutrients from deeper soil layers, we expected the most significant changes in the pool of available nutrients to occur in the upper soil layers. Therefore, laboratory findings on soil salinity, SOM and N were included in the statistical analyses for the 0–30 cm depth only. However, we complement these findings with a soil profile description over the total depth of these parameters. The statistical level of significance was defined to be $p < 0.1$. Statistical analyses were conducted with SPSS and SAS (SAS Institute 2005; SPSS 2005).

13.3 Results

Soil profile: Throughout the experiment, soil salinity, SOM and N contents in lower soil strata were always lower than in the topsoil. Soil salinity was significantly lower by approximately 20% down to 80 cm depth. In the 80–100 cm layer, salinity was slightly higher, although insignificantly. Comparing the two upper soil layers (0–30 and 30–50 cm), SOM and N contents were significantly lower over depth by circa 20% (data not shown).

Crop residue (CR): Irrespective of soil tillage treatment, the residue cover (+CR) significantly increased SOM and soil N content, whereas its effect on soil salinity was insignificant (Table 13.2). During these first years, cotton and winter wheat yields were insignificantly affected by the residue treatments (Table 13.2).

Soil tillage and crop residue interaction: Both soil salinity and SOM content under −CR were significantly different between CT and the three CA treatments (results not shown), but residue retention (+CR) had no effect on soil salinity,

Table 13.2 Effect of crop residue retention on percent changes of selected soil parameters between 2003 and 2005

CR	Salinity [%]	SOM [%]	N [%]	Cotton yield [kg ha^{-1}]	Winter wheat yield [kg ha^{-1}]
+	0.44 (±0.20) a	0.66 (±0.11) a	0.067 (±0.011) a	2 265.4 (±961.2) a	5 265.4 (±916.3) a
−	0.45 (±0.21) a	0.63 (±0.11) b	0.065 (±0.010) b	2 299.6 (±1 258.2) a	5 404.2 (±1 106.3) a

Note: Means with the same letters in one column are not significantly different at $p < 0.1$ according to the Bonferroni t-test. These findings exclude the results of the first soil sampling due to an insufficient number of repetitions

Numbers in brackets indicate standard deviations

Table 13.3 Means of soil parameters with crop residue retention (+CR) and removal (−CR) per treatment, pooled over the whole experimental period (± standard deviation)

Treatment	Salinity [%] −CR	Salinity [%] +CR	SOM [%] −CR	SOM [%] +CR	N [%] −CR	N [%] +CR
CT	0.44 (±0.19) b	0.51 (±0.21) a	0.63 (±0.13) a	0.64 (±0.10) a	0.067 (±0.010) a	0.066 (±0.014) a
IT	0.49 (±0.25) a	0.40 (±0.17) b	0.62 (±0.11) b	0.65 (±0.10) a	0.066 (±0.009) b	0.069 (±0.009) a
PB	0.44 (±0.19) a	0.44 (±0.23) a	0.63 (±0.10) b	0.66 (±0.15) a	0.067 (±0.008) a	0.063 (±0.010) b
ZT	0.42 (±0.18) a	0.40 (±0.19) a	0.64 (±0.12) b	0.71 (±0.09) a	0.060 (±0.010) b	0.071 (±0.008) a

Note: Means with the same letters in one row of the respective variable are not significantly different at $p < 0.1$ according to the Bonferroni t-test

Table 13.4 Means (± standard deviation) of cotton (2004) and winter wheat (2005) yields for soil tillage treatments with crop residue retention (+CR) and removal (−CR)

Treatment	Cotton (kg ha^{-1}) −CR	Cotton (kg ha^{-1}) +CR	Winter wheat (kg ha^{-1}) −CR	Winter wheat (kg ha^{-1}) +CR
CT	2 830.2 (±874.6) a	3 422.2 (±780.2) a	5 542.0 (±1 541.3) a	4 277.0 (±1 758.2) a
IT	1 859.8 (±765.4) a	1 302.5 (±406.7) a	5 403.3 (±615.1) a	5 630.8 (±252.9) a
PB	1 957.7 (±1 168.3) a	2 840.3 (±1446.2) a	5 616.7 (±822.5) a	6 052.5 (±415.8) a
ZT	2 550.5 (±1 015.2) a	1 496.8 (±893.7) a	5 055.0 (±720.9) a	5 100.8 (±670.3) a

Note: Means with the same letters in one row of the respective variable are not significantly different at $p < 0.1$ according to the Bonferroni t-test

SOM and soil N contents, irrespective of the soil tillage treatment (Table 13.3). Cotton yields in 2004 and winter wheat yields in 2005 were insignificantly affected by the residue treatment, irrespective of tillage treatment (Table 13.4).

Tillage: Confounded over the CR management practices, only soil salinity and SOM were significantly influenced by soil tillage, whereas soil N contents were not (Table 13.5). At the end of 2005, soil salinity was highest under CT with 0.47% and lowest under ZT with 0.41% ($p < 0.1$). At an assumed bulk density of 1.4 g cm^{-3} (averaged over the entire 2.85 ha experimental area), this corresponds to a difference of 2,520 kg salts ha^{-1} in the 0–30 cm top soil. SOM contents peaked with 0.67%

Table 13.5 Means (± standard deviation) of soil salinity, soil organic matter (SOM) and soil nitrogen (N) content at the end of the study period 2005 at 0–30 cm depth

Treatment	Salinity [%]	SOM [%]	N [%]
CT	0.47 (±0.20) a	0.63 (±0.11) b	0.067 (±0.010) a
IT	0.45 (±0.22) ab	0.63 (±0.10) b	0.068 (±0.010) a
PB	0.44 (±0.21) ab	0.65 (±0.13) ab	0.065 (±0.010) a
ZT	0.41 (±0.18) b	0.67 (±0.11) a	0.065 (±0.010) a

Note: Means with the same letters in one column are not significantly different at $p < 0.1$ according to the Bonferroni t-test

Table 13.6 Means (± standard deviation) of cotton yield in 2004 and winter wheat yield in 2005 according to soil tillage treatment

Treatment	Cotton (kg ha^{-1})	Winter wheat (kg ha^{-1})
CT	3 126.2 (±830.0) a	4 909.6 (±1,673.2) a
IT	1 581.1 (±641.7) b	5 517.1 (±452.1) a
PB	2 399.0 (±1,305.3) ab	5 834.6 (±646.8) a
ZT	2 023.6 (±1,049.4) ab	5 077.9 (±644.9) a

Note: Means with the same letters in one column are not significantly different at $p < 0.1$ according to the Bonferroni t-test

under treatment ZT, which was insignificantly different from PB (0.65%), but significantly higher than the SOM contents under CT and IT (both 0.63%).

In the first season after the conversion to CA practices, the highest cotton yields were obtained with CT (3,126 kg ha^{-1}). These differed significantly only from IT (Table 13.6), which had the lowest yields (1,581 kg ha^{-1}). Yet, the effect of soil tillage on winter wheat yields in the following season was already insignificant, irrespective of soil treatment (Table 13.6).

Hence, in summary, although CT had the highest cotton yields (Table 13.6), the highest soil salinity was also observed with this soil tillage practice (Table 13.5). In contrast, SOM content was highest and salinity lowest under ZT. The results of the treatments IT and PB were always similar. Although IT was implemented in 2004 on previous PB trials with higher bed height, this does not seem to have affected the yields. If the previous PB trial had influenced yields in the later IT trial, they should have been more similar to PB than to ZT.

13.4 Temporal Dynamics

Crop residue (CR): The ANOVA findings show that CR retention influenced SOM and N contents, but not soil salinity level. In general, +CR resulted in higher SOM and N contents than −CR. The SOM and soil N contents under IT and PB did not significantly differ.

Tillage: Soil salinity, SOM and soil N content were only significantly influenced by tillage method until March 2004. The monitored differences for the final three growing periods (October 2004–October 2005) were not significant. The soil N content in July and October 2005 showed significant differences as a result of soil tillage method. For both periods, CT yielded highest N values, which differed significantly from ZT where the soil N concentration was lowest. Hence, soil N followed the order CT, IT > PB, ZT during these two periods.

When analyzing the differences at each individual measuring point separately, the ANOVA revealed that +CR had only influenced SOM and N content. Tillage operations with +CR increased SOM and N amounts significantly compared to tillage operations without CR retention (−CR). Differences in SOM and soil N contents between IT and PB were not significant.

Since the selected variables had been measured over time, they were subjected also to repeated-measures ANOVA. The findings from this not only confirm the results of the standard ANOVA that the effect of soil tillage varied between the measuring periods, but they also reveal the development of soil salinity, SOM and soil N contents. In the following, these temporal dynamics are presented for these selected soil parameters only. Based on the findings at the onset of the experiment (March 2002) and at the end (October 2005), the repeated measures ANOVA revealed that soil salinity increased significantly over the years for all treatments (Fig. 13.3), and that SOM also increased significantly over time (Fig. 13.4), whereas soil N content did not (Fig. 13.5). In fact, soil N content declined under PB and ZT (Table 13.7) over the experimental period.

Soil salinity showed the highest increase under CT (67%) and the lowest under PB (28%) (Table 13.8). When considering all tillage treatments, soil salinity increased on average by about 44%, and the SOM content by about 31%. The average soil N content decreased by about 2% (Table 13.8). In contrast to the changes in SOM and soil N, changes in soil salinity were not influenced by CR. The increase in SOM was significantly higher in the +CR treatments.

Based on these findings, the levels of the soil parameters at the onset (March 2002) and end (October 2005) were compared with a standard ANOVA to identify the influence of CR retention and soil tillage on these soil parameters (Table 13.9).

Crop residue: In contrast to SOM and soil N, the changes in soil salinity were not influenced by crop residue retention (data not shown). The increase in SOM was, however, significantly higher under soil tillage treatments with +CR.

Soil tillage: Soil tillage alone significantly affected soil salinity and soil N content over time, but this was not the case with SOM. Hence, it can be assumed that the observed increase in SOM in most treatments was not related to soil tillage *per se* but rather to the CR treatment and the interaction between tillage and CR. Regarding soil salinity and soil N content, CT always showed highest values (0.22% and 0.004%, respectively) that differed significantly from those of PB, which had the lowest values. The order of these two parameters was generally CT > ZT, IT > PB. Soil N content decreased under PB and ZT, but increased slightly under CT and IT (Table 13.9). In general, the highest increase in soil salinity, SOM and soil N occurred under CT, followed by IT > PB > ZT.

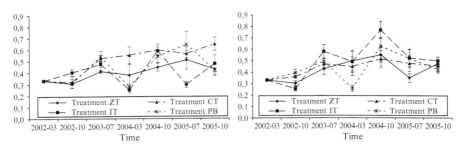

Fig. 13.3 Soil salinity (%) changes according to treatment (mean and standard error of 0–30 cm depth) over seven vegetation cycles 2002–2005; +CR (*left*) and −CR (*right*)

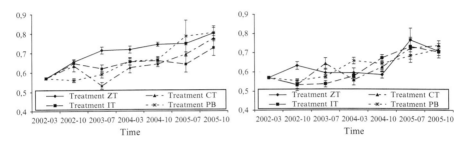

Fig. 13.4 SOM content (%) changes according to treatment (mean and standard error of 0–30 cm depth) over seven vegetation cycles 2002–2005; +CR (*left*) and −CR (*right*)

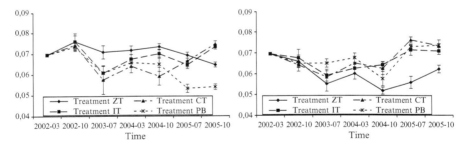

Fig. 13.5 N content [%] changes according to treatment (mean and standard error of 0–30 cm depth) over seven vegetation cycles 2002–2005; +CR (*left*) and −CR (*right*)

13.5 Location Effect

A regression analysis was used to identify whether or not the plot location (expressed as the distance to the main channel) on the 2.8 ha sized experimental area had led to changes in the selected soil parameters (Table 13.10). This analysis showed that initially (till July 2003) SOM content was influenced by the treatment factors and

13 Introducing Conservation Agriculture on Irrigated Meadow Alluvial Soils... 207

Table 13.7 Average (± standard deviation) changes in soil parameters between the onset (March 2002) and end of the experiment (October 2005) according to soil tillage treatment

Time	Treatment	Soil salinity (%)	Soil organic matter (%)	Soil N (%)
March 2002	CT	0.33 (±0.00) b	0.57 (±0.00) b	0.070 (±0.000) b
October 2005	CT	0.55 (±0.21) a	0.76 (±0.07) a	0.074 (±0.008) a
March 2002	IT	0.33 (±0.00) b	0.57 (±0.00) b	0.070 (±0.000) b
October 2005	IT	0.49 (±0.19) a	0.71 (±0.13) a	0.072 (±0.005) a
March 2002	PB	0.33 (±0.00) b	0.57 (±0.00) b	0.070 (±0.000) a
October 2005	PB	0.42 (±0.14) a	0.76 (±0.12) a	0.064 (±0.010) b
March 2002	ZT	0.33 (±0.00) b	0.57 (±0.00) b	0.070 (±0.000) a
October 2005	ZT	0.45 (±0.17) a	0.75 (±0.08) a	0.064 (±0.009) b

Note: Means with the same letters in one column are not significantly different at $p < 0.1$ according to the Bonferroni t-test

Table 13.8 Soil salinity, organic matter and nitrogen content increase/decrease according to soil tillage treatment at end of experiment (October 2005) compared to onset (March 2002)

Treatment	Soil salinity (%)	Soil organic matter (%)	Soil N (%)
CT	67.1	32.5	5.7
IT	47.4	25.3	2.9
PB	27.8	33.3	−8.6
ZT	36.6	32.3	−8.6

Table 13.9 Changes in soil salinity, organic matter and nitrogen content according to soil tillage treatment at end of experiment (October 2005) compared to onset (March 2002) (± standard deviation)

Treatment	Soil salinity (%)	Soil organic matter (%)	Soil N (%)
CT	0.22 (± 0.21) a	0.19 (± 0.07) a	0.004 (± 0.008) a
IT	0.16 (± 0.19) ab	0.14 (± 0.12) a	0.002 (± 0.005) a
PB	0.09 (± 0.14) b	0.19 (± 0.12) a	−0.006 (± 0.011) b
ZT	0.12 (± 0.13) ab	0.18 (± 0.08) a	−0.006 (± 0.005) b

Note: Means with the same letters in one column are not significantly different at $p < 0.1$ according to the Bonferroni t-test

not by the location of the plots. However, in July 2005, plot location significantly influenced SOM, nevertheless, this was not confirmed after this period. In contrast, in October 2002 and July 2005 when the model produced significant results, soil salinity was influenced by the location rather than by the treatment factors tillage method and crop residue management (Table 13.10).

The spatial distribution of SOM, soil salinity and N contents for the different plots and measurement periods allowed distinguishing location-related from treatment-related differences (Figs. 13.5–13.7). The analyses showed a high heterogeneity among the plots over the entire experimental field with regard to soil salinity and N contents, but this heterogeneity was less distinct for SOM. Furthermore, the distribution of these parameters varied according to observation period. Soil salinity

Table 13.10 Significances of the regression model and the independent variables (p<0.1) according to Bonferroni *t*-test

Variable	Time	Model significance	Treatment significance	Distance significance
SOM	October 2002	0.68	0.94	0.92
	July 2003	0.02	0.02	0.19
	March 2004	0.09	0.11	0.17
	October 2004	0.37	0.19	0.63
	July 2005	0.05	0.27	0.03
	October 2005	0.62	0.39	0.61
Salinity	October 2002	0.03	0.82	0.01
	July 2003	0.56	0.42	0.51
	March 2004	0.95	0.77	0.89
	October 2004	0.21	0.57	0.10
	July 2005	0.04	0.49	0.01
	October 2005	0.20	0.51	0.09

concentrations in the southeastern part of the field increased after July 2003 and exceeded those in most other plots (Fig. 13.6). In contrast and also indicated by the statistical analysis, SOM content increased for the entire field during the experimental period irrespective of plot location (Fig. 13.7). Nevertheless, according to the Uzbek classification, the SOM contents remained low (0.4–0.8%) to moderate (0.8–1.2%). Soil N contents seemed to have gradually decreased, irrespective of plot location (Fig. 13.8).

The yields per plot of cotton in 2004 (Fig. 13.9) and winter wheat in 2005 (Fig. 13.10) are visualized based on the ANOVA results. Cotton yield was lowest in the IT treatment compared to all other soil tillage methods (Fig. 13.9). Soil salinity seems to have influenced winter wheat yields, e.g., the highly saline plots located in the southeastern corner gave lower yields.

13.6 Discussion

13.6.1 Dynamics of Soil Parameters Under Conservation Agriculture Practices

Soil organic matter: Despite the relative, but significant, increase in SOM content over time in the CA practices from about 0.57% to about 0.75% (i.e., about 32% of increase), the absolute increase remained moderate, as expected. An increase in SOM is generally a long-term process, in particular in semi-arid environments with high turn-over rates of SOM. Sanchez et al. (2004) therefore argued that under semi-arid agro-climatic conditions, an increase in SOM is usually proportional to the annual amount of organic matter added irrespective of whether mulch is applied or residues are incorporated in the soil. To accurately evaluate the effect of soil tillage on SOM, long-term experimental data is hence required (Ding et al. 2002).

13 Introducing Conservation Agriculture on Irrigated Meadow Alluvial Soils... 209

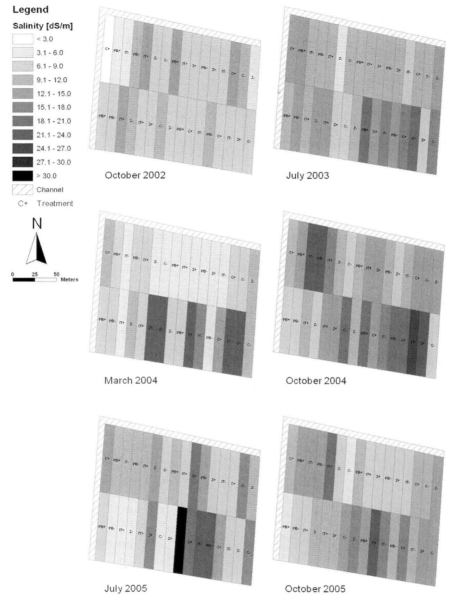

Fig. 13.6 Soil salinity changes over experimental period March 2002–October 2005. For classification of measured salinity in the FAO salt-tolerant classification, percentage values were converted into dS m^{-1} according to Forkutsa (2006)

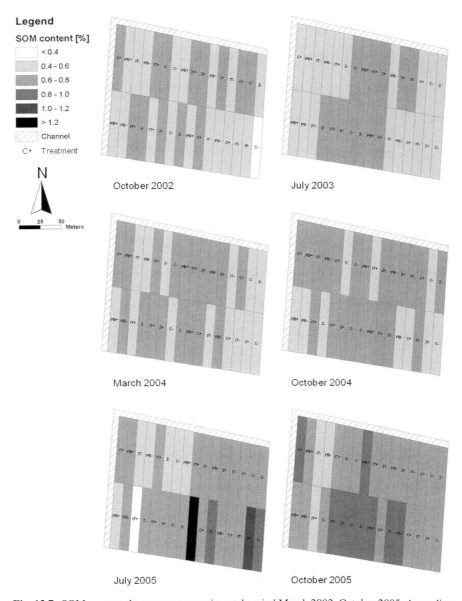

Fig. 13.7 SOM content changes over experimental period March 2002–October 2005. According to Uzbek classification, most SOM amounts were in the category 'low' (0.4–0.8%), a few were 'moderate' (0.8–1.2%)

This is especially important since it has been postulated that over shorter observation periods, SOM may vary naturally due to changing climatic conditions (Scheffer and Schachtschabel 2002) or according to conditions that cannot be controlled, e.g., groundwater movement.

13 Introducing Conservation Agriculture on Irrigated Meadow Alluvial Soils... 211

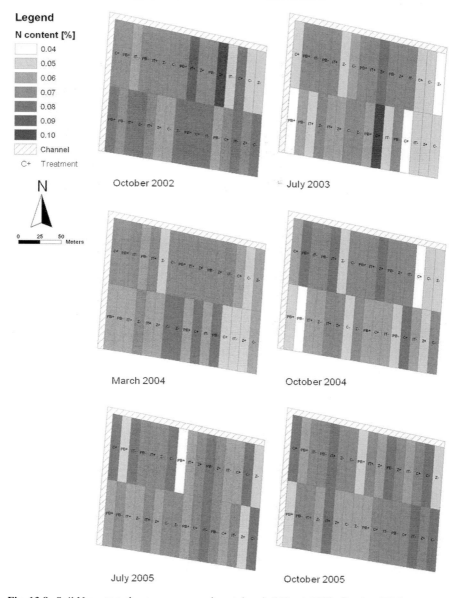

Fig. 13.8 Soil N content changes over experimental period March 2002–October 2005

The experimental period of seven growing seasons is nevertheless sufficiently long to obtain indications that the observed SOM accumulation was associated with tillage treatments. These results are also in line with previous findings that showed that SOM content increased more under ZT than under CT (Baldesdent et al. 2000).

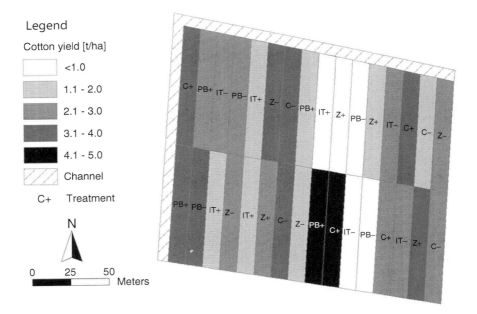

Fig. 13.9 Cotton yields per plot (t ha^{-1}) in 2004

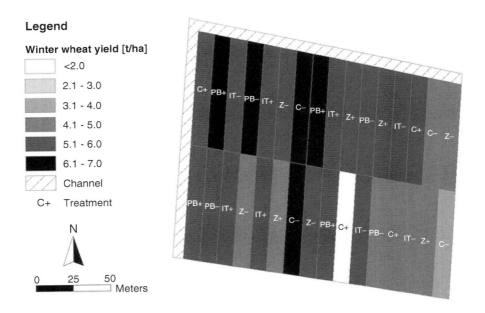

Fig. 13.10 Winter wheat yields per plot (t ha^{-1}) in 2005

The data set, however, does not allow clearly identifying those treatments that had led to a higher accumulation of SOM. Assuming that this accumulation is a result of the time needed to reach a certain SOM value, one can look at the point in time where this level is reached. For instance, the arbitrarily set value of 0.7% SOM in the topsoil in this experiment was reached at the end of the observation period in all +CR treatments, but was reached much earlier under ZT (Fig. 13.4). In contrast, without crop residue retention (−CR), the SOM content under ZT never reached this value of 0.7%, while in the other treatments the differences between the −CR and +CR treatments were insignificant. These findings indicate that ZT practices should be combined with some level of residue retention in order to increase the SOM content. This aspect was investigated in subsequent years (2007–2010) in studies on CA practices in the same region.

Soil salinity: Secondary soil salinization, which is a major cause of soil degradation in the study region (cf. Akramkhanov et al. 2011; Tischbein et al. 2011), depends on the degree of soil-water evaporation. In general, both soil mulching and tillage operations influence water evaporation, but to a different extent. Soil tillage operations can either increase or decrease evaporation depending on the tillage depth, soil-water content, or soil texture (Unger and Cassel 1991). Reduced tillage does therefore not necessarily reduce evaporation. However, one would expect that, when combined with a sufficient soil cover (mulch), it reduce capillary rise, so that less salts are moved to and accumulate in the topsoil. However, no systematic reduction in soil salinity with CR retention was observed (Fig. 13.3). This could have been due either to the high salinity levels in the irrigation water or to high initial soil salinity. Nonetheless, in the long run, mulching the soil surface is favorable. This was also confirmed by subsequent CA studies in the same region.

Soil tillage may have contributed in different ways to the observed differences in soil salinity, e.g., the larger total soil surface area exposed to the air in the bed-and-furrow systems (PB) than in the flat area under ZT may have increased the potential evaporative surface. On well-leveled, flood-irrigated treatments (e.g., ZT), irrigation water could have leached out the salt, which would have decreased topsoil salinity. Although the present data set does not allow far-reaching conclusions, recent research findings in the same area show that crop residue retention on permanent beds increased soil moisture and reduced the increase in soil salinity rates already in the second growing season, thus indicating that CA practices can be beneficial. Also others (Tursunov 2009) concluded that PB practices are a good alternative to CT in salt-affected irrigated drylands when an appropriate crop residue management is introduced.

Soil N: Various previous findings indicate that total topsoil N content is expected to increase under CA (no-tillage) compared to CT practices (Agenbag and Meree 1989; Hao et al. 2000; Thomas et al. 2007), since soil tillage accelerates mineralization and short-term N availability to plants. However, N mineralization with ZT plus crop residue retention may take longer because microorganisms are less active and in turn less plant-available N is present. Furthermore, ZT sometimes tends to entail a lower N_{min} content, possibly caused by N immobilization in crop residues. Based on this, to counterbalance the expected N immobilization some authors advocated for increasing the N applications at the onset of the conversion of CT to CA practices (Hickmann 2006; Sommer et al. 2007). Nitrogen immobilization occurs especially when microorganisms consume all available N to decompose the high carbon portion of the SOM in the crop residues. A high C/N ratio of the residues will enhance N fixation (Bosshardt 1984). The consequently lower mineralization rates result in lower N_{min} contents in the soil. Hence, the general decrease in topsoil N content observed in the CA practices (Fig. 13.5) contradicts these previous findings. A decrease in SOM content could have explained the decline in soil N content, but this was not the case here. We suggest therefore future studies to monitor both N_{min} and C/N ratio rather than the total soil N content alone. Should the decreasing trend persist over several years, higher N applications than previously concluded may be required to counterbalance N losses (Hickmann 2006; Sommer et al. 2007). On the other hand, recent findings in the study region also indicate high N emissions with conventional cotton, wheat and rice cultivation (Scheer et al. 2008), and these N emissions peaked when irrigation events immediately followed N-fertilizer applications. Hence, generally increasing N-fertilizer applications may turn out to be less efficient if present farmer's practice is followed. Moreover, on-going work on CA practices in the same region indicates that the response of N applications was higher in PB than under CT, i.e., increased total N use efficiency was observed. Proper N fertilization schemes under CA in semi-arid environments such as in Uzbekistan are therefore certainly crucial for successful crop management and need further detailed research.

Reduced or no-tillage is perhaps interesting for farmers because of the cost reduction due to a much lower number of machine operations, which decreases fuel consumption and workload (Knowler et al. 2001). However, immediately after introduction of CA practices, especially ZT, and sometimes for long periods, yields may not be higher than under CT (Gomez et al. 1999). In the present study, cotton yields were lowest with ZT. Thus CA practices may not pay off immediately, and this may be a major obstacle for the acceptance of such practices by the farmers in rural Khorezm because of the farmers' high dependence on income generation from cropping activities (cf. Djanibekov et al. 2011). Financial estimates of the CA treatments using cumulative gross margin (GM) and dominance analyses (over 3 years) showed, however, higher GM values with CA practices compared to those of CT (Tursunov 2009). The highest GM was found under IT with crop residue retention. A dominance analysis based on the accumulated results of three consecutive growing seasons clearly showed the advantages of CA over conventional practices, owing to the higher total variable costs and lower GMs of the latter.

13.7 Summary and Conclusions

With the intention to assess options for introducing CA practices under irrigated conditions in an arid environment, it was demonstrated that under the continental climate of Khorezm, higher SOM levels and soil N contents could be secured when mulching the soil surface with crop residues (+CR). In addition, significant differences in SOM dynamics between conventional and reduced tillage operations were found during the seven cropping cycles. Highest SOM levels were observed under ZT and PB, but both treatments showed simultaneous losses in topsoil N contents. Furthermore, the soil salinity increase was 25% less in PB than under ZT. Only the reduced soil N content reduced the otherwise positive evaluation of PB and ZT, which requires more long-term research. Among the three CA practices, the yields of cotton and (not significantly) winter wheat were highest with PB and lowest with ZT. Although the cotton yields remained highest under CT, and thus higher than with CA practices, CT, as the present farmers' practice, also increased soil salinity. One cannot therefore exclude the possibility that CT yields will decrease in the long run –an often observed phenomenon in the region. Increasing soil salinity levels are counterbalanced by increasing amounts and events of leaching (cf. Tischbein et al. 2011). With IT, the lowest SOM contents and the lowest cotton yields were observed. Although this CA practice was designed as an intermediate step to facilitate the conversion from CT to CA practices, based on the short-term and preliminary findings, it seems reasonable to opt for a continuation of PB and for a discontinuation of IT and ZT practices. The current practice CT is recommendable under yield but not under ecological considerations.

References

Agenbag GA, Meree PCJ (1989) The effect of tillage on soil carbon, nitrogen and soil strength of simulated surface crusts in two cropping systems for wheat (*Triticum aestivum*). Soil Tillage Res 14:53–65

Akramkhanov A, Kuziev R, Sommer R, Martius C, Forkutsa O, Massucati L (2011) Soils and soil ecology in Khorezm. In: Martius C, Rudenko I, Lamers JPA, Vlek PLG (eds) Cotton, water, salts and *Soums*: economic and ecological restructuring in Khorezm, Uzbekistan. Springer, Dordrecht/Berlin/Heidelberg/New York

ASSRI (1975) Agro-chemistry and soil science research institute of ministry of agriculture and water resources. Soils of Uzbekistan. Fan, Tashkent, p 222

ASSRI (2003) Agro-chemistry and soil science state research institute. Soils of Khorezm region. Fan, Tashkent, p 188

Baldesdent J, Chenu C, Balabane M (2000) Relationship of soil organic matter dynamics to physical protection and tillage. Soil Tillage Res 53:215–230

Bosshardt U (1984) Einfluss der Stickstoffdüngung und der landwirtschaftlichen Bewirtschaftungsweise auf die Nitratauswaschung in das Grundwasser. Dissertation, Kassel University

Breiman L, Friedman JH, Olshen RA, Stone CJ (1984) Classification and regression trees. Chapman & Hall/CRC, Boca Raton, (Reprint 1998), 358 pp

Cart Salford Systems (2000–2006) CART extended edition version 6.0. California Statistical Software Inc, California

Choudhari A, Gill M, Pulatov A (2000) Prospects of no-till wheat in rotation with rice or cotton in Central and South Asia. In: Karabaev M, Satybaldin A, Benites J, Friedrich T, Pala M, Payne T (eds) Conservation tillage: a viable option for sustainable agriculture in Central Asia. CIMMYT and ICARDA, Aleppo, pp 125–131

Conrad C, Schorcht G, Tischbein B, Davletov S, Sultonov M, Lamers JPA (2011) Agro-meteorological trends of recent climate development in Khorezm and implications for crop production. In: Martius C, Rudenko I, Lamers JPA, Vlek PLG (eds) Cotton, water, salts and

Soums: economic and ecological restructuring in Khorezm, Uzbekistan. Springer, Dordrecht/Berlin/Heidelberg/New York

Craswell ET, Lefroy RDB (2001) The role and function of organic matter in tropical soils. In: Martius C, Tiessen H, Vlek PLG (eds) Managing organic matter in tropical soils: scope and limitations. Springer, Dordrecht, p 235

Ding G, Noak JM, Amarasiriwardena D, Hunt PG, Xing B (2002) Soil organic matter characteristics as affected by tillage management. Soil Sci Soc Am J 66:421–429

Djanibekov N, Bobojonov I, Djanibekov U (2011) Prospects of agricultural water service fees in the irrigated drylands, downstream of Amudarya. In: Martius C, Rudenko I, Lamers JPA, Vlek PLG (eds) Cotton, water, salts and *Soums*: economic and ecological restructuring in Khorezm, Uzbekistan. Springer, Dordrecht/Berlin/Heidelberg/New York

Egamberdiev O (2005) Resource conservation technologies. Agric Uzbekistan 4:29

Egamberdiev O (2007) Changes of soil characteristics under the influence of resource saving and soil protective technologies within the irrigated meadow alluvial soil of the Khorezm region. PhD dissertation, Tashkent

Egamberdiev O, Pulatov A, Tursunov L (2006) Soil aggregate condition. Agric Uzbekistan 6:33–34

Ekboir J (2002) Developing no-till packages for small-scale farmers. In: Ekboir J (ed) CIMMYT 2000–2001 world wheat overview and outlook: developing no-till packages for small-scale farmers. CIMMYT, Mexico, pp 1–37

FAO (2006a) Spotlight: conservation agriculture. Agriculture and consumer protection department. Rome, Italy. http://www.fao.org/ag/magazine/0110sp.htm. Accessed Aug 2008

FAO (2006b) World reference base for soil resources 2006. A framework for International classification, correlation and communication. FAO, Rome, p 128

FAO (2007) Conservation agriculture. http://www.fao.org/ag/ca/index.html. Accessed 15 May 2007

FAO/WFP (2000) Global Information and early warning system on food and agriculture (GIEWS): crop and food supply assessment mission to the Karakalpakstan and Khorezm Regions of Uzbekistan. www.fao.org/WAICENT/faoinfo/economic/giews/english/alertes/2000/SRuzb12.htm. Accessed 15 May 2007

FAO (2007). Project FAO/TCP/UZB/3001 Enhanced productivity of cotton-wheat systems through the adoption of conservation agriculture practices in Uzbekistan. Pulatov A. (ed) Final report

FAO (2009) Conservation agriculture in Uzbekistan. Working paper 2, FAO, Rome, pp.48

Forkutsa I (2006) Modeling water and salt dynamics under irrigated cotton with shallow groundwater in the Khorezm region of Uzbekistan. ZEF Ser Ecol Dev 37:158

Gomez JA, Giraldez JV, Pastor M, Fereres E (1999) Effects of tillage method on soil physical properties, infiltration and yield in an olive orchard. Soil Tillage Res 52:167–175

Hao X, Chang C, Larney FJ, Nitschelm J, Regitnig P (2000) Effect of minimum tillage and crop sequence on physical properties of irrigated soil in southern Alberta. Soil Tillage Res 57:53–60

Hickmann S (2006) Conservation agriculture in northern Kazakhstan. FAO – Agricultural and Food Engineering working document 4: 33

Karimov A (2003) Analysis of soil conservation technologies in wheat production in Khorezm Province. MS dissertation, Tashkent Institute of Irrigation and Melioration

Knowler D, Bradshaw B, Gordon D (2001) The economics of conservation agriculture. FAO report

Kooistra K, Termorshuizen A (2006) The sustainability of cotton: consequences for man and environment. Science Shop Wageningen University & Research Centre, Report no. 223

Lamers J, Egamberdiev O, Pulatov A, Tursunov L (2006a) Soil salinity changes under soil and resource conservation technologies. Uzbekistan Nat Univ J 1:102–106

Lamers J, Egamerdiev O, Martius C, Tursunov L, Pulatov A (2006b) New agrobiological approaches to improve soilfertility. Agric Uzbekistan 11:29

MAWR (2004) Annual report for crop production in Uzbekistan. Ministry of Agriculture and Water resources of the Republic of Uzbekistan, Tashkent

Pulatov A (1999) Development of environmentally sustainable agricultural systems in Uzbekistan. In: Banskota M, Karki A, Croon F (eds) Strategic considerations on the development of central Asia. Council for Sustainable Development of Central Asia, Urumqi, pp 21–23

Pulatov A (2002) Results of zero tillage in wheat production in Uzbekistan. In: International workshop on conservation agriculture for sustainable wheat production in rotation with cotton in water resource areas, Tashkent, 14–18 October 2002

Pulatov A, Akramkhanov A, Aldabergenov M, Fakhrutdinova M, Kanwar R, Kumar A (1997) Soil and water quality under alternative crop production systems in Uzbekistan. Paper No MC97-129, Mid -Central conference of ASAE, St. Joseph, MO, 11–12 April 1997, p 111

Pulatov A, Akramkhanov A, Choudhary A (2001a) Pioneering conservation tillage seeds are established in Uzbekistan Conservation Agriculture. In: A worldwide challenge, I world congress on conservation agriculture, Madrid, 1–5 October 2001

Pulatov A, Choudhary A, Akramkhanov A (2001b) Status of conservation tillage practices in Uzbekistan. In: International workshop on conservation agriculture for food security and environment protection in rice-wheat cropping systems, Lahore, 2–9 February 2001

Rudenko I, Lamers JPA (2006) The comparative advantages of the present and future payment structures for agricultural producers in Uzbekistan. Cent Asian J Manag Econ Soc Res 5(1–2):106–125

Rudenko I, Nurmetov K, Lamers JPA (2011) State order and policy strategies in the cotton and wheat value chains. In: Martius C, Rudenko I, Lamers JPA, Vlek PLG (eds) Cotton, water, salts and *Soums*: economic and ecological restructuring in Khorezm, Uzbekistan. Springer, Dordrecht/Berlin/Heidelberg/New York

Sanchez JE, Harwood RR, Wilson TC, Kizilkaya K, Smeenk J, Parker E, Paul EA, Knezek BD, Robertson GP (2004) Managing soil carbon and nitrogen for productivity and environmental quality. Agron J 96:769–775

SAS Institute (2005) SAS 9.1 TS Level 1 M2 for Windows. SAS Institute Inc. Cary

Scheer C, Wassmann R, Kienzler K, Ibraghimov N, Lamers JPA, Martius C (2008) Methane and nitrous oxide fluxes in annual and perennial land-use systems of the irrigated areas in the Aral Sea Basin. Glob Chang Biol 14:1–15

Scheffer F, Schachtschabel P (2002) Lehrbuch der Bodenkunde, 15th edn. Spektrum Akademischer Verlag, Berlin/Heidelberg, 593 p

Sommer R, Wall PC, Govaerts B (2007) Model-based assessment of maize cropping under conventional and conservation agriculture in highland Mexico. Soil Tillage Res 94:83–100

SPSS (2005) SPSS Version 14.0 for Windows. SPSS Inc

Thomas GA, Dalal RC, Standley J (2007) No-till effects on organic matter, pH, cation exchange capacity and nutrient distribution in a Luvisol in the semi-arid subtropics. Soil Tillage Res 94-2:295–304

Tischbein B, Awan UK, Abdullaev I, Bobojonov I, Conrad C, Forkutsa I, Ibrakhimov M, Poluasheva G (2011) Water management in Khorezm: current situation and options for improvement (hydrological perspective). In: Martius C, Rudenko I, Lamers JPA, Vlek PLG (eds) Cotton, water, salts and *Soums*: economic and ecological restructuring in Khorezm, Uzbekistan. Springer, Dordrecht/Berlin/Heidelberg/New York

Tursunov M (2009) Potential of conservation agriculture for irrigated cotton and winter wheat production in Khorezm, Aral Sea Basin. PhD dissertation, Bonn University, Bonn

Unger PW, Cassel DK (1991) Tillage implement disturbance effects on soil properties related to soil and water conservation: a literature review. Soil Tillage Res 19:363–382

Chapter 14
Crop Diversification in Support of Sustainable Agriculture in Khorezm

Ihtiyor Bobojonov, John P.A. Lamers, Nodir Djanibekov,
Nazirbay Ibragimov, Tamara Begdullaeva, Abdu-Kadir Ergashev,
Kirsten Kienzler, Ruzumbay Eshchanov, Azad Rakhimov,
Jumanazar Ruzimov, and Christopher Martius

Abstract Escalating soil degradation caused by soil salinity and rising saline groundwater tables, limits crop production in the irrigated lowlands of arid Uzbekistan. Crop diversification is one option for obtaining more stable farm incomes while improving natural resource use and environmental sustainability. Although the agro-climatic conditions in the country allow growing a wide variety of crops, few crops (cotton, winter wheat, rice, maize) dominate the crop portfolio, which also reflects the restrictions imposed by the state. In the Khorezm region in northwest Uzbekistan, we examined the economic and ecological suitability of

---

I. Bobojonov (✉)
Department of Agricultural Economics, Farm Management Group, Humboldt-Universität zu Berlin, Philippstr. 13 – Building 12 A, D-10115 Berlin, Germany
e-mail: ihtiyorb@yahoo.com

J.P.A. Lamers • N. Djanibekov • N. Ibragimov • T. Begdullaeva
ZEF/UNESCO Khorezm Project, Khamid Olimjan Str., 14, 220100 Urgench, Uzbekistan
e-mail: j.lamers@uni-bonn.de; nodir79@gmail.com; nazar@zef.uznet.net; tamarabeg@mail.ru

A.-K. Ergashev
UNESCO office in Tashkent, Amir Temur Str., 95, 100084 Tashkent, Uzbekistan
e-mail: ake_ergashev@unesco.org.uz

K. Kienzler
International Bureau of the Federal Ministry of Education and Research at the Project Management Agency c/o German Aerospace Center (DLR), Heinrich-Konen-Str. 1, 53227 Bonn, Germany
e-mail: kirsten.kienzler@dlr.de

R. Eshchanov • A. Rakhimov
Urgench State University, Khamid Olimjan Str., 14, 220100 Urgench, Uzbekistan
e-mail: ruzimboy@mail.ru; ozod_eko@mail.ru

C. Martius et al. (eds.), *Cotton, Water, Salts and Soums: Economic and Ecological Restructuring in Khorezm, Uzbekistan*, DOI 10.1007/978-94-007-1963-7_14,
© Springer Science+Business Media B.V. 2012

alternative crops in a stepwise approach. A literature review resulted in a list of about 30 crops that would theoretically fit the agro-climatic conditions in this region. For field research, five crops with a high potential were selected based on socio-economic (potential income) and bio-physical (potential yield, crop quality, options for soil improvement, water use efficiency) criteria. The crops included sorghum (*Sorghum bicolor* (L.) Moench), potato (*Solanum tuberosum*), the cash crop indigo (*Indigofera tinctoria*), and the food and feed crops mung bean (*Vigna radiata*) and sweet maize (*Zea Mays L.*). Field experiments were complemented with laboratory analyses and mathematical modeling for estimating the potential economic and ecological benefits from these crops. Three potato varieties from Germany out-yielded the local variety by at least 50%. Sorghum, indigo, maize and mung bean grew well on marginal lands and obtained very high revenues. Findings from the simulation runs demonstrate that crops such as maize for grain, potato and fodder crops could play an important role in coping with risks in drought years and for securing farm income. Field experiments and modeling results based on this extensive data set from Khorezm allow upscaling to regions in Central Asia with similar agro-climatic conditions.

Keywords Alternative crops • Economic and ecological benefits • Central Asia • Income security • Indigo • Maize • Mung bean • Potato • Sorghum

14.1 Introduction

Agricultural production plays an important role in the Central Asian countries (CAC) Kazakhstan, Kyrgyzstan, Tajikistan, Turkmenistan and Uzbekistan. The contribution of agriculture to the GDP is with 11% lowest in Kazakhstan, but as high as 38% in Kyrgyzstan (Bucknall et al. 2003). While the livelihoods of more than 22,000,000 people depend on irrigated agriculture in this region (Bucknall et al. 2003), the farming population in all CAC has to permanently cope with escalating land degradation. This is caused by soil salinization, which has already affected ca 12% of the total irrigated area in Kyrgyzstan, 50–60% in Uzbekistan and almost 96% in Turkmenistan (Saigal 2003). With the exception of Kazakhstan and Turkmenistan, agriculture is a major source of export revenues for the CAC. During

---

J. Ruzimov
ZEF/UNESCO Khorezm Project/Urgench State University, Khamid Olimjan Str., 14, 220100 Urgench, Uzbekistan
e-mail: jumanazar@zef.uznet.net

C. Martius
Center for Development Research (ZEF), Walter-Flex-Str. 3, 53113 Bonn, Germany
e-mail: gcmartius@gmail.com

14 Crop Diversification in Support of Sustainable Agriculture in Khorezm

the Soviet era, over 60% of all cotton produced in the region originated in the Aral Sea Basin, and the export of cotton still contributes substantially to the revenues from agriculture. For instance, exports of cotton fiber accounted for about 18% of the GDP in 2004 in Uzbekistan (Center for Effective Economic Policy 2005). However, aside from the economic advantages, the almost complete mono-cropping of cotton and winter wheat, crops mandated by the Uzbek government, threatens sustainable agricultural development. At present, farmers do not have to make crucial crop choice decisions themselves, nor need they respond to market signals. But this means also that they do not cultivate crops for which a comparative advantage exists in the region (Bobojonov and Lamers 2008).

The agro-climatic conditions in Uzbekistan in general, and in the Khorezm region in particular, allow the production of a variety of crops (FAO 2003; Kohlschmitt et al. 2007) aside from cotton, winter wheat, rice and fodder crops, which are the four crops/crop types annually cultivated on virtually 87% of all farmland (cf. Djanibekov et al. 2011b). A wide variety of crops such as potatoes, tomatoes, onions, cucumbers, melons, sunflowers, and beans are cultivated in kitchen gardens, albeit on a small area (Müller 2006), but nonetheless indicating the scope for crop diversification.

It is often suggested that crop diversification will contribute to better land use, reduced water use and soil deterioration, and secure profits (Prohens et al. 2003). Crop diversification would thus not only allow farmers to respond more flexibly to market changes and stabilize their income, but also allow a more judicious crop rotation, and in turn a more environment-friendly land resource management. Therefore, diversifying the cropping pattern could accelerate the achievement of a greatly needed economic, ecological and social sustainability – three pillars of sustainable agricultural development (Kassie and Zikhali 2009). For the selection of alternative crops with a high potential to cope with the socio-economic and bio-physical conditions prevailing in the Khorezm region, criteria such as opportunities for improving the ecological situation, increasing farm profits and improving life quality should be considered. However, it cannot be recommended to only identify a single crop to cope with these diverse demands, and a selection of crops needs to be examined to find out which crops are suitable for generating and improving farmers' income, securing livelihoods and food security, and which crops have comparative advantages on degraded areas and the potential to reverse the ecological deterioration in the region. In this overview, our investigations of crops fulfilling these specific criteria are summarized.

14.2 A Step-by-Step Approach for Selecting Suitable Alternative Crops

To identify the range of potentially suitable crops for specific purposes, a stepwise procedure was implemented. In a first step, the boundaries set by the agro-ecological conditions of the region were formulated and used for identifying crops and suitable varieties (Fig. 14.1).

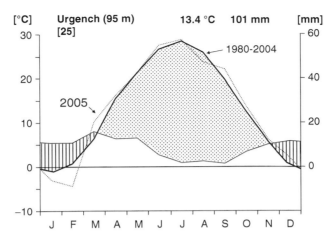

Fig. 14.1 1980–2004 and 2005 monthly mean air temperature and monthly precipitation in Khorezm

The literature review yielded a list of 30 crops (Table 14.1) with a potential to be produced under the harsh environmental conditions of the region (Kohlschmitt et al. 2007). However, the review also revealed difficulties in identifying single crops that would concurrently support all three pillars of sustainable development.

In a second step, several crops were short-listed based on their potential for ecological and/or livelihood benefits. The selected crops included sorghum (*Sorghum bicolor* (L.) Moench) (for growing on degraded land patches), potato (*Solanum tuberosum* L.) (for increasing food security), and indigo (*Indigofera tinctoria*) (an example of a cash crop). Mung bean (*Vigna radiata*), as a potential leguminous crop for enriching the cotton-wheat dominated crop portfolio and the diets of the rural population, and maize (*Zea Mays* L.), as a representative of potential fodder and grain crops, were chosen because of a growing local interest in fodder crops. In the following, the findings of the screening for the suitability for these specific purposes in the Khorezm region are appraised.

14.3 *Sorghum bicolor* – A Multipurpose Staple for Saline Areas

Sorghum was selected as a crop with the potential to grow on degraded soils, uptake salt from the soil and generate income. This C4 species is recognized as moderately tolerant to soil salinity (Fransois et al. 1984), is highly water-use efficient (Rai et al. 2004), capable of assimilating water from deeper soil layers and tolerates drought (FAO 2002). The crop is recommended for arid regions, where the growth of other staples is restricted (Michaelis 1984; FAO 1995), and also for feed (Pedersen et al. 2000).

Table 14.1 Alternative crops for the Khorezm region

Botanic name	Temperature requirements				Drought tolerance	Frost tolerance	Salt tolerance	Soil requirements
	Opt.	Max.	Min.	Winter min.				
Prunus dulcis	12–35°C	35°C	10°C	−17°C	Resistant			pH 6–7; sandy loams
Pistacia vera	Hot, dry				Quite	Cool winter	Relatively	Modest
Punica granatum	Hot			−11°C	Extremely		Tolerant	
Ribes uva-crispa	15–25°C	40°C	3°C	−28°C				pH 5.5–7; deep loam
Phoenix dactylifera	32°C				Quite	Moderate	Tolerant	Sand, sandy loam, others
Prunus cerasus	15–25°C	30 °C	4°C	−29°C			Sensitive	pH 6–7; loamy
Ziziphus jujuba	Hot			−29°C	Tolerant	Tolerant	High	Sandy, well drained
Ficus carica	16–26°C	38°C	4°C	−12°C	Resistant		Moderate	pH 6–7.5
Morus rubra	12–25°C	32°C	12°C	−38°C				pH 5.5–7.5
Pyrus communis	20–35°C	37°C	10°C	−28°C				pH 5.1–6.7; sandy. clay loams
Eragrostis tef	17–27°C	35°C	7.5°C				Tolerant	pH 6–6.5
Eleusine coracan	17–27°C	35°C	7.5°C		Very adaptable		Tolerant	pH 6–6.5; very daptable
Pennisetum americanum					Resistant		Tolerant	
Secale cereale	10–20°C	30°C	5°C	−40°C				pH 5.5–7.5; several types
Hordeum vulgare					Tolerant		Resistant	
Helianthus tuberosus	15–27°C	30°C	7°C	−42°C	Needs irrigation			pH 4–7;several types
Cyamopsis tetragonoloba	25–35°C	45°C	10°C		Quite tolerant		Tolerant	pH 7.5–8; sandy loam
Acacia karroo					Resistant	Resistant		
Fagopyrum esculentum	17–27°C	40°C	7°C		Temp., drought	To cold		pH 5–6.5; Sand, silt loam

(continued)

Table 14.1 (continued)

Botanic name	Temperature requirements				Drought tolerance	Frost tolerance	Salt tolerance	Soil requirements
	Opt.	Max.	Min.	Winter min.				
Solanum tuberosum	15–25°C	30°C	7°C	1°C	Likes high temp.		Some	pH 5–6.5; sandy loams
Sorhum bicolor	40°C	8°C			Needs heat, moderate water	Sensitive to cold		pH 5.5–7.5; loams, heavy clays
Indigofera suffruticosa	22–28°C	32°C	7°C		Resistant	Suffers from frost		pH 6–7; sandy
Isatis tinctoria								Calcareous loess, limestone
Corchorus olitorius	27–32°C							pH 6–7; loams, sandy loams
Nicotiana tabacum							High	pH 5–5.6; sandy, sandy loams
Glycine max	20–32°C	38°C	10°C	0°C				5.5–7.3; fertile loams
Vicia faba	18–28°C	32°C	5°C	−10°C	Not tolerant			
Scorzonera hispanica						Tolerant		Moderate on lean soil
Carthamus tinctorius	20–32°C	45°C	5°C	−14°C	Moderate	Some varieties	Tolerant	5.4–8.2

Source: Kohlschmitt et al. (2007)

14 Crop Diversification in Support of Sustainable Agriculture in Khorezm

Table 14.2 Dry weight of grain, stover of four sorghum varieties at soils with low, medium and high salinity

	Dry weight (t ha⁻¹)			
	Grain	Stover (Stem and leaves)	Grain	Stover (Stem and leaves)
Varieties	2003		2004	
Low saline soil				
S. vulgare	2.41c[a] (±0.21[b])	3.62a (±0.32)	3.44b (±0.24)	7.20a (±0.99)
S. cernuum	1.83c (±0.49)	2.75a (±0.73)	5.13a (±0.14)	6.75a (±0.43)
S. durra	3.00ba (±0.54)	2.81a (±037)	3.73b (±0.40)	7.40a (±0.83)
S. technicum	3.56a (±0.23)	3.62a (±0.55)	2.64c (±0.09)	4.98b (±0.53)
Medium saline soil				
S. vulgare	2.12b (±0.28)	3.18a (±0.42)	4.49b (±0.36)	7.22b (±0.72)
S. cernuum	2.08b (±0.39)	3.12a (±0.59)	6.05a (±0.08)	8.18a (±0.92)
S. durra	2.75ba (±0.45)	3.12a (±0.47)	3.70c (±0.51)	7.55ba (±0.63)
S. technicum	3.37a (±0.59)	3.31a (±1.14)	3.32c (±0.28)	5.87c (±0.47)
Highly saline soil				
S. vulgare	0	0.87a (±0.62)	2.23c (±0.29)	5.17a (±0.85)
S. cernuum	0	0.72a (±0.60)	3.33a (±0.26)	5.22a (±0.80)
S. durra	0.56b (±0.48)	1.78a (±0.43)	2.69b (±0.16)	5.28a (±0.15)
S. technicum	0.42c (±0.21)	2.12a (±1.49)	2.01c (±0.37)	4.83a (±0.80)

Source: Begdullaeva et al. (2009)
[a]Means with the same letter are not significant different at $P < 0.1$ by the Fischers' comparison test
[b]Numbers in brackets indicate standard deviation

Field experiments were conducted on three sites differing in soil salinity: low saline (0.2–0.5% water-soluble salts), medium saline (0.5–1.0%), and highly saline (1.0–2.0%) with four local sorghum varieties including *S. vulgare* Pers., *S. cernuum* (Ard.) Host, *S. durra* Battand et Trab. and *S. technicum* (Koern.) Roshev (Begdullaeva et al. 2007) during 2003 and 2004. Grain yield produced in 2003 was used in the following experimental years as seeds were now adapted to local conditions.

The results show that the total water soluble salt (TDS) accumulation in all four sorghum varieties increased significantly with the advancing growing season. The lowest TDS values were determined at emergence, and increased further in the order of booting<maturity<blooming (Begdullaeva et al. 2007). The TDS accumulation varied between 317 and 152 kg ha⁻¹ in 2003, and 406 and 185 kg ha⁻¹ in 2004 depending on variety and soil salinity. The values among the four varieties did not significantly differ in 2004, but showed a decreasing total uptake with increasing salinity for *S. vulgare* and *S. durra*. The highest TDS values were found at the medium saline site in *S. cernuum* and *S. technicum*.

Grain and stover on the three salt-affected soils differed significantly among the four sorghum varieties (Table 14.2). In 2003, the highest grain yields and harvest indices (ratio of grain yield to total above ground plant mass; HI) were obtained from *S. technicum* and *S. durra* at the low and the medium saline sites. Yields and HI of *S. vulgare* and *S. cernium* were lowest and statistically did not differ among each other. Irrespective of soil salinity, stover yields of all four varieties did not differ significantly from each other, but in the highly saline soil, grain production and HI of sorghum was substantially reduced.

In 2004, grain and total dry matter (DM) production of *S. cernuum* was highest and of *S. technicum* lowest at all sites. Stover production of *S. vulgare* and *S. durra* did not differ significantly and that of grain only at the medium saline site. Although statistically not always significant, average stover, grain and in particular total DM production of all four varieties decreased in the order medium soil salinity > low soil salinity > high soil salinity.

The results of the experiments show that sorghum production is feasible even on highly saline soils, although total DM, grain and stover yields were clearly reduced compared to the other sites. With regard to the growing interest in sorghum as a food and feed crop, the feed quality of sorghum stover on the saline sites ranked between that of alfalfa hay and wheat stalks and could represent a good alternative if harvested during the maturation period. Yet the baking quality of all sorghum varieties assessed was extremely low, irrespective of soil salinity level. Targeted applications of nitrogenous fertilizers, in particular, may upgrade the baking quality, as was experienced with winter wheat experiments in the same region (Kienzler 2010). Yet this needs further in-depth research.

Begdullaeva et al. (2009) analyzed the economic potential of 16 sugar sorghum varieties including local and imported varieties from India in saline soils of Karakalpakstan, the autonomous republic of Uzbekistan that borders the Khorezm region. The sugar content and yield of these varieties varied from variety to variety, but all had potential to generate income (from 230,000 to 1,166,000 UZS ha^{-1}) on saline soils.

14.4 The Multi-faceted Potato Tuber – Income Improvement and Food Security

Potato was selected as a representative crop to generate income and improve food security at household level. Potato tubers are appreciated for their multi-purpose use. Worldwide, annually more than 300,000,000 tons of potato is harvested, and potato is, after wheat and maize, the third most cropped staple in the world. Potato gives more volume than any other crop per hectare, with less water and more food from less land, i.e., "more crop per drop". It gains stable prices, and the time till harvest is must shorter than, for instance, for rice. Potato produces about 500–1,000 kg of protein per year per hectare in contrast to 164–500 kg of protein from soybeans, 98–300 kg of protein from wheat, and only 33 kg of protein from cow milk. Potato production for human consumption is not a new food source in the Khorezm region, since it used to be widely cropped during the Soviet era. However, production declined substantially after the 1990s, owing to a deficiency in high quality seed

14 Crop Diversification in Support of Sustainable Agriculture in Khorezm

Table 14.3 Yield of potato tubers

Year	Variety	Mean	Standard deviation
2003	Koretta	30.5	0.6
	Delikat	32.2	0.6
	Beluga	36	0.5
	Local variety	23.5	0.1
2004	Koretta	31.3	0.3
	Delikat	31.8	0.5
	Beluga	36.1	0.5
	Local variety	23.3	0.3

potatoes, which used to be imported (Olimjanov and Mamarasulov 2006). Food potato production then became limited to smaller areas within household plots and the use of seed potatoes of low quality.

The saline soil conditions in Khorezm, which is located in the floodplain terraces of the Amudarya river (~100 m a.s.l.), were generally judged as a natural obstacle, and the region was thus classified as an area ill-suited for seed potato. Potato seed production is recommended for the mountain terrains of Uzbekistan with altitudes of >1 500 m a.s.l. However, the findings of Ruzimov et al. (2005) show that seed potatoes of three, early-maturing varieties (namely Koretta, Beluga and Delikat) imported from Germany produced high-quality (meaning virus-free) seed potatoes under the agro-climatic conditions in the study region. Seed potatoes were distributed to local farmers by the German Agro-Action project (GAA). These activities were flanked by a field experiment in which these three varieties were compared with the locally used variety, roughly translated as "40 Days", in a jointly farmer-researcher managed experiment (Ruzimov et al. 2005).

The findings show that germination, sprouting and flowering of all three varieties did not differ from the local variety. But all out-yielded the local variety each study year (Table 14.3). The potato yield of Koretta, Delikat and Beluga varieties ranged from 30 to 36 t dry matter (DM) ha^{-1}, compared to an average production of 23 t DM ha^{-1} by the local variety. The imported tubers also had significantly higher starch contents than the local variety. Koretta and Beluga had a much higher tolerance to various diseases than Delikat and the local variety.

14.5 Producing "the King of Colors" – Indigo as a High-Value Cash Crop

Khorezm is currently experiencing a modest revival of the use of natural fibers and dyes for carpet and textile manufacturing, mainly oriented at catering to the tourism industry. Therefore, we examined the agronomic and economic feasibility of indigo (*Indigofera tinctoria* L.) in Khorezm. Indigo is a tropical plant, which mostly is cultivated in India, but also in Ecuador and Ghana. The species belongs to the family of legumes and is known for producing a highly valued natural blue dye (the plant is called sometimes the "king of colors"). This natural dye has a promising local market value, and hence a potential for income generation. Indigo is not native to the region, and currently the dye is typically imported from India (Shimoyama 1991) at high prices for dying silk and wool. Given the declared national policy in Uzbekistan to intensify the local processing of cotton, silk and wool, several enterprises have emerged in a short time also in the Khorezm region that need higher amounts of dyes. Furthermore, as a leguminous crop that fixes atmospheric nitrogen, indigo is rich in nitrogen, and thus can be considered for crop rotations, since it can grow as a green manure crop; it can furthermore be used as a feed source for livestock. Moreover, following dye extraction, the leaf mass provides excellent compost and can be ploughed back into the soil.

Fields experiments were conducted under two conditions: (1) main planting in spring as a first crop and (2) after harvesting winter wheat as a second crop. These pilot plantations showed that indigo germinated without problems, and provided satisfactory leaf and seed yields in Khorezm, albeit at lower levels than those reported from India. Preliminary laboratory tests showed that the dye could be successfully extracted. Growing indigo after winter wheat accumulated on average 59 g of green mass (per one plant), 57.6 g of seeds and 4.6 g of roots. After drying the samples at 70°C for 48 h, almost 50% of the mass weight was lost. The plants yielded a fresh weight of about 15 t biomass ha^{-1} (about 300 kg of pigment). At an estimated price of this dye of about 20 US\$ kg^{-1} (which is at the lower end of prices paid internationally that can be as high as 250 US\$ kg^{-1} depending on quality and purity), profits were in the range of 4,374,970 UZS or about 3,000 US\$ ha^{-1}, which is higher than the gross margin (GM) of about 2,000,000 UZS or about 1,500 US\$ ha^{-1} of rice, currently the most profitable crop in the region (Djanibekov 2008). Therefore, indigo could be considered as an alternative cash crop to rice since it could generate higher income. However, the corresponding regional demand of this cash crop and export opportunities need to be investigated before recommending its full-scale adoption. Yet, in 2011 the government of Uzbekistan decided to encourage the establishment of an indigo farmer association as part of the national farmer association.

14.6 Maize and Mung Bean – Improving Crop Rotation and Easing Food Insecurity

In a series of both researcher- and jointly researcher-farmer managed on-farm experiments conducted at different locations in Khorezm, Djumaniyazova et al. (2010) and Kienzler (2010) could show that nitrogen (N) is the most limiting nutrient in the cultivation of cotton, winter wheat, maize and other crops. Especially for maize and wheat, the N rates presently typically applied by the farmers seem insufficient for a good quality production. Therefore, diversifying the traditional cotton-wheat rotation by introducing a legume crop was expected to benefit livelihoods, food security and the environment alike. Mung bean was selected as the most suitable

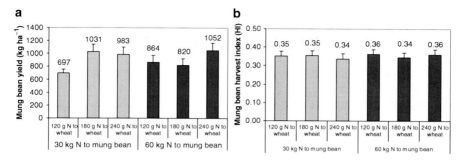

Fig. 14.2 Yield (**a**; kg ha^{-1}) and harvest indices (**b**) for mung bean following differently fertilized winter wheat. Means of 2 locations and 3 replications+SE

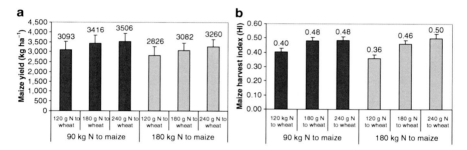

Fig. 14.3 Yield (**a**; kg ha^{-1}) and harvest indices (**b**) for maize following differently fertilized winter wheat. Means of 2 locations and 3 replications+SE

legume for local conditions. Another examined alternative was the summer crop sweet corn (maize for human consumption), also with a good potential for increasing the income of farmers.

Mung bean and maize were planted as a second crop after winter wheat, which had three fertilizer rates (120, 180 and 240 kg N ha^{-1}). The influence of fertilization rate on maize (90 and 180 kg N ha^{-1}) and mung bean (30 and 60 kg N ha^{-1}) was also analyzed.

For maize and mung bean (Figs. 14.2 and 14.3), the yields were 4,242 kg DM ha^{-1} and 868 kg DM ha^{-1}, and HI was 0.45 and 0.35, respectively. The quality of the maize grain reached the quality thresholds (Kienzler 2010) for classification as ready for human consumption, although at present maize is still considered mainly as a fodder crop. In contrast to the irrigation-intensive maize, mung beans were found to be less water demanding and showed favourable effects on the yields of subsequent crops. Therefore, these two crops could become major income-generating crops with the potential to improve the nutrition of the rural population in the area where availability of irrigation water is threatened.

14.7 Scope for Alternative Crops for Household Security and Regional Welfare

Different studies investigated the potential of alternative crops with the help of mathematical programming models in the framework of the ZEF/UNESCO project. The outcomes of the KhoRASM model (Djanibekov 2008) emphasized the profits of potato production, since the GM from potato production was 812 US$ ha⁻¹, which was the second highest GM after rice (ca 1,100 US$ ha⁻¹; both data from household plots in Khazarasp and Bagat districts). However, paddy rice cultivation demands at least 2,600 mm (26,000 m³ ha⁻¹) of water, whereas potato production uses on average only 800 mm with the available irrigation methods (flood and furrow). It is unlikely that rice will maintain high GMs with the introduction of water pricing, a policy measure anticipated to take place in the near future, as confirmed by analysis of various scenarios (cf. Djanibekov et al. 2011a). When introducing a price for water of at least 2.5 US$ cent m⁻³, the production of potato is likely to become the most important cash crop to maintain farm income. Similarly, maize can become a very important crop if a water price is introduced or water availability in the region further declines. Gross margins of fodder crops such as maize are not very high when compared with rice, but these crops are crucial for securing feed production for poultry and animal husbandry in the region, which in turn is an essential part of the income and household security of the rural population (Djanibekov 2008).

Findings by Bobojonov et al. (2010) show that under the assumption of a reduction in irrigation water supply, irrespective of whether or not due to climate change or to an increasing water demand in upstream countries, cropping patterns are likely to change drastically, and a shift towards potato and vegetable production is to be expected at the expense of other crops. In compliance with the Uzbek saying, "where water ends, there ends the land", under an increasingly uncertain water supply, it is to be expected that farmers may abandon the cultivation of the high-value, high-water-demanding and risky rice if they want to secure their profits (Bobojonov et al. 2010); this was confirmed by household-based studies (cf. Oberkircher et al. 2011). But alternative crops allow going beyond the Uzbek saying, and in particular those crops with a considerably lower demand for irrigation water are likely to increase their share in the downstream locations of the irrigation systems. Crops such as mung beans, maize and sorghum, despite their presently low GMs, would become suitable in situations where low availability of irrigation water would not allow planting high-water-demanding crops. These three crops would not only offer the option to maintain farm income during water-scarce years in the region, but their wide-scale cultivation would also allow increasing irrigation efficiency in the entire irrigation system, which could create the basis for environmental improvement (Bobojonov 2009; Bekchanov et al. 2010).

Unfortunately, this potential change is far from taking place, as it demands first a shift in governmental policies with respect to increasing farm-level decision-making by the producers and land users. As long as the state order to grow cotton and winter wheat still stands, expanding the area of alternatives to these crops is not viable. At present, farmers in the region have to allocate more than 70% of their arable land to cotton and winter wheat production, irrespective of the lower GMs from these crops and in turn of lower profits. The government's policy is to maintain a high share of cotton production, which provides currency inflow to the country, together with the winter wheat introduced in the early independence years, which provides food security and self-sufficiency for grains in Uzbekistan. Therefore, the area available for high-value alternative crops is limited, despite their high economic and ecological potential (Bobojonov 2009). Nevertheless, Müller (2006) has shown that the sustainability of cotton production in the region may be threatened under declining world market prices for cotton fibre, which was regularly the case in the last two decades. Thus, future policies should account for the long-term suitability of cotton production considering a value chain approach (cf. Rudenko et al. 2011), export alternatives and the water footprint of crops (Rudenko et al. 2009).

Finally, alternative crops such as fruits and vegetables are not part of the state order, and their profits are for a larger part determined by market prices. At present, these crops are used for household consumption due to the lack of processing capacities in the region. Consequently, a surplus production cannot be absorbed by the domestic and regional markets and often leads to price declines and decreasing profits, as demonstrated by the annual sharp decline in prices after harvest, and price rises during the winter and spring periods (Bobojonov and Lamers 2008). Storage facilities established during Soviet times have deteriorated (Bobojonov and Lamers 2008), and this limits further development of fruit and vegetable production. Surplus production could be exported to other regions of Uzbekistan, but export of agricultural products other than cotton, meat and rice currently rarely occurs, being limited by transport facilities and export constraints in Khorezm (Bobojonov and Lamers 2008), and is also due to less research on the export potential of other agricultural products in Uzbekistan.

An adequate and functioning processing sector, as well as storage and export channels, needs to be available before promoting the wide-scale cultivation of alternative crops. However, as the findings presented here clearly support a wider crop diversification, care should be taken to prevent a shift from the present dominating crops cotton and winter wheat to a monoculture of any of the recommended alternative crops. For example, if the economic benefits of potato production were to lead to a monoculture of this crop, future crashes in the potato production can be foreseen, as that caused by late blight (*Phytophthora infestans*) in the early 1920s in Ireland. Without crop diversity, or at least the introduction of different varieties of the same crop, food security remains threatened. Uzbek farmers, on the other hand, have wide experience with different crops in their home gardens, and this raises hopes that future agricultural policies will support diversification of agricultural production to the benefit of the land users that have to make a decent living, and the environment, that needs to be preserved for future generations.

14.8 Conclusions

The results of a series of field experiments show five crops selected among 30 to be theoretically suitable for Khorezm. Not only is plant growth successful, but also yields are acceptable, and sometimes excellent. Thus, there is scope for diversification of the agricultural production system, making it ecologically more resilient, economically more reliable and socially acceptable. Each of the five examined crops has comparative advantages for different purposes. The production of sorghum, mung bean and maize was found suitable mainly on areas with high soil salinity and unstable irrigation water availability. Despite these harsh circumstances, these crops could maintain a certain level of revenue and provide income security in water-scarce years. In contrast, potato and indigo could be considered as high-value crops, but demand better soil quality and a more reliable water supply during the vegetation period.

Sustainable development from crop diversification will depend on structural and policy adjustments. Alternative crops have to be integrated into the common farming practices in the region, and state regulations must become conducive for this diversification. Furthermore, the processing and marketing sectors need to be revitalized in order to allow channelling over-production into markets near and distant, to improve food security, and to improve sustainable agricultural development in the region.

References

Begdullaeva T, Kienzler KM, Kan E, Ibragimov N, Lamers JPA (2007) Response of Sorghum bicolor varieties to soil salinity for feed and food production in Karakalpakstan, Uzbekistan. Irrig Drain Syst. doi:10.1007/s10795-10007-19020-10798

Begdullaeva T, Orel M, Rudenko I, Ibragimov N, Lamers JPA, Toderich K, Khalikulov Z, Martius C (2009) Productivity of sugar sorghum varieties imported from India under the conditions of Karakalpakstan. Vestnik 215:20–22

Bekchanov M, Lamers JPA, Martius C (2010) Pros and cons of adopting water-wise approaches in the lower reaches of the Amu Darya: a socio-economic view. Water 2:200–216

Bobojonov I (2009) Modeling crop and water allocation under uncertainty in irrigated agriculture: a case study on the Khorezm Region, Uzbekistan. PhD dissertation, Bonn University, Bonn

Bobojonov I, Lamers JPA (2008) Analysis of agricultural markets in Khorezm, Uzbekistan. In: Wehrheim P, Schoeller-Schletter A, Martius C (eds) Continuity and change: land and water use reforms in rural Uzbekistan, Socio-economic and legal analyses for the Region Khorezm. IAMO, Halle/Saale, pp 167–186

Bobojonov I, Franz J, Berg E, Lamers JPA, Martius C (2010) Improved policy making for sustainable farming: a case study on irrigated dryland agriculture in Western Uzbekistan. J Sustain Agric 34(7):800–817

Bucknall L, Klytchnikova I, Lampietti J, Lundell M, Scatasta M, Thurman M (2003) Irrigation in Central Asia: social, economic and environment considerations. World Bank, Washington, DC

Center for Effective Economic Policy (2005) Uzbekistan economy statistical and analytical review for the year 2004. www.bearingpoint.uz

Djanibekov N (2008) A micro-economic analysis of farm restructuring in Khorezm region, Uzbekistan. PhD dissertation, Bonn University, Bonn

Djanibekov N, Bobojonov I, Djanibekov U (2011a) Prospects of agricultural water service fees in the irrigated drylands, downstream of Amudarya. In: Martius C, Rudenko I, Lamers JPA, Vlek

PLG (eds) Cotton, water, salts and *Soums*: economic and ecological restructuring in Khorezm, Uzbekistan. Springer, Dordrecht/Berlin/Heidelberg/New York

Djanibekov N, Bobojonov I, Lamers JPA (2011b) Farm reform in Uzbekistan. In: Martius C, Rudenko I, Lamers JPA, Vlek PLG (eds) Cotton, water, salts and *Soums*: economic and ecological restructuring in Khorezm, Uzbekistan. Springer, Dordrecht/Berlin/Heidelberg/New York

Djumaniyazova Y, Sommer R, Ibragimov N, Ruzimov J, Lamers JPA, Vlek PLG (2010) Simulating water use and N response of winter wheat in the irrigated floodplains. Field Crop Res 116:239–251

FAO (1995) Sorghum and millets in human nutrition, FAO food and nutrition series no. 27. Food and Agriculture Organization, Rome, p 184

FAO (2002) Crop water management: sorghum. AGLW water management group. FAO, Rome. http://www.fao.org/AG/AGL/aglw/cropwater/sorghum.stm. Last accessed 12 Feb 2007

FAO (2003) Fertilizer use by crop in Uzbekistan. FAO Land and Water Development Division, Rome

Fransois J, van Schaftingen E, Hers H-G (1984) The mechanism by which glucose increases fructose 2,6-bisphosphate concentration in Saccharomyces cerevisiae. A cyclic-AMP-dependent activation of phosphofructokinase 2 Eur. J Biochem 145:187–193

Kassie M, Zikhali P (2009) The contribution of sustainable agriculture and land management to sustainable development. United Nations Sustainable development Innovation Brief Issue, 7

Kienzler KM (2010) Improving the nitrogen use efficiency and crop quality in the Khorezm region, Uzbekistan. Ecology and Development Series no. 72, Rheinische Friedrich-Wilhelms-Universität Bonn, Bonn

Kohlschmitt S, Eshchanov R, Martius C (2007) Alternative crops for Khorezm (Uzbekistan) and their sales opportunities as well as risks on the European market. ZEF Work Papers for Sustainable Development in Central Asia 11:42

Michaelis S (1984) Sorghum- und Millet-Hirsen: anbau. Nutzpflanzen der Tropen und Subtropen. Franke G. Leipzig, Band II.4. Auflage. S. Hirzel Verlag. pp 101–105

Müller M (2006) A general equilibrium approach to modeling water and land use reforms in Uzbekistan. PhD dissertation, Rheinische Friedrich-Wilhelms-Universität Bonn, Bonn

Oberkircher L, Haubold A, Martius C, Buttschardt T (2011) Water patterns in the landscape of Khorezm, Uzbekistan. A GIS approach to socio-physical research. In: Martius C, Rudenko I, Lamers JPA, Vlek PLG (eds) Cotton, water, salts and *Soums*: economic and ecological restructuring in Khorezm, Uzbekistan. Springer, Dordrecht/Berlin/Heidelberg/New York

Olimjanov O, Mamarasulov K (2006) Economic and social context of the vegetable system in Uzbekistan. In: Kuo CG, Mavlyanova RF, Kalb TJ (eds) Increasing market-oriented vegetable production in Central Asia and the Caucasus through collaborative research and development. AVRDC – The World Vegetable Center, Shanhua, pp 91–95

Pedersen JF, Milton T, Mass RA (2000) Crop quality & utilization: a twelve-hour in vitro procedure for sorghum grain feed quality assessment. Crop Sci 40:204–208

Prohens J, Ruiz JJ, Nuez F (2003) Vegetable crop diversification in areas affected by salinity: the case of pepino (solanum muricatum). Acta Hortic ISHS 618:267–273

Rai M, Varma A, Pandey AK (2004) Antifungal potential of spilanthes calva after inoculation of piriformospora indica. Mycosis 47(112):479–481

Rudenko I, Djanibekov U, Lamers JPA (2009) Cotton water footprint. Is rational water use possible in Khorezm? Vestnik. The Journal of Uzbek Academy of Science, # 3(216):44–48

Rudenko I, Nurmetov K, Lamers JPA (2011) State order and policy strategies in the cotton and wheat value chains. In: Martius C, Rudenko I, Lamers JPA, Vlek PLG (eds) Cotton, water, salts and *Soums*: economic and ecological restructuring in Khorezm, Uzbekistan. Springer, Dordrecht/Berlin/Heidelberg/New York

Ruzimov J, Ibragimov N, Lamers J (2005) Potatoes in Khorezm. Agric Uzbekistan 10:20

Saigal S (2003) Combating desertification in Central Asia: Kyrgyz republic issues and approaches to combat desertification. ADB, Manila

Shimoyama A (1991) A "Devil's" Dye, indigo: the Indian craze and establishment of indigo plantations in colonial America and the West Indies; its technology and slavery. J Mark Hist Shijoushi Kennkyu 9:29–48

Chapter 15
Conversion of Degraded Cropland to Tree Plantations for Ecosystem and Livelihood Benefits

Asia Khamzina, John P.A. Lamers, and Paul L.G. Vlek

Abstract This section summarizes the findings of a multidisciplinary, long-term research program that assessed the potential of plantation forestry to rehabilitate a degraded, irrigated-agriculture ecosystem. Bio-physical and socio-economic studies conducted during 2002–2009 in Khorezm assessed the suitability of afforestation with multipurpose species as an alternative land-use option, by examining the ecosystems services and opportunities for income generation. The initial phase involved inventory, evaluation, and selection of suitable tree species and determination of the irrigation demand for establishing plantations on highly salinized, nutrient-poor soils with a shallow, saline groundwater table. Next, the environmental services were investigated including biological drainage for soil salinity control, improvement of soil nutrient stocks, and carbon sequestration into soil and tree biomass. Potential income generation from timber and non-timber products such as fuelwood, leaf fodder, and fruits was compared to that from a continued cropping of the degraded cropland. Sociological surveys evaluated farmers' perceptions and current silvicultural practices to determine the prerequisites for introducing farm forestry in the area. Overall, evidence on ecosystem rehabilitation and financial benefits suggest that converting degraded cropland to long-term forestry use is an attractive option. Socio-economic obstacles such as legislative aspects of retiring degraded cropland and related land tenure issues, poor market conditions for tree products, lack of incentives and under-appreciation of the benefits of tree-based systems need to be addressed to ensure farmer and governmental support for afforestation.

A. Khamzina (✉) • P.L.G. Vlek
Center for Development Research (ZEF), Walter-Flex-Str. 3, 53113 Bonn, Germany
e-mail: asia.khamzina@uni-bonn.de; p.vlek@uni-bonn.de

J.P.A. Lamers
ZEF/UNESCO Khorezm Project, Khamid Olimjan Str., 14, 220100 Urgench, Uzbekistan
e-mail: j.lamers@uni-bonn.de

C. Martius et al. (eds.), *Cotton, Water, Salts and Soums: Economic and Ecological Restructuring in Khorezm, Uzbekistan*, DOI 10.1007/978-94-007-1963-7_15,
© Springer Science+Business Media B.V. 2012

Keywords Afforestation • Biodrainage • Carbon sequestration • Farmers' preference and motivation • Multipurpose tree species • Net Present Value • Nitrogen fixation • Salinity • Tree fodder • Wood for fuel

15.1 Afforestation as an Alternative Use of Degraded Cropland

Rising water scarcity, cropland degradation due to soil salinization, and their threat to sustainable economic development are topics that have been discussed in Uzbekistan for several decades (UNEP and Glavgidromet 1999). The climate change issue has recently been added to the country's list of environmental challenges (Chub and Ososkova 2008). In this context, adaptive land- and water-use strategies that aim to use the dwindling resources more efficiently are required. As one of such strategies, in the ZEF/UNESCO's landscape restructuring project in the Khorezm Region of Uzbekistan we have investigated the option of converting to tree plantations the degraded cropland areas, abandoned from cropping or where yields hardly justify investments. The resources saved can be used in productive agricultural areas (Martius et al. 2004). In several parts of the world, afforestation has proven successful in re-vegetating saline agricultural landscapes and providing environmental benefits and valuable products to land users (Heuperman et al. 2002; Marcar and Crawford 2004). In response to global climate change, afforestation has also been suggested as an important mitigating and adaptive land-use strategy (FAO 2000; Katyal and Vlek 2000; Garrity 2004; Scherr and Sthapit 2009).

In Khorezm, about 15–20% of arable lands are considered poorly suited or unsuitable for cropping, mostly due to soil salinization, and some of them could be considered for alternative uses, such as plantation forestry (Martius et al. 2004). As these degraded lands are not traditionally used as forestry areas, there was little published research on silvicultural strategies appropriate for this environment. To fill this information gap, the following issues were addressed:

- Screening of available tree species and selection of most promising candidates;
- Establishment of plantations on degraded cropland under deficit irrigation;
- Evaluation of ecosystem services provided by afforestation;
- Analysis of financial returns from investment in afforestation;
- Survey of current farmers' perceptions and silvicultural practices.

15.2 Species Screening

To ensure effective and sustainable outcomes, afforestation of marginal lands must be preceded by a comprehensive evaluation of available species (Harrington 1999). The ZEF/UNESCO project made an inventory of regional tree and forest resources by interpreting locally available aerial photographs with a resolution 1:20,000 m (Tupitsa 2009). The assessment revealed a tree cover of about 4% of the total surveyed area, with forests accounting for only 0.3%. These forests were primarily the remnants of riparian *Tugai* forests and consisted mostly of *Populus euphratica* Oliv. In addition, "desert forests", desert and xeric shrub lands, including *Haloxylon persicum* Bunge ex Boiss. and Buhse and *H. ammodendrom* (C. A. Mey.) Bunge as well as various shrub species covered about 2% of the surveyed area, but were characterized by a low standing volume and a sparse canopy cover. Windbreak plantings on agricultural land accounted for some 0.8% of the area and were dominated by *Morus alba* L. and various *Salix* and *Populus* spp. Fruit plantations constituted about 1% of the area and, for the most part, were intercropped with annual plants. The share of timber plantations of hybrid poplars was only 0.02% (Tupitsa 2009).

These and other promising candidate species were assessed using multiple physiological and socio-economic criteria in a field trial conducted on two soil types (Lamers et al. 2005; Khamzina et al. 2006b). Given the predominance of highly salinized soils, salt-tolerance was a compulsory criterion, while nitrogen (N) fixation and phreatophytic characteristics were viewed advantageous for growth on saline, infertile soils with a shallow groundwater table (Danso et al. 1992; Heuperman et al. 2002; Khamzina et al. 2009a). Local species, known and valued by the land users, were also preferred to avoid competition with indigenous plant community (IUCN 2004) and to facilitate their adoption by farmers. Rapid establishment capability and high growth rate were also considered desirable as they shorten the "waiting" period before forestry activities generate appreciable revenues (Harrington 1999).

A judicious choice of species would also maximize associated environmental services such as biodrainage and soil salinity control, improvement of the soil nutrient stocks, carbon sequestration, biodiversity, and amenities (shadow, shelter for livestock, bee foraging, scenic beauty) contributing to overall ecosystem health. Multipurpose species that provide more than one useful product, i.e. fuelwood, high-protein livestock fodder, edible fruits and, in the long-term, timber, were also desirable for optimal economical use of the otherwise unproductive cropland (Galiana et al. 2004; Khamzina et al. 2006b, 2008). Mixed-species plantations also help diversify risks and ensure stable production rates due to different rotation periods.

Considering all these criteria, neither the fruit species *Prunus armeniaca* L. and *M. alba* nor *Populus* and *Salix,* species that are frequently planted on agricultural lands, showed high potential for afforestation of the degraded cropland (Khamzina et al. 2006a, b). Instead, the species ranking singled out the currently underutilized *Elaeagnus angustifolia* L., *Ulmus pumila* L. and *P. euphratica.* This selection was based on the evaluation of trees in their early growth stage but was consistent with the presence of mature individuals of these species on abandoned land patches, where no other species were found to survive (Khamzina 2003).

15.3 Irrigation Demand

In an arid climate, afforestation success initially depends on the availability of sufficient water for tree establishment. In Khorezm, furrow irrigation with seasonal water application of 400–800 mm throughout the lifespan of plantations is a

Fig. 15.1 Overview of the experimental site showing (**a**) severe secondary soil salinization in the spring of 2004, 2 years after tree planting; (**b**) tree plantations in the spring of 2006 (Photos by J. Lamers and A. Khamzina)

standard recommendation (Makhno 1962; Fimkin 1972; MAFRI 1972). However, such amounts of water are unlikely to be consistently available for the marginal lands.

Thus the establishment and growth of *E. angustifolia*, *P. euphratica* and *U. pumila* were field-tested under deficit irrigation applied via drip and traditional furrows in a long-term field trial during 2003–2009 (Fig. 15.1). Irrigated with 80–160 mm year^{-1}, the tested species successfully established on highly saline soils with the root-zone electrical conductivity (EC) over 20 dS m^{-1}, underlain by shallow (0.9–2.0 m) groundwater with an EC ranging between 1 to 5 dS m^{-1}. Following the cessation of irrigation after 2 years, the trees effectively used the groundwater and produced 10–60 t ha^{-1} year^{-1} of above-ground biomass (Khamzina et al. 2008, 2009a, 2009b).

Compared to furrow irrigation, drip irrigation enhanced the initial growth of *P. euphratica* which, known for poor survival and slow onset (Wang and Chen 1996; Khamzina et al. 2006b, 2008), benefited from the constant and homogeneous root-zone moisture conditions provided by the high-frequency drip water supply. At a later stage, once the trees gained sufficient access to the groundwater table, the drip irrigation no longer provided an advantage over the furrow technique. *E. angustifolia* and *U. pumila*, more robust during the early growth, were insensitive to the mode of irrigation (Khamzina et al. 2008). Even at these low application rates, the traditional furrow irrigation was sufficient to meet the initial water demand of plantations. Thus, applying the costly drip irrigation system is deemed unnecessary for forest establishment at sites where a shallow, slightly-to-moderately saline groundwater table prevails throughout the growing season. Moreover, by drawing on relatively untapped groundwater resources (Ibrakhimov et al. 2007) afforestation can contribute to water saving as "unused" irrigation water from afforested plots would become available for use on productive cropland.

15.4 Ecosystem Services

Worldwide, awareness is growing of the importance of ecosystems services, the benefits provided to humans by natural ecosystem functions, which were popularized and their definitions formalized by the United Nations Millennium Ecosystem Assessment (MEA 2005).

15.4.1 Biodrainage and Salinity Control

Within the irrigated land-use systems, where secondary soil salinization is chiefly responsible for the cropland degradation, mitigating dryland salinization by reducing elevated groundwater tables via biodrainage (Marcar and Crawford 2004) is one of the important ecosystem services that could be provided by afforestation. Biodrainage uses the transpirative capacity of trees to control the recharge or enhance the discharge of the shallow groundwater. In this case, profligate tree groundwater use is an asset (Heuperman et al. 2002).

To assess the potential of the selected species for biodrainage purposes, Khamzina et al. (2009b) examined the transpiration of established, 2–4-year-old tree plantations (years 2003–2005) on marginalized irrigated cropland. In the course of the growing season, the plantations transpired 0.1–7 mm day^{-1} in 2003 and 1–13 mm day^{-1}

in 2004–2005, as determined with the Penman-Monteith model. In the absence of irrigation in 2005, the annual stand transpiration averaged 1,250, 1,030, and 670 mm for *E. angustifolia*, *P. euphratica* and *U. pumila*, respectively (Khamzina et al. 2009b).

The transpiration of *E. angustifolia*, the most water consuming species in this study, ranged from 16 to 23 Lday^{-1} tree^{-1} in 2005. This is at the lower end of the summer transpiration range of evergreen *Eucalyptus* spp. of similar age in semi-arid regions of Australia (Lima 1984). There, the effective biodrainage function of *Eucalyptus* species has been widely acknowledged (Heuperman et al. 2002). On the basis of land area, transpiration rates (9–13 mm day^{-1}) in the dense plantation in Khorezm were similar or superior to that of the eucalypts. Such high (ground)water use despite the highly saline environment is explained by the salt-tolerance of the selected species, a high atmospheric evaporative demand, and availability of the groundwater with a tolerable salinity level, to satisfy this demand (Khamzina et al. 2009b).

Despite the ample water use and vigorous juvenile growth, the groundwater drawdown effect under this 2 ha tree plantation was less than 1 m over 5 years of plantation growth (Khamzina et al. 2008, 2009a). Ibrakhimov et al. (2007) and Khamzina et al. (2005; 2009b) suggested that a significant subsidence of the groundwater table under small-scale plantations within the Khorezm irrigated area was unlikely due to a continuous refill from the surrounding irrigated cropped fields. On the other side, this underground inflow might allow the plantation growth even under non-irrigated conditions.

Of concern to plantation long-term viability was an observed rise in root-zone soil salinity from 4 to 12 dS m^{-1} already during the early growth (2003–2005) and also after irrigation was stopped and groundwater uptake increased (Khamzina et al. 2009b). Modeling studies assessing the sustainability of tree plantations elsewhere predicted salt accumulation due to exclusion of salts from the groundwater uptake. This phenomenon might endanger the long-term success of tree planting in the groundwater discharge areas (Thorburn et al. 1995; Morris et al. 1998; Thorburn 1999; Paydar et al. 2005), which prevail in Khorezm (Ibrakhimov et al. 2007). Nevertheless, the significant biomass production on the degraded land in our study over the first 7 years suggests a great tolerance of the selected species. Even though the trees do not "cure" the salinity problem, they do thrive and make better use of land with otherwise a little productive value. Continued observations of soil salinity dynamics and simulation analyses should quantify the growth response to increasing salinity over time, and any ameliorating leaching required for long-term plantation sustainability.

15.4.2 Improving Soil Nutrient Stocks

Planting nitrogen-fixing tree species on nutrient-exhausted fallow lands or interplanting with valuable (tree) crops has been a widely acknowledged silvicultural practice. It utilizes the ability of N-fixers to replenish nutrient stocks and increase the productivity of agroforestry and silvopastoral systems, particularly in the tropics (Dawson 1986;

Danso et al. 1992; Dommergues 1995; Paschke 1997; Galiana et al. 2004). To assess the potential for improving soil fertility by afforestation in the arid region, N_2 fixation of *E. angustifolia* was examined using the ^{15}N natural abundance method (Khamzina et al. 2009a; Djumaeva et al. 2010).

Drought, salinity and P and micronutrient deficiency—conditions that predominate in degraded soils in Khorezm—can suppress nitrogen fixation by reducing plant growth or by directly affecting the symbiosis (Dommergues 1995). In our study, despite elevated root-zone salinity and deficiency in plant-available P (4–15 mg kg^{-1}), the proportion of N derived from the atmosphere by *E. angustifolia* increased from an initial value of 20% to almost 100% over 5 years following planting. Due to the shallow groundwater table, the soil water content was not a limiting factor (Khamzina et al. 2008, 2009b) whereas the high soil salinity did not inhibit N_2 fixation either, implying a high salt tolerance of both, the host plant and the *Frankia* strains, naturally occurring in the soil. The high N_2 fixation despite P deficiency might be attributed to the *mycorrhizal* associations on roots stimulating efficient P uptake (Gardner 1986).

Nitrogen fixation initially averaged 0.02 t ha^{-1} year^{-1}, peaked at 0.5 t ha^{-1} year^{-1} during the next 2 years, and thereafter stabilized at 0.3 t ha^{-1} year^{-1} (Khamzina et al. 2009a). According to the classification of Dommergues (1995), species with a N_2-fixing potential of 0.1–0.3 t ha^{-1} year^{-1} are regarded as highly efficient. *Elaeagnus angustifolia* would thus fit this category.

The conversion of degraded cropland to tree plantations increased soil total N stocks in the upper 20 cm layer by 6–30% in 5 years. The increase in plant-available soil N was significantly higher in *E. angustifolia* plots than in *P. euphratica* and *U. pumila* plots. Increases in the concentrations of plant-available P of up to 74% were significant irrespective of tree species, suggesting an efficient nutrient pump (Khamzina et al. 2009a). This improvement in soil fertility via including N_2-fixing trees is further evidence that afforestation with mixed-species plantations can be a sustainable land-use option for the degraded cropland.

15.4.3 Carbon Sequestration

Planting trees on the degraded lands provides an opportunity to combine the efforts of combating land degradation and reducing CO_2 emissions (FAO 2000). Moreover, payments are offered to land users implementing such bio-sequestration projects through mandatory (operated under Kyoto Protocol) and, in a larger share, voluntary markets. The Kyoto Protocol allows non-Annex I countries, such as Uzbekistan, to participate in the global carbon (C) sequestration effort by selling C units gained from re- and afforestation under the Clean Development Mechanism (CDM). These are the only land uses currently eligible under the CDM, but their potential in arid regions has not yet been fully recognized, let alone exploited.

Our results in Khorezm showed that 5 years after afforestation, the soil organic C (SOC) stocks rose by 10–35%, adding 2–7 t C ha^{-1} to the upper 0–20 cm soil layer, with *E. angustifolia* being the most effective tree species in soil C sequestration.

The SOC concentrations remained below 1%, reflecting the low soil fertility, but the observed increase is impressive given the short time period. A meta analysis by Guo and Gifford (2002) showed a mean of 18% increase in soil C stocks following conversion of cropland to tree plantation. Furthermore, results of a chronosequence study in Khorezm suggested that soil C sequestration continued at the rate of $0.15\,t\,ha^{-1}\,year^{-1}$ in the first 20 years and $0.09\,t\,ha^{-1}\,year^{-1}$ thereafter (Hbirkou et al. 2011). According to these results, the SOC stocks in tree plantation systems in the long-term exceed those in the native *Tugai* forest.

Depending on tree species, C sequestration in woody biomass ranged from 11 to $23\,t\,ha^{-1}$ already in the fifth year after afforestation. When such C sequestration in the biotic and soil pools occurs in an afforestation project certified under the CDM, the resulting C payments could encourage this alternative land use in degraded areas (Smith and Scherr 2003). In the study region, the cost effectiveness of forestry activities under the CDM and the magnitude of potential C payments remain to be determined.

15.5 Provisional Ecosystem Services – Non-timber Products

The ecological benefits alone are insufficient to incentivize the adoption of farm forestry on degraded land. The challenge is to enhance ecosystem services while meeting farmers' domestic and income generation needs. Non-timber tree products can provide annual cash flows while waiting to harvest and market timber, the most valuable tree product, or, in a CDM project, before the C offsets become significant. Few studies from Central Asia reported on appropriate management practices to receive non-timber benefits (Fisher et al. 2004; Kan et al. 2008).

15.5.1 Wood for Fuel

Fuelwood is an important source of energy even in countries well-endowed with fossil fuels as an access to gas and electricity is often limited in rural areas due to difficulties in linking remote regions to national grids or due to interruptions of energy supplies. In this context, tree (residual) wood biomass is a cheap, additional energy option in Uzbekistan where over 50% of the rural population has insecure or reduced access to gas supplies[1] (UNFCC 2001).

In pilot plantations, the observed calorific value of wood from *E. angustifolia, U. pumila* and *P. euphratica* varied little over time, within $18–19\,MJ\,kg^{-1}$. Thinning the 5-year-old plantations by a half of their initial density ($2,300\,stems\,ha^{-1}$), to make room for growing trees, generated an energy value varying from 6 tons of oil energy equivalent (toe) ha^{-1} (*U. pumila*) to $10\,toe\,ha^{-1}$ (*P. euphratica*). This would satisfy the

---

[1] http://www.ieguzexpo.com/page/exhibition/clean_energy/71/res

15 Conversion of Degraded Cropland to Tree Plantations... 243

average annual per capita energy needs of 55–90 people in Uzbekistan and exceeds by 400% the energy value gained over the same period from cotton stalks, commonly used in rural households (Lamers and Khamzina 2008). The new fuelwood source could also reduce the illegal cutting of natural *Tugai* and desert forests.

15.5.2 Supplementary Fodder

The importance of livestock rearing to rural livelihood has been growing in Khorezm, but fodder crop production remains restricted, as the state-ordered cotton and wheat cultivation occupies ca. 70–80% of cropland (Djanibekov 2006, 2008; Müller 2006). Livestock owners in Uzbekistan do crop forages on privately owned plots, but these are too small in size and number to produce the desired fodder quantity (Djanibekov 2006). Introducing fodder trees on degraded croplands could contribute to forage availability provided the nutritional quality of the planted species is appropriate.

Recent surveys in Khorezm showed that farmers feed their dairy cows with wheat bran and residues of maize and sorghum (Djanibekov 2008). With a ratio of crude protein to metabolizable energy content (CP:ME) of ~11 $g\,MJ^{-1}$, these feedstuffs are insufficient for milk production of at least 15–20 $L\,day^{-1}$ (Close and Menke 1986). In contrast, due to high CP and relatively low ME values, CP:ME ratio in tree leaves was at least 13 $g\,MJ^{-1}$, a well-balanced ratio for milk production (Close and Menke 1986). In fact, the CP:ME ratio of *E. angustifolia* was close to that of CP-rich cottonseed cake (CP:ME = 29.9 $g\,MJ^{-1}$), an expensive supplement which, when affordable, is added to the diet of dairy cows (Djanibekov 2008). Given the high CP concentrations (up to 260 g DM kg^{-1}), the tree foliage should be mixed with roughages, which are protein-poor but would yield the appropriate ME content to digest the proteins in the mixture. Supplementing the livestock diets with protein-rich leaves can thus contribute to forage-saving by reducing the amount of basic feed needed (Djumaeva et al. 2009; Lamers and Khamzina 2010). Moreover, using tree leaves from afforestation plots to enrich and increase feed production can ease the pressure on natural pastures without competing for prime agricultural land.

An overall *in-vitro* evaluation of the leaves, considering the organic matter digestibility and the effects of tannins, indicated a medium-to-good feed quality of *E. angustifolia, U. pumila, M. alba*, and *Gleditchia triacantos* (Khamzina et al. 2006b; Djumaeva et al. 2009; Lamers and Khamzina 2010). These results need to be confirmed by *in vivo* studies that access the intake and palatability of tree foliage.

15.6 Potential Financial Gains from the Alternative Land Use

The biomass data of the *E. angustifolia, U. pumila*, and *P. euphratica* plantings (Khamzina et al. 2009a) was supplemented with that of mature trees already growing on marginal land (Khamzina 2003). This combined dataset formed the basis for elaborating 20-year growth functions for these species. Potential financial returns were assessed by considering annual fuelwood, fodder and fruit production, plus the stumpage value after 20 years. At a 16% discount rate (base case), the estimated Net Present Value (NPV, US\$ ha^{-1}) was greatest for *E. angustifolia* (ca. 13,900), followed by *P. euphratica* (ca. 4,100), and *U. pumila* (ca. 1,700) showing a benefit to cost ratio (BCR) of 7.8, 2.2 and 1.8, respectively (Lamers et al. 2008). Between 47% (*E. angustifolia*) and 89% (*U. pumila*) of the total revenues accrued 20 years after planting, stemmed from the estimated timber production (Lamers et al. 2008). This high-value commodity is at present largely imported in Uzbekistan.

A comparative analysis of tree-based land use with the conventional annual cropping of cotton, winter wheat and rice on the land of low fertility showed that, except for the first year, gross margins of tree plantations exceeded those of the annual crops, owing to the annually recurring benefits from fuelwood and fodder. Among the crops, particularly cotton persistently caused losses due to high expenses for labor and machinery and the low yields on marginal lands (cf. Rudenko et al. 2011).

The estimated opportunities for positive returns to investment in afforestation and the comparative advantages of perennial *vs.* annual vegetation on the degraded cropland suggest that conversion to tree plantations could bring significant financial and social benefits to the local agriculture-based economy (Lamers et al. 2008).

15.7 Social Settings for Introduction of Forestry on Degraded Land

Farmer acceptance is essential for the successful adoption of afforestation practices. However, until recently farmers' knowledge of, and motivation for practicing the tree-based land-use systems hardly received research attention.

Therefore, a survey conducted by Kan et al. (2008) among 133 households during 2003–2005 in Khorezm found that farmers rarely invested in pure tree plantations. Instead, various agroforestry systems were managed, with 97% of all sites including fruit tree species. The preference for these species was motivated by the opportunity their fruits presented for additional income and enrichment of the family food basket. Annual crop components i.e. cereals, vegetables, fodder and cash crops were still considered commercially more important than the trees and were given the highest priority in these agroforestry systems.

The large share of trees below 10 years of age (over 40% of cases) was evidence of the emerging interest in fruit tree planting and appeared linked to the recent land reforms (Djanibekov 2008). The reforms included tax exemptions for land users who commit themselves to horticultural crops (Kan et al. 2008) and encouraged the inclusion of perennial crops in the land-use systems. However, farmers seemed to capture the tax exemption while maintaining a bias for annual crops rather than combining the two components in an optimal production system (Kan et al. 2008).

The potential for growing trees on degraded cropland was underutilized, and, more often than not, the farmers' knowledge about the environmental value of this land use and direct benefits (e.g., tree fodder) was either superficial or absent (Kan et al. 2008). The lack of knowledge may be linked to the only recent legalization of land ownership and thus limited experience in land management, and to the absence of suitable farmer training and education. Nevertheless, the boost in tree planting following the governmental support shows an emerging interest in perennial crops and is an encouraging step toward enhancing forestry activities with appropriate species on marginal agricultural land.

15.8 Outlook

Soil salinity, the primary cause of on-going cropland degradation in Khorezm, was not eliminated by conversion to tree plantations. Hence a return to cropping on these lands in the long-run seems impractical unless more sustainable irrigation and drainage practices are adopted to maintain salts at levels acceptable for common crops. Increasing regional water scarcity could render such technical solutions uneconomical for unproductive croplands. Thus highly degraded cropland parcels should likely be permanently converted to small-scale forests to assist in ecosystem rehabilitation rather than the restoration of the land to annual cropping.

However, the legislative aspects of retiring degraded croplands, which are currently under the state-ordered land use, for use in forestry, and related land tenure issues need to be addressed. Afforestation can also be constrained by socio-economic factors, such as poor market conditions for tree products, lack of infrastructure or an under-appreciation of the benefits of tree-based systems. Economic obstacles will need to be removed by the provision of incentives and access to capital (Kan et al. 2008; Lamers et al. 2008). The annual cash flows predicted from non-timber products are promising, but a commercialization of these products would be necessary to actually influence the cash income of rural Uzbek households. Processing locally gathered non-timber products, e.g., turning traditional fuelwood into wood briquettes for fuel could add value, create jobs, and exploit new markets. Agricultural extension services and transfer of technical and ecological know-how may help increase the motivation of farmers to plant ecologically appropriate tree species.

Environmental services from afforestation, when translated in monetary terms, can significantly increase the value of degraded land used for forestry (Costanza et al. 1997). The existing international C market offers an opportunity for small-scale forestry participation in the CDM thus linking the local and global interest via participatory afforestation. In Uzbekistan, the bio-sequestration forestry projects have been under-represented on the country's CDM agenda due to presumably low cost effectiveness of such projects. We hope that the diversity and magnitude of benefits from afforestation demonstrated in our studies could create awareness of its financial and ecological advantage in degraded croplands.

References

Chub V, Ososkova T (2008) Problems of Aral: impact on the gene pool of the population, flora, fauna and international cooperation for mitigating consequences. Tashkent, Uzbekistan

Close W, Menke KH (1986) Selected topics in animal nutrition. Deutsche Stiftung Fur Internationale Entwicklung (DSE), Feldafing

Costanza R, Arge R, de Groot R, Farberk S, Grasso M, Hannon B, Limburg K, Naeem S, O'Neill RV, Paruelo J, Raskin RG, Suttonkk P, van den Belt M (1997) The value of the world's ecosystem services and natural capital. Nature 387:253–260

Danso SKA, Bowen GD, Sanginga N (1992) Biological nitrogen fixation in trees in agroecosystems. Plant Soil 141:177–196

Dawson JO (1986) Actinorhizal plants: their use in forestry and agriculture. Outlook Agric 15:202–208

Djanibekov N (2006) Cattle breeding system in dekhkan households. Agric J Uzbekistan 4:25–26

Djanibekov N (2008) A micro-economic analysis of farm restructuring in Khorezm region, Uzbekistan. PhD dissertation, Bonn University, Bonn

Djumaeva DM, Djanibekov N, Vlek PLG, Martius C, Lamers JPA (2009) Options for optimizing dairy feed rations with foliage of trees grown in the irrigated drylands of Central Asia. Res J Agric Biol Sci 5:698–708

Djumaeva D, Lamers JPA, Martius C, Khamzina A, Vlek PLG (2010) Quantification of symbiotic nitrogen fixation by *Elaeagnus angustifolia L.* on salt-affected irrigated croplands using two ^{15}N isotopic methods. Nutr Cycl Agroecosyst 88:329–339

Dommergues YR (1995) Nitrogen fixation by trees in relation to soil nitrogen economy. Fertil Res 42:215–230

FAO (2000) Carbon sequestration options under the clean development mechanism to address land degradation. World soil resources reports, 92. FAO, Rome

Fimkin VP (1972) Use of mineralized water for managing protective forest plantations on salt-affected soils. Final report N68001131. Middle-Asian Forestry Research Institute, State Committee of Forest Management at the Council of Ministers of USSR, Tashkent, Uzbekistan, p 149

Fisher RJ, Schmidt K, Steenhof B, Akenshaev N (2004) Poverty and forestry: a case study of Kyrgyzstan with reference to other countries in West and Central Asia. Livelihood Support Programme (LSP). Working paper 13. FAO, Rome

Galiana A, Bouillet JP, Ganry F (2004) The importance of biological nitrogen fixation by trees in agroforestry. In: Rachid S (ed) Symbiotic nitrogen fixation: prospect for enhanced application in tropical agriculture. International workshop on biological nitrogen fixation for increased crop productivity, enhanced human health and sustained soil fertility, Science Publishers, Enfield

Gardner IC (1986) Mycorrhizae of actinorhizal plants. MIRCEN J 2:147–160

Garrity DP (2004) Agroforestry and the achievement of the millennium development goals. Agrofor Syst 61:5–17

Guo LB, Gifford RM (2002) Soil carbon stocks and land use change: a meta analysis. Glob Change Biol 8:345–360

Harrington CA (1999) Forests planted for ecosystem restoration or conservation. New For 17:175–190

Hbirkou C, Martius C, Khamzina A, Lamers JPA, Welp G, Amelung W (2011) Reducing topsoil salinity and raising carbon stocks through afforestation in Khorezm, Uzbekistan. J Arid Environ 75:146–155

Heuperman AF, Kapoor AS, Denecke HW (2002) Biodrainage – principles, experiences and applications. International Programme for Technology and Research in Irrigation and Drainage (IPTRID) Secretariat, Food and Agriculture Organization of the United Nations (FAO), Italy

Ibrakhimov M, Khamzina A, Forkutsa I, Paluasheva G, Lamers JPA, Tischbein B, Vlek PLG, Martius C (2007) Groundwater table and salinity: spatial and temporal distribution and influence

15 Conversion of Degraded Cropland to Tree Plantations...

on soil salinization in Khorezm region (Uzbekistan, Aral Sea Basin). Irrig Drain Syst 21:219–236

IUCN (2004) Afforestation and reforestation for climate change mitigation: potentials for pan-European action. The World Conservation Union (IUCN) and Foundation IUCN Poland (IUCN Programme Office for Central Europe), Warsaw

Kan E, Lamers JPA, Eschanov R, Khamzina A (2008) Small-scale farmers' perceptions and knowledge of tree intercropping systems in the Khorezm region of Uzbekistan. For Trees Livelihoods 18:355–372

Katyal JC, Vlek PLG (2000) Desertification-concept, causes and amelioration, ZEF-discussion papers on development policy 33. Center for Development Research (ZEF), Bonn

Khamzina A (2003) Root development of tree species under different soil and water conditions in irrigated and marginal areas of Khorezm Region. Midterm report. Center for Development Research (ZEF), Bonn

Khamzina A, Lamers JPA, Wickel B, Jumaniyazova Y, Martius C (2005) Evaluation of young and adult tree plantations for biodrainage management in the lower Amu Darya river region, Uzbekistan. In: Proceedings of ICID 21st European regional conference integrated land and water resources management: towards sustainable rural development, Frankfurt (Oder) and Slubice – Germany/Poland

Khamzina A, Lamers JPA, Martius C, Worbes M, Vlek PLG (2006a) Potential of nine multipurpose tree species to reduce saline groundwater tables in the lower Amu Darya River region of Uzbekistan. Agrofor Syst 68:151–165

Khamzina A, Lamers JPA, Worbes M, Botman E, Vlek PLG (2006b) Assessing the potential of trees for afforestation of degraded landscapes in the Aral Sea basin of Uzbekistan. Agrofor Syst 66:129–141

Khamzina A, Lamers JPA, Vlek PLG (2008) Tree establishment under deficit irrigation on degraded agricultural land in the lower Amu Darya River region, Aral Sea Basin. For Ecol Manag 255:168–178

Khamzina A, Lamers JPA, Vlek PLG (2009a) Nitrogen fixation by *Elaeagnus angustifolia* L. in the reclamation of degraded croplands of Central Asia. Tree Physiol 29:799–808

Khamzina A, Sommer A, Lamers JPA, Vlek PLG (2009b) Transpiration and early growth of tree plantations established on degraded cropland over shallow saline groundwater table in Northwest Uzbekistan. Agric For Meteorol 149:1865–1874

Lamers JPA, Khamzina A (2008) Fuelwood production in the degraded agricultural areas of the Aral Sea Basin, Uzbekistan. Bois et Forets des Tropiques 297:47–57

Lamers JPA, Khamzina A (2010) Seasonal quality profile and production of foliage from trees grown on degraded cropland in arid Uzbekistan, Central Asia. J Anim Physiol Anim Nutr 94:77–85

Lamers JPA, Khamzina A, Worbes M (2005) The analyses of physiological and morphological attributes of 10 tree species for early determination of their suitability to afforest degraded landscapes in the Aral Sea Basin of Uzbekistan. For Ecol Manag 221:249–259

Lamers JPA, Bobojonov I, Khamzina A, Franz J (2008) Financial analysis of small-scale forests in the Amu Darya lowlands of rural Uzbekistan. For Trees Livelihoods 18:373–386

Lima WP (1984) The hydrology of eucalypt forests in Australia – a review. IPEF 28:11–32

Makhno G (1962) Elaboration of methods for forest management on salt-affected soils in the Khorezm Region. Middle-Asian Forestry Research Institute, Tashkent

Marcar NE, Crawford DF (2004) Trees for saline landscapes. Rural Industries Research and Development Corporation, Canberra

Martius C, Lamers JPA, Wehrheim P, Schoeller-Schletter A, Eshchanov R, Tupitsa A, Khamzina A, Akramkhanov A, Vlek PLG (2004) Developing sustainable land and water management for the Aral Sea Basin through an interdisciplinary research. In: Seng V, Crasswell E, Fukai S, Fischer K (eds) Water in agriculture. ACIAR proceedings, Australian Centre for International Agricultural Research, Canberra

Middle-Asian Forestry Research Institute (MAFRI) (1972) State forestry committee under the cabinet of ministers of the USSR. Recommendations on growing tree plantations of valuable deciduous, coniferous and fruit species on salt-affected land in the Golodnaya Steppe, Khorezm and Ferghana Valley, Tashkent

Millennium Ecosystem Assessment (MEA) (2005) Ecosystems and human well-being: synthesis report. Island Press, Washington, DC

Morris J, Mann L, Collopy J (1998) Transpiration and canopy conductance in a eucalypt plantation using shallow saline groundwater. Tree Physiol 18:547–555

Müller M (2006) A general equilibrium approach to modeling water and land use reforms in Uzbekistan. PhD dissertation, Rheinischen Friedrich-Wilhelms-Universität Bonn, Bonn

Paschke MW (1997) Actinorhizal plants in rangelands of the western United States. J Range Manag 50:62–72

Paydar Z, Huth N, Snow V (2005) Modelling irrigated eucalyptus for salinity control on shallow watertables. Aust J Soil Res 43:587–597

Rudenko I, Nurmetov K, Lamers JPA (2011) State order and policy strategies in the cotton and wheat value chains. In: Martius C, Rudenko I, Lamers JPA, Vlek PLG (eds) Cotton, water, salts and *Soums*: economic and ecological restructuring in Khorezm, Uzbekistan. Springer, Dordrecht

Scherr SJ, Sthapit S (2009) Farming and land use to cool the planet. In: Starke L (ed) State of the world: into a warming world. Worldwatch institute report on progress toward a sustainable society. W.W. Norton & Company, New York

Smith J, Scherr S (2003) Capturing the value of forest carbon for local livelihoods. World Dev 31:2143–2160

Thorburn PJ (1999) Interactions between plants and shallow, saline water tables: implications for the management of salinity in Australian agriculture. Agric Water Manag 39(2/3):234

Thorburn PJ, Walker GR, Jolly ID (1995) Uptake of saline groundwater by plants: an analytical model for semi-arid and arid areas. Plant Soil 175:1–11

Tupitsa A (2009) Photogrammetric techniques for the functional assessment of tree and forest resources in Khorezm, Uzbekistan. PhD dissertation, Rheinischen Friedrich-Wilhelms-Universität Bonn, Bonn

UNEP, Glavgidromet (1999) National action programme to combat desertification in republic of Uzbekistan. Tashkent

UNFCC (2001) Initial national communication of the republic of Uzbekistan under the UNFCC. Uzbekistan: Country study on climate change, phase 2. Expedited financing of climate change enabling activities. Project implementation report. In: Main administration under the Cabinet of Ministers, Uzbekistan

Wang S, Chen B (1996) Euphrates poplar forest. China Environmental Science Press, Beijing

Chapter 16
Abundance of Natural Riparian Forests and Tree Plantations in the Amudarya Delta of Uzbekistan and Their Impact on Emissions of Soil-Borne Greenhouse Gases

Clemens Scheer, Alexander Tupitsa, Evgeniy Botman,
John P.A. Lamers, Martin Worbes, Reiner Wassmann,
Christopher Martius, and Paul L.G. Vlek

Abstract Through a forest inventory in parts of the Amudarya river delta, Central Asia, we assessed the impact of ongoing forest degradation on the emissions of greenhouse gases (GHG) from soils. Interpretation of aerial photographs from 2001, combined with data on forest inventory in 1990 and field survey in 2003 provided comprehensive information about the extent and changes of the natural *Tugai* riparian forests and tree plantations in the delta. The findings show an average annual deforestation rate of almost 1.3% and an even higher rate of land use change from *Tugai* forests to land with only sparse tree cover. These annual rates of deforestation and forest degradation are higher than the global annual forest loss. By 2003, the *Tugai* forest area had drastically decreased to about 60% compared to an inventory in 1990.

C. Scheer (✉)
Institute for Sustainable Resources, Queensland University of Technology,
2 George Street, QLD 4001 Brisbane, Australia
e-mail: clemens.scheer@qut.edu.au

A. Tupitsa • P.L.G. Vlek
Center for Development Research (ZEF), Walter-Flex-Str. 3, 53113 Bonn, Germany
e-mail: atupitsa@uni-bonn.de; p.vlek@uni-bonn.de

E. Botman
Republican Scientific Production Center for Decorative Gardening and Forestry,
Darkhan, 111104 Zangiota District, Tashkent province, Uzbekistan
e-mail: darhanbek@yandex.ru

J.P.A. Lamers
ZEF/UNESCO Khorezm Project, Khamid Olimjan Str., 14, 220100 Urgench, Uzbekistan
e-mail: j.lamers@uni-bonn.de

M. Worbes
Department of Crop Sciences, Division Agronomy in the Tropics,
Georg-August University of Göttingen, Grisebach Str. 6, 37077 Göttingen, Germany
e-mail: mworbes@gwdg.de

C. Martius et al. (eds.), *Cotton, Water, Salts and Soums: Economic and Ecological Restructuring in Khorezm, Uzbekistan*, DOI 10.1007/978-94-007-1963-7_16,
© Springer Science+Business Media B.V. 2012

Significant differences in soil GHG emissions between forest and agricultural land use underscore the impact of the ongoing land use change on the emission of soil-borne GHGs. The conversion of *Tugai* forests into irrigated croplands will release 2.5 t CO_2 equivalents per hectare per year due to elevated emissions of N_2O and CH_4. This demonstrates that the ongoing transformation of *Tugai* forests into agricultural land-use systems did not only lead to a loss of biodiversity and of a unique ecosystem, but substantially impacts the biosphere-atmosphere exchange of GHG and soil C and N turnover processes.

Keywords Aral Sea Basin • Forest degradation • Deforestation • Climate change impact • Global warming • Nitrous oxide • Methane

16.1 Introduction

The conversion of natural, unmanaged vegetation to cropland releases substantial amounts of carbon dioxide (CO_2) to the atmosphere and in turn reduces C storage in soil and vegetation (Robertson and Grace 2004). Soil microbial transformations (e.g., nitrification, denitrification and methanogenesis) can produce significant amounts of the greenhouse gases (GHGs) methane (CH_4) and nitrous oxide (N_2O) (Mosier 1998). Recent estimates show that agriculture accounts for about 60% of N_2O and about 50% of CH_4 global anthropogenic emissions (Smith et al. 2007); whereas deforestation accounts for about 20% of global CO_2 emissions, larger than the entire global transportation sector (Meridian Institute 2009). Land use change and deforestation are very acute processes also ongoing in the Amudarya river delta, Central Asia, where both direct and indirect anthropogenic causes are responsible for the degradation of the riparian (*Tugai*) forests.

Prior to the expansion of the area for irrigated agriculture during the Soviet Union epoch, the natural vegetation in the Amudarya delta consisted mainly of *Tugai* forests and widespread reed communities occurring as narrow belts flanking the Amudarya river on its course towards the Aral Sea through the Karakum and Kyzylkum deserts. *Tugai* forests are fast-growing deciduous trees mainly of poplar (*Populus euphratica* Oliv. and *P. pruinosa*), but also including Russian olive

---

R. Wassmann
Institute for Meteorology and Climate Research, Atmospheric Environmental Research, Kreuzeckbahn Str. 19, 82467 Garmisch-Partenkirchen, Germany
e-mail: r.wassmann@cgiar.org

C. Martius
Center for Development Research (ZEF), Walter-Flex-Str. 3, 53113 Bonn, Germany
e-mail: gcmartius@gmail.com

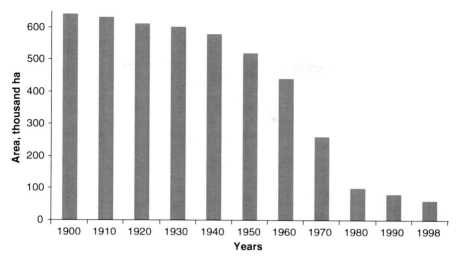

Fig. 16.1 Dynamics of *Tugai* forest area along the Amudarya river from 1900 to 1998 (Treshkin 2001a)

(*Elaeagnus angustifolia*) and willow (*Salix* spp.), in the floodplains and river deltas along the Amudarya, Syrdarya, Zeravshan and Vaksh rivers (Novikova 2001; Treshkin 2001b). Their growth requires regular flooding, which explains their occurrence mainly on sand banks, islands and low terraces (Kuzmina and Treshkin 1997). With more than 230 plant species, the *Tugai* forest is considered to be one of the most diverse vegetation types of arid regions in Central Asia (Novikova 2001).

At the beginning of the last century, the narrow belts of *Tugai* forest stretched for hundreds of kilometers along rivers. They covered about 600,000 ha in the floodplains of the Amudarya alone, including 300,000 ha in its lower reaches and delta (Fig. 16.1). Nowadays, this area is reportedly reduced to less than 30,000 ha in the delta, having lost about 90% in the past 60 years (Treshkin 2001b). Varying from a few hundred meters up to several kilometers length, narrow belts of *Tugai* forest flank also the river reaches and canals in the irrigated areas of Khorezm, an administrative region in the northwest of Uzbekistan, Central Asia.

Uncontrolled tree cutting, overgrazing and bushfires, which transformed the tree-shrub communities into pasture ecosystems, are direct anthropogenic factors that reduced the area of *Tugai*. Many former *Tugai* forest areas have been converted also into agricultural croplands, in particular flooded rice fields, or were replaced by permanent forest plantations. Also, the intensive irrigation agriculture introduced during the Soviet era for the production of cotton (cf. Tischbein et al. 2011), has reduced river flow, lowered the groundwater table along the river and increased soil salinity, which are indirect factors having caused the disappearance of the *Tugai* forests (Kuzmina and Treshkin 1997; Treshkin 2001a). Furthermore, the Tuyamuyun reservoir upstream of Khorezm (cf. Tischbein et al. 2011) regulates river flow and reduces the regular floods that previously filled up the *Tugai* groundwater reservoirs.

The impact of the manifold direct and indirect causes of land use change and deforestation of *Tugai* forests on the emission of soil-borne GHGs has not yet been quantified, although changes in land-use contribute to the anthropogenic emissions of GHGs to the atmosphere (IPCC 2007). Lacking, however, are both an up-to-date inventory of the *Tugai* forests in the Amudarya delta (including those areas on the left bank of the river where most irrigated agriculture occurs) and the quantification of GHG emission levels caused by the conversion of the *Tugai* forests into cropland or tree plantations. The objectives of this study therefore were to (i) evaluate the extent and changes of the *Tugai* forests and tree plantations in a transect through Khorezm and (ii) assess their impact on emissions of soil-borne GHGs in this region.

16.2 Materials and Methods

16.2.1 Study Region

The study was conducted in the Khorezm region, a 610,000 ha area which is part of the Turan Lowland of the Aral Sea Basin, a vast low-lying desert basin stretching from southern Turkmenistan through Uzbekistan to Kazakhstan. The region is situated between 41°08′ and 41°59′ N latitude and 60°03′ and 61°24′ E longitude at 90–138 m above sea level. It is bordered by the Karakum and Kyzylkum deserts to the south and east and is about 250–350 km south from the remains of the Aral Sea (cf. Conrad et al. 2011). Khorezm belongs to the lower reaches of the Amudarya and is one of the areas most intensively used for agriculture with roughly 270,000–300,000 ha irrigated land (cf. Conrad et al. 2011).

16.2.2 Forest Inventory of Khorezm

Two arbitrarily selected transects crossing Khorezm in NS (320 km^2) and WE (230 km^2) directions were delineated as the study area (Fig. 16.2). They covered ca. 10% of the entire region in which four *Tugai* forest compartments and tree plantations were analyzed. Two datasets were used for the forest assessment: (i) information of the last Khorezm forest inventory conducted in 1990 (KhFI 1990), which was provided by the Khorezm Forestry Department, and (ii) black-and-white aerial photographs, scale 1:20,000, taken in November 2001 by the Land and Geodesy Cadastre Center of Uzbekistan (Ergeodezkadastr 2001) complemented with field studies in 2003 (Tupitsa 2009).

The aerial photographs were analyzed with a VISOPRET 10-DIG analytical stereo plotter and AutoCAD software. Prior to the analyses, 70 stereo pairs of aerial photographs were prepared for the consecutive procedures of interior, relative and absolute orientations. ESRI ArcGIS 9.0 and ESRI ArcView 3.2 software and their extensions were used for the analyses. The photogrammetric technique included

16 Abundance of Natural Riparian Forests and Tree Plantations...

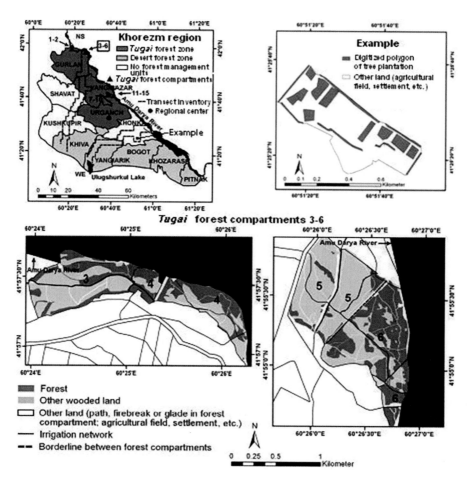

Fig. 16.2 Extent of forest and other wooded land in the *Tugai* forest compartments and example of an area with tree plantations in the transects, Khorezm

ocular photo-interpretations with the plotter to identify forests and tree plantations according to the developed thematic classes (Table 16.1). Following the identification of each forest patch or tree plantation, its polygon was plotted and an output file (digitized polygons) was produced (Fig. 16.2). The application of GIS-based tools allowed computing the area to the nearest 0.01 ha of the digitized polygons. Based on the FAO (2005) threshold values of the classification of a forest canopy of >10% (forest) and 5–10% (other wooded land), four *Tugai* forest compartments were classified. Ocular estimates of crown closure to the nearest 0.05 coefficient were carried out with the plotter. Finally, aerial photography was complemented with field surveys to identify species composition, vitality and age classes. Therefore, randomly selected plots of 0.01–0.1 ha representing a 5% area in each forest patch or tree plantation were surveyed. For rating insect and disease problems (Rohrmoser 1984),

Table 16.1 Developed forest and tree plantation thematic classes in Khorezm, based on aerial photographs of 2001

	Photographic indicator				Other criterion[a]	
		Crown shadow profile				Minimal crown closure (coef.)
Thematic class	Horizontal crown projection	At edge of projection	Between projections (tree plantations) or in projection (forests)	Site condition	Minimal area (ha)	
Tugai forest	Various-shaped and clumped (individual crowns cannot be identified) projection corresponding to a forest stand in an irregularly shaped area	Visible long and clumped profile (individual crowns cannot be identified)	Long, but not visible profile due to the clumped projection	Floodplains along margins of the Amudarya river	≥0.5	≥0.1
Other wooded land in *Tugai* forest	Single tree projection corresponding to single trees in an irregularly-shaped area	Yong trees: poorly visible profile due to short length; mature trees: visible long and un-clumped (individual crowns can be identified) profile	Young trees: poorly visible profile due to short length; mature trees: visible long and un-clumped (individual crowns can be identified) profile	Floodplains along margins of the Amudarya river	≥0.5	<0.1
Populus spp. plantation	Line-shaped and clumped projection (individual crowns/rows cannot be identified) corresponding to triple to sextuple rows in a line-shaped cluster. There are several clusters or projections in a square- or rectangular-shaped area and where spaces between the projections are not visible due to a crown shadow profile exceeding those spaces	Visible long and clumped profile (individual crowns/rows cannot be identified)	Long, but not visible profile due to the not visible spaces	Any space in agricultural fields	≥0.5	≥0.1

[a]According to FRA 2005 (FAO 2005)

simplified versions of international conventions were used. The locally used age classes, young (≤10 year), pre-mature (11–25 year) and mature (≥26 year) were determined with a Suunto 300 mm tree increment borer.

16.2.3 Emission of GHGs from Soils

In 2005, emissions of soil-borne GHG were measured in the *Tugai* forest reserve located in the Amudarya floodplain, North of Khorezm. The reserve was set aside in 1971 and spans ca 6,400 ha (UNDP/GEF 1998). The forest in this reserve had suffered severely from deforestation and for years experienced decreased natural flooding caused by the diversion of the Amudarya water for irrigation. The fluxes of N_2O and CH_4 were measured in the *Tugai* forest vegetation with the closed chamber technique (Mosier 1989) in four chambers (replications). In the experimental area, poplars (*Populus ariana* L. and *P. pruinosa*) dominated the woody species. In 2005 and 2006, N_2O and CH_4 fluxes were monitored also in a forest plantation on a gleyic solonchak soil (FAO 1998) with a sandy loam texture. The entire plantation covered 75 ha, including 70 ha of 10-year-old black poplar (*P. nigra*) and 5 ha of 6-year-old Siberian elm (*Ulmus pumila* L.) trees. Fluxes were primarily measured in the *P. nigra* plantation.

Additional details about the method and the experimental set-up were presented previously (Scheer et al. 2008a, b).

16.3 Results and Discussion

16.3.1 Forestry Inventory of Khorezm

Tugai forest covered less than 1% of the study area (Table 16.2), which is classified as a low forest cover according to the classification by Lund (1999). No forest patches were found in the WE transect. Comparing the results of the inventories in 1990 and 2003 showed that about 40% of the forest area had been turned into 'other wooded land' (thinned out and strongly degraded forest) resulting in an annual forest degradation rate of 3.1% (Fig. 16.3). However, these areas could still regenerate naturally or under rehabilitation efforts. If forests and other woodland are lumped together, then 17% of their combined area had been converted into cropland in that period, corresponding to an annual deforestation rate of 1.3% (Fig. 16.3). Due to the intensive agricultural land use, these areas cannot be rehabilitated to *Tugai* forest anymore. Thus, the latter value should be considered as a conservative deforestation estimate.

The estimated annual rates of forest degradation and deforestation in the *Tugai*, which are higher than the global annual forest loss, are due to land use change caused by numerous factors. The water shortages of 2000 and 2001 significantly affected the *Tugai* forests (Treshkin 2001a). An exposure to frequent grazing, particularly of young trees, forest fires and illegal cuttings has substantially reduced

Table 16.2 Spatial extent of the *Tugai* forest compartments in the transects, Khorezm

Transect	Transect area (ha)		Forest classification[a] Forest	Other wooded land	Total
NS	32,020	Area (ha)	139.1	177.8	316.9
		Ratio to the transect area (%)	(0.4)	(0.6)	(1.0)
WE	23,020	Area (ha)	–	–	–
Total	55,040	Area (ha)	139.1	177.8	316.9
		Ratio to the transect area (%)	(0.3)	(0.3)	(0.6)

[a]FRA 2005 classification (FAO 2005)

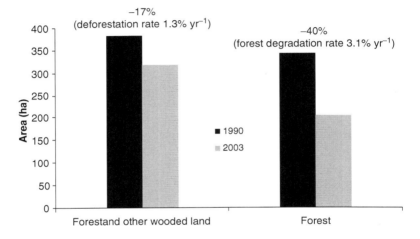

Fig. 16.3 Changes in forest area in the *Tugai* forest compartments in the transects Khorezm

the potential regeneration of the vulnerable *Tugai* (Treshkin et al. 1998). Unless these low regeneration rates are increased by protective measures implemented by the national and regional authorities, it is highly likely that more, maybe all, of the unique *Tugai* forests will be lost. Future forest management must be based on comprehensive, reliable and near real-time information which, as Tupitsa (2009) has adopted and verified for *Tugai* forests, can be extracted from aerial photographs (regularly taken by local authorities) combined with photogrammetry.

Since the mid 1990s, a nationwide program pushed the establishment of 50,000 ha (including 7,300 ha in Khorezm) of fast-growing hybrid poplar (*Populus* spp.) plantations to satisfy the national demand for construction wood and pulp, which used to be imported from Russia (Government of Uzbekistan 1994; FAO 2006). Yet, due to environmental constraints and disturbances affecting tree health, the recent tree-planting initiatives failed (FAO 2006), and poplar now only covers a small part of the forest area. The transect inventory in Khorezm confirmed the low extent of <0.1% of *Populus* spp. plantations as compared to other plantings (Table 16.3).

Table 16.3 Spatial extent of tree plantations in the transects, Khorezm

Transect	Transect area (ha)		Tree plantations (ha)		Orchards				Vineyards (*Vitis* L.)	Total
			Populus spp.	Mulberry (*M. alba*)	Apple (*Malus* spp.)	Apricot (*P. arme-niaca*)	Other spp.	Total		
NS	32,020	Area (ha)	11	48	178	93	17	288	30	377
		Ratio (%)[a]	(<0.1)	(0.2)	(0.6)	(0.3)	(<0.1)	(0.9)	(0.1)	(1.2)
WE	23,020	Area (ha)	6	54	136	55	11	202	24	286
		Ratio (%)[a]	(<0.1)	(0.2)	(0.6)	(0.2)	(<0.1)	(0.9)	(0.1)	(1.2)
Total	55,040	Area (ha)	17	102	314	148	28	490	54	663
		Ratio (%)[a]	(<0.1)	(0.2)	(0.6)	(0.3)	(<0.1)	(0.9)	(0.1)	(1.2)

[a]Ratio to the transect area (%)

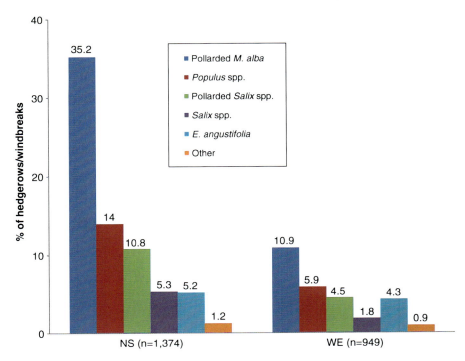

Fig. 16.4 The overall presence of dominant species of hedgerows/windbreaks in the transects, Khorezm

However, in hedgerows/windbreaks along fields, canals and roads, the share of *Populus* spp. was about 20% (Fig. 16.4). Furthermore, the tree plantations were relatively young (≤10 year) and *Populus* spp. had no pre-mature and mature (>10 year) age classes. About 60% area of *Populus* spp. plantations was infested by pests and diseases (Tupitsa 2009). The environmental constraints and disturbances affecting tree health suppressed the recent initiatives on planting *Populus* spp. plantations in Uzbekistan (FAO 2006). Furthermore, although forest logging is strictly forbidden in Uzbekistan, except for sanitary felling and clearings, wood removal between 1990 and 2003 amounted to about 30,000 m^3 annually of which about 30% was for industrial use for sawn wood and hardwood for producing wooden cases, matches, pulp and paper. During the past years, wood has generally been supplied by the private sector and consumed by the small-scale wood-processing businesses for manufacturing small furniture (FAO 2006). Wood is also commonly used as fuel (UNECE 2005).

Given the growing need for timber by the local population, the extension of appropriate tree plantations on degraded, salt effected cropland could be an attractive option in the near future (Khamzina et al. 2006b). In that case, it seems judicious to plant not only softwood species, such as *Populus* spp., but also other species such as *U. pumila* (Khamzina et al. 2006b).

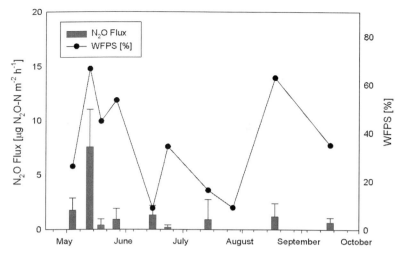

Fig. 16.5 N$_2$O flux rates and water-filled pore space (WFPS) of the *Tugai* forest plot over the April-September 2005 measurement period. Error bars indicate the standard error

16.3.2 Soil-Borne GHG Emissions

Emissions of N$_2$O and CH$_4$ in the *Tugai* forest were measured at regular intervals from May to September 2005. During the entire sampling period, no CH$_4$ fluxes and only very low N$_2$O fluxes (<2 μg N$_2$O-N m^{-2} h^{-1}) were observed. Maximum fluxes occurred in one replication (chamber) on May 12, which reached 17.6 μg N$_2$O-N m^{-2} h^{-1} (Fig. 16.5). The arithmetic mean of the N$_2$O flux for all measurements in the four measuring chambers was 1.4 ± 0.75 μg N$_2$O-N m^{-2} h^{-1}, corresponding to a cumulative annual emission of 0.12 kg N$_2$O-N ha^{-1} year^{-1}.

Throughout summer 2005, natural flooding did not occur in this riparian zone and the sole water source was the shallow groundwater table. However, the topsoil water content remained low throughout the measurement period (<40% WFPS). Soil N$_2$O emissions are largely regulated by soil water content (Simojoki and Jaakkola 2000) and are disproportionately reduced with low soil moisture (Zheng et al. 2000). The very low N$_2$O emissions in the *Tugai* land-use system can thus largely be explained by the low water content in the upper layer of the soil. However, it should be noted that the *Tugai* forests used to be regularly inundated by the rivers. As a result of the extensive use of river water for irrigation, this natural flooding has ceased. Under a natural flooding regime, which coincides with depositions of sediments and nutrients, it is very likely that N$_2$O emissions would peak after inundation events. Hence, the presently observed 'natural' N$_2$O flux rates of the *Tugai* forest are likely to be lower than previous rates, owing to the diversion of river water for irrigated agriculture (Scheer et al. 2008b).

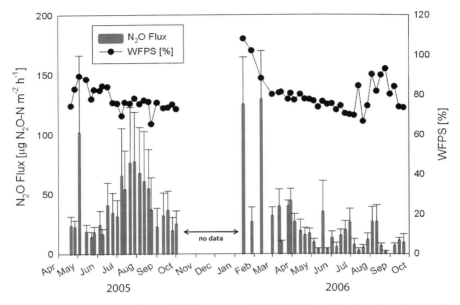

Fig. 16.6 N$_2$O flux rates and water-filled pore space (WFPS) of the poplar plantation plot over the April 2005-September 2006 measurement period. Error bars indicate the standard error (Modified after Scheer et al. 2008b)

In the poplar plantations, "substantial N$_2$O emissions" were observed throughout the entire measurement period (Fig. 16.6), while significant CH$_4$ fluxes could not be detected. The N$_2$O fluxes showed clear inter-annual differences. In 2005, N$_2$O fluxes were measured from April to October and high emissions were observed over the entire period (39.8 µg N$_2$O-N m^{-2} h^{-1}) with a clear peak during the summer months (July-August). Since no fertilizer was added and soil moisture had stayed at a constant level, soil temperature can be considered the main regulating parameter of these N$_2$O emissions. In 2006, the mean flux rates were significantly lower (22.9 µg N$_2$O-N m^{-2} h^{-1}) with highest emissions in winter and lower flux rates in summer. Extraordinarily high N$_2$O emissions were observed on two occasions in January and February, which corresponded to thawing after a frost period in the upper soil. Such emission peaks have previously been described as freeze-thaw events that are known to induce an emission pulse of N$_2$O at, or shortly after, thawing during winter and spring (Papen and Butterbach-Bahl 1999; Morkved et al. 2006). After these initial events in the mid-winter months, the flux rates stayed below 50 µg N$_2$O-N m^{-2} h^{-1} and followed no clear temporal trend. Compared to N$_2$O emissions from forest plantations elsewhere (Robertson et al. 2000; Ferré et al. 2005), the measured flux rates were unexpectedly high, particularly when realizing that N-fertilizers had not been added. It seems that the main reason for these high flux rates at this study site was the joint impact of constantly high soil water content and high soil temperatures during the summer months. Due to the shallow groundwater table of 1–2 m below ground level throughout the observation period, the soil water content remained

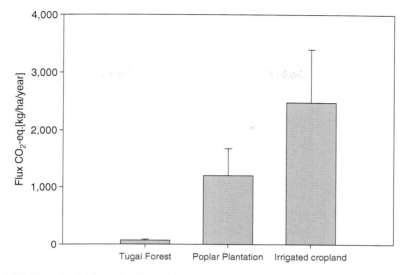

Fig. 16.7 Mean Global Warming Potential in CO_2 equivalents of N_2O and CH_4 fluxes for different land-use systems of the Amudarya delta (Modified after Scheer et al. 2008b)

high in these poplar plantations, even in the absence of regular irrigation events. The WFPS has been described in various studies as the determining factor regulating N_2O emissions (Simojoki and Jaakkola 2000). During the whole measuring campaign, the soil water content was in the range of 65–100% WFPS and largest fluxes occurred at high WFPS (70–100%), indicating that denitrification was the major factor responsible (Dobbie et al. 1999).

A third factor influencing the N emissions is the N source. No N-fertilizer was added at the examined site and topsoil N (0.1%) and organic matter (0.6%) contents were rather low, as is common in the entire region (cf. Akramkhanov et al. 2011). Hence, the N_2O emissions should have been limited due to the low mineral-N content in the soil. A possible explanation for the elevated N_2O emissions is that a significant amount of N was transported via groundwater (cf. Kienzler et al. 2011) or atmospheric deposition from adjacent agricultural fields with high fertilizer application rates into this site (Janzen et al. 2003). However, this could not be confirmed because measurements were not taken.

The results reflect the potential impact of *Tugai* forests conversion (into degraded forests and agricultural land) on the emission of soil borne GHGs from these systems. GHG emissions previously reported from irrigated agricultural fields in the Amudarya delta were up to 100 times higher than the emissions observed in the *Tugai* forests (Scheer et al. 2008b). Consequently, the reduction of the *Tugai* forests will lead to elevated emissions of GHGs. Conversion of *Tugai* forests into irrigated cropping land will release 2.5 t CO_2 equivalents per hectare per year (Fig. 16.7). Accordingly, the ca. 50,000 ha of *Tugai* forests that have been converted into irrigated agricultural land will continuously be releasing an equivalent of 125,000 t CO_2 eq. every year in emissions of N_2O and CH_4 only, due to fertilization and flooding. This figure does not take into account the amount of CO_2 that was lost from the aboveground biomass and C stored in the soils, so the absolute amount of CO_2 emitted is even higher.

16.4 Conclusions

Air photography coupled with photogrammetry provided comprehensive information about the extent and changes of forested land in the Amudarya delta. The findings confirmed both the ongoing reduction of the *Tugai* forest and the degradation of forests. Furthermore, significant differences in GHG emissions from soil under different land uses were found. Degradation and the conversion of natural *Tugai* forests into cropland will result in elevated GHG emissions via enhanced N_2O and CH_4 fluxes from the modified land-use systems. Hence, the conservation of the *Tugai* forests would simultaneously help protect this unique ecosystem and mitigate emission of GHGs. However, this would require setting aside water for ecological purposes, which currently is not the case. Approaches to water saving in agriculture as discussed in several other chapters in this book could therefore have clear ecological benefits if properly administered.

To reduce the impact of the ongoing deforestation, sustainable forest management plans should urgently be developed and implemented. Given the growing need for timber and firewood, an extension of appropriate tree plantations in the near future will be needed. Removing land from irrigated agricultural production, or changing land use from an annual cropping system to perennial forest plantations, in particular on marginal lands where annual crops are no longer profitable (Khamzina et al. 2006a; Khamzina et al. 2008), could help reducing land degradation and, at the same time, mitigate N_2O and CH_4 fluxes. In addition, such plantations would offer mitigation options through carbon sequestration, owing to a high rate of soil C storage and the accumulation of C in non-harvested wood, given the high biomass production which can reach 20–32 $t\,ha^{-1}$ per year over 5 years, depending on the species (cf. Khamzina et al. 2011). The creation of a market for trading carbon dioxide emissions could thus provide an additional income source for farmers.

References

Akramkhanov A, Kuziev R, Sommer R, Martius C, Forkutsa O, Massucati L (2011) Soils and soil ecology in Khorezm. In: Martius C, Rudenko I, Lamers JPA, Vlek PLG (eds) Cotton, water, salts and *Soums*: economic and ecological restructuring in Khorezm, Uzbekistan. Springer, Dordrecht/Berlin/Heidelberg/New York

Conrad C, Schorcht G, Tischbein B, Davletov S, Sultonov M, Lamers JPA (2011) Agro-meteorological trends of recent climate development in Khorezm and implications for crop production. In: Martius C, Rudenko I, Lamers JPA, Vlek PLG (eds) Cotton, water, salts and *Soums*: economic and ecological restructuring in Khorezm, Uzbekistan. Springer, Dordrecht/Berlin/Heidelberg/New York

16 Abundance of Natural Riparian Forests and Tree Plantations...

Dobbie KE, McTaggart IP, Smith KA (1999) Nitrous oxide emissions from intensive agricultural systems: variations between crops and seasons, key driving variables, and mean emission factors. J Geophys Res Atmos 104:26891–26899

Ergeodezkadastr (2001) Land and Geodesy Cadastre Center of Uzbekistan. Aerial photography of Khorezm taken by the Ergeodezkadastr in 2001

FAO (1998) World reference base for soil resources. ISSS-ISRIC-FAO, FAO, Rome

FAO (2005) Global forest resources assessment, FRA 2005. Uzbekistan country report. Forestry department, FAO, Rome, p 42

FAO (2006) Forest and forest products. Country profile Uzbekistan. Geneva timber and forest discussion paper 45. ECE/TIM/DP/45, p 68

Ferré C, Leip A, Matteucci G, Previtali F (2005) Impact of 40 years poplar cultivation on soil carbon stocks and greenhouse gas fluxes. Biogeosci Discuss 2:897–931

Government of Uzbekistan (1994) Development of industrial poplar plantations and other fast-growing tree species. Decision of the Government of Uzbekistan, 62. 68, Feb 1994

IPCC (2007) Summary for Policymakers. In: Salomon S, Quin D, Manning N, Chen Z, Marquis N, Averyt KB, Tignor M, Miller HL (eds) Climate change 2007. Cambridge University Press, Cambridge/New York

Janzen HH, Beauchemin KA, Bruinsma Y, Campbell CA, Desjardins RL, Ellert BH, Smith EG (2003) The fate of nitrogen in agroecosystems: an illustration using Canadian estimates. Nutri Cycl Agroecosyst 67(1):85–102

Khamzina A, Lamers JPA, Martius C, Worbes M, Vlek PLG (2006a) Potential of nine multipurpose tree species to reduce saline groundwater tables in the lower Amu Darya River region of Uzbekistan. Agrofor Syst 68:151–165

Khamzina A, Lamers JPA, Worbes M, Botman E, Vlek PLG (2006b) Assessing the potential of trees for afforestation of degraded landscapes in the Aral Sea Basin of Uzbekistan. Agrofor Syst 66:129–141

Khamzina A, Lamers JPA, Vlek PLG (2008) Tree establishment under deficit irrigation on degraded agricultural land in the lower Amu Darya River region, Aral Sea Basin. For Ecol Manag 255:168–178

Khamzina A, Lamers JPA, Vlek PLG (2011) Conversion of degraded cropland to tree plantations for ecosystem and livelihood benefits. In: Martius C, Rudenko I, Lamers JPA, Vlek PLG (eds) Cotton, water, salts and *Soums*: economic and ecological restructuring in Khorezm, Uzbekistan. Springer, Dordrecht/Berlin/Heidelberg/New York

Khorezm Forest Inventory (KhFI) (1990) The complete report by Lesproekt and Khorezm leskhoz, Tashkent 1–3:400

Kienzler K, Ibragimov N, Lamers JPA, Vlek PLG (2011) Groundwater contribution to N fertilization in irrigated cotton and winter wheat in the Khorezm region, Uzbekistan. In: Martius C, Rudenko I, Lamers JPA, Vlek PLG (eds) Cotton, water, salts and *Soums*: economic and ecological restructuring in Khorezm, Uzbekistan. Springer, Dordrecht/Berlin/Heidelberg/New York

Kuzmina Z, Treshkin S (1997) Soil salinazation and dynamics of tugai vegetation in the Southeastern Caspian Sea Region and in the Aral Sea Coastal Region. Eurasian Soil Sci 30(6):642–649

Lund HG (1999) Definition of Low Forest Cover (LFC). Report prepared for IUFRO. Manassas, VA. Forest Information Services, p 22. http://home.comcast.net/~gyde/LFCreport.html. Verified Nov 2008

Meridian Institute (2009) Reducing emissions from deforestation and forest degradation (REDD): an options assessment report. http://www.REDD-OAR.org. Verified Nov 2009

Morkved PT, Dorsch P, Henriksen TM, Bakken LR (2006) N^2O emissions and product ratios of nitrification and denitrification as affected by freezing and thawing. Soil Biol Biochem 38:3411–3420

Mosier AR (1989) Chamber and isotope techniques. In: Andrae MO, Schimel DS (eds) Exchange of trace gases between terrestrial ecosystems and the atmosphere. Wiley, Chichester, pp 175–187

Mosier AR (1998) Soil processes and global change. Biol Fertil Soils 27:221–229

Novikova NM (2001) Ecological basis for botanical biodiversity conservation within the Amu Darya and Syr Darya. In: Breckle S-W, Veste M, Wuchere W (eds) Sustainable land use in deserts. Springer, Berlin/New York, pp 95–103

Papen H, Butterbach-Bahl K (1999) A 3-year continuous record of nitrogen trace gas fluxes from untreated and limed soil of a N-saturated spruce and beech forest ecosystem in Germany 1. N_2O emissions. J Geophys Res Atmos 104:18487–18503

Robertson GP, Grace PR (2004) Greenhouse gas fluxes in tropical and temperate agriculture: the need for a full-cost accounting of global warming potentials. In: Wassmann R, Vlek PLG (eds) Tropical agriculture in transition-opportunities for mitigating greenhouse gas emissions? Kluwer Academic, Dordrecht, pp 51–63

Robertson GP, Paul EA, Harwood RR (2000) Greenhouse gases in intensive agriculture: contributions of individual gases to the radiative forcing of the atmosphere. Science 289:1922–1925

Rohrmoser K (1984) Compendium for field research, 2 edn. Federal Republic of Germany: Deutsche Gesellschaft für Technische Zusamenarbeit (GTZ)

Scheer C, Wassmann R, Kienzler K, Ibragimov N, Eschanov R (2008a) Nitrous oxide emissions from fertilized, irrigated cotton (*Gossypium hirsutum L.*) in the Aral Sea Basin, Uzbekistan: influence of nitrogen applications and irrigation practices. Soil Biol Biochem 40:290–301

Scheer C, Wassmann R, Kienzler K, Ibragimov N, Lamers JPA, Martius C (2008b) Methane and nitrous oxide fluxes in annual and perennial land use systems of the irrigated areas in the Aral Sea Basin. Glob Change Biol 14:2454–2468

Simojoki A, Jaakkola A (2000) Effect of nitrogen fertilization, cropping and irrigation on soil air composition and nitrous oxide emission in loamy clay. Eur J Soil Sci 51:413–424

Smith P, Martino D, Cai Z, Gwary D, Janzen H, Kumar P, MacCarl B, Ogle S, O'Mara F, Rice C, Scholes B, Sirotenko O (2007) Agriculture. In: Metz B, Davidson OR, Bosch PR, Dave R, Meyer LA (eds) Climate change 2007: mitigation. contribution of working group III to the fourth assessment report of the Intergovernmental Panel on Climate Change. Cambridge University Press, Cambridge/New York

Tischbein B, Awan UK, Abdullaev I, Bobojonov I, Conrad C, Forkutsa I, Ibrakhimov M, Poluasheva G (2011) Water management in Khorezm: current situation and options for improvement (hydrological perspective). In: Martius C, Rudenko I, Lamers JPA, Vlek PLG (eds) Cotton, water, salts and *Soums*: economic and ecological restructuring in Khorezm, Uzbekistan. Springer, Dordrecht/Berlin/Heidelberg/New York

Treshkin SE (2001a) Transformation of tugai ecosystems in the floodplains of the lower reaches and delta of the Amu-Darya and their protection. Ecological research and monitoring of the Aral Sea Deltas. A basis for restoration. Book 2. UNESCO Aral Sea Project, 1997–2000. Final scientific reports, UNESCO, Paris, pp 1189–1216

Treshkin SE (2001b) The tugai forests of floodplain of the Amu Darya River: ecology, dynamics and their conservation. In: Breckle S-W, Veste M, Wuchere W (eds) Sustainable land use in deserts. Springer, Berlin/New York, pp 95–103

Treshkin SE, Kamalov SK, Bachiev A, Mamutov N, Gladishev AI, Aimbetov I (1998) Present status of tugai forests in the lower Amu Darya Basin and problems of their protection and restoration. Ecological research and monitoring of the Aral Sea Deltas. UNESCO Aral Sea Project 1992–1996, Final Scientific Report, UNESCO, Paris

Tupitsa A (2009) Photogrammetric techniques for the functional assessment of tree and forest resources in Khorezm, Uzbekistan. PhD dissertation, Rheinische Friedrich-Wilhelms-Universität Bonn, Bonn

UNDP/GEF (1998) (United Nations Development Program/Global Environmental Facility). Biodiversity conservation. National strategy and action plan, international network. Tashkent, Uzbekistan

UNECE (2005) (United Nations Economic Commission for Europe). Capacity building in sharing forest and market information. Country statement. Uzbekistan. Overview of forestry. Workshop, Prague and Krtiny, Czech Republic, 24–28 Oct 2005

Zheng X, Wang M, Wang Y, Shen R, Gou J, Li J, Jin J, Li L (2000) Impacts of soil moisture on nitrous oxide emission from croplands: a case study on the rice-based agro-ecosystem in Southeast China. Chemosphere Glob Change Sci 2:207–224

Part V
Land and Water Management Improvement Tools

Chapter 17
Economic-Ecological Optimization Model of Land and Resource Use at Farm-Aggregated Level

Rolf Sommer, Nodir Djanibekov, Marc Müller, and Omonbek Salaev

Abstract A Farm-Level Economic Ecological Optimization Model (FLEOM) was developed in the ZEF/UNESCO Khorezm project as a land-use planning and decision-support tool at the level of farms and Water Users Associations (WUAs) to couple ecological and economic optimization of land allocation. The agronomic database for cotton, winter wheat and maize that underlines the model was established with the cropping system simulation model CropSyst using data sets, field experience and knowledge of a range of agronomic and hydrological studies on irrigation and fertilizer response, planting dates, tillage and residue management. Potential users of this tool are medium-level stakeholders such as representatives of WUAs and the local water authority. Besides, the model is intended to be a tool for scientists and for university education. The features of FLEOM are presented through simulation of four different management scenarios, each with different sets of assumptions relating to changes in socio-economic conditions: (i) business-as-usual, (ii) commodity market liberalization, (iii) ecological commodity market liberalization, and (iv) dry-year scenario. The evaluation of the scenarios demonstrates that FLEOM produces consistent and plausible outputs, and that it can be used for quite complex scenario simulations. The scenario results reveal that under conditions of a liberalized commodity market, cotton production had no comparative advantages and would completely disappear, and with it the state income from cotton exports. However,

---

R. Sommer (✉)
International Center for Agricultural Research in the Dry Areas (ICARDA),
P.O. Box 5466, Aleppo, Syrian Arab Republic
e-mail: r.sommer@cgiar.org

N. Djanibekov • O. Salaev
ZEF/UNESCO Khorezm Project, Khamid Olimjan str., 14, 220100 Urgench, Uzbekistan
e-mail: nodir79@gmail.com; omonbek@gmail.com

M. Müller
Center for Development Research (ZEF), Walter-Flex-Str. 3, 53113 Bonn, Germany
e-mail: molinero@uni-bonn.de

C. Martius et al. (eds.), *Cotton, Water, Salts and Soums: Economic and Ecological Restructuring in Khorezm, Uzbekistan*, DOI 10.1007/978-94-007-1963-7_17,
© Springer Science+Business Media B.V. 2012

simulations also highlighted that the state procurement of cotton seems to indirectly mitigate excessive use of irrigation water and that without the state procurement system, scarcity and conflicts over irrigation water even in normal years are likely to occur. Double cropping of rice and maize as summer crops after wheat was constrained mainly by the availability of water and/or by the obligation of fulfilling the state procurement production of cotton. Despite the model being normative, the simulation results are very reasonable and thus enable a better understanding of the impacts of different cotton policies on the farm economy as well as on farmers' decisions with respect to land and water use in Khorezm. The results of this study can further contribute to the discussion on what policy options are available for promoting income and food resilience of rural producers in other areas of Uzbekistan that are prone to water scarcity, and with agronomic and economic conditions closely resembling those observed in the Khorezm region.

Keywords Bio-economic modeling • Farm-aggregate model • Natural resource use efficiency • CropSyst • Central Asia

17.1 Introduction

The current practice of agronomic land and resource management by small- and medium-scale farmers in Khorezm (Uzbekistan) is subject to a variety of environmental challenges. For instance, while the overall irrigation water use is excessively high, water use efficiency on system as well as on field level is notoriously low. Irrigation water use is highly unsustainable on a regional scale. The immense withdrawal of on average 5 km^3 water per annum for the Khorezm region alone from the Amudarya river has significantly contributed to the desiccation of the Aral Sea, bringing about a large-scale ecological catastrophe with severe impacts on human health and livelihoods (UNESCO 2000; Roll et al. 2005; Micklin 2007). Furthermore, the inefficient use of irrigation water triggers soil degradation by secondary salinization, negatively affecting the soil fertility of the desert soils of Khorezm (cf. Tischbein et al. 2011).

However, rural livelihoods in Khorezm mainly build on agriculture. The agricultural sector accounts for roughly 67% of the total regional GDP, 75% of the regional population lives in rural areas, and almost 40% of the labor force is employed in the agricultural sector (Djanibekov 2008). For various socio-economic reasons, and also because land and resource use is not optimal, income from agriculture is comparably low (cf. Djanibekov et al. 2011a). In particular, an economic-ecological restructuring of land and water use in the Khorezm region has considerable potential to improve farmers' incomes while at the same time maintaining the natural resource base.

To address this issue, the so-called integrated Farm-Level Economic-Ecological Optimization Model (FLEOM) was developed, a site-specific integrated model capable of optimizing land and resource use at the micro-scale, while at the same time assessing the respective economic and environmental impacts. FLEOM captures the basic features of the regional agriculture, as well as the interrelations of production activities most prevalent to the local farmers. It aggregates, integrates and optimizes field-level management decisions on the allocation of water, resources and labor, and reports the respective ecological and economic consequences. Additionally, FLEOM operates in a spatially explicit way. It furthermore relates farm-level decisions with constraints or (optimization) goals of networks at the next higher level, such as Water Users Associations (WUAs) or farmers' associations. To meet these requirements, the size of a target area for FLEOM lies in the range of a WUA of around 1,000 ha, but it can also handle individual farms of different sizes.

FLEOM is a tool for scenario analysis (what-if) and for presenting possibilities, potentials and constraints for alternative land management, and is tailored to the characteristics of the irrigated agriculture predominating in the Khorezm region. The overall objectives to develop FLEOM were to:

- understand options for optimal and sustainable land and resource allocation;
- explore options to increase the income of farms and farmers' associations while maintaining or even increasing crop and animal production with medium-term sustainable land management;
- assess opportunities to promote an efficient use of irrigation water;
- analyze the effects of various external "shocks" on farm income, crop and animal production, cropping pattern, water and resource use, and
- develop and suggest optimal land use under alternative environmental conditions, e.g., water scarcity, to stakeholders.

User friendliness was one of the priorities when developing FLEOM, with the aim of enabling potential users with a less comprehensive understanding of the particular agronomic or socio-economic background of the model details to use this tool, such as medium-level stakeholders (WUA representatives, local water authorities and extension agents). Besides, FLEOM was developed as a tool for scientists in the respective fields and for education at universities, for workshops, etc. A detailed description of characteristics of the FLEOM model is given in Sommer et al. (2010).

This chapter briefly describes FLEOM, and the results of four different management scenarios are presented and discussed in detail. The scenarios were set up in particular to demonstrate the possibility of simulating and optimizing land use with FLEOM on the basis of economic and agro-ecological land-use drivers or constraints.

17.2 Model Description

FLEOM comprises five sub-components (Fig. 17.1): a graphical user interface (FLEOM-GUI) programmed in Java, a GIS visualization component realized with the Open Systems Mapping Technology (OpenMap, http://openmap.bbn.com/), and an economic-ecological optimization routine written in GAMS. This optimization component is connected to a MS-Access agronomic database. The database was established with the cropping system simulation model, CropSyst (Stöckle et al. 2003), using data sets, field experience and knowledge of a range of agronomic and hydrological studies. At present, the datasets contain the four crops cotton, wheat, rice and maize.

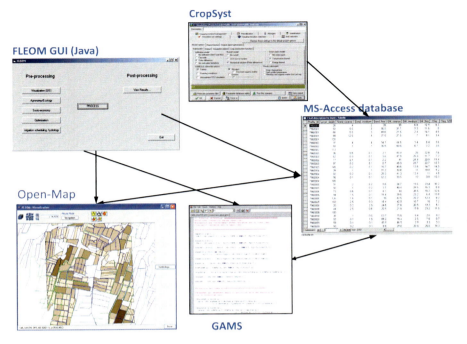

Fig. 17.1 Components of FLEOM and their links

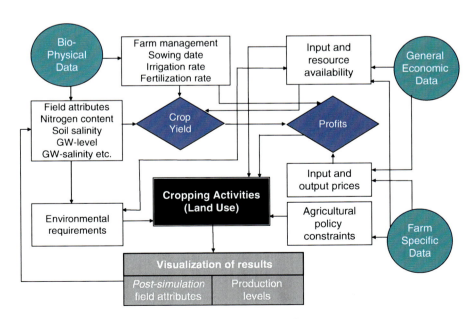

Fig. 17.2 Workflow of the optimization sub-routine of FLEOM

After setup and start of the optimization analysis, the GUI launches the optimization routine in GAMS (Fig. 17.2). This sub-model receives all necessary data from the databases. The GAMS routine itself is an economic farm-household linear-programming (LP) optimization routine developed to describe, simulate and optimize the farm-household economy in Khorezm. Settings and results of FLEOM are visualized in tables and figures or via the GIS environment.

17.3 Management Scenario Analysis

17.3.1 Management Scenario Description

Four different management scenarios (henceforth called scenarios 1 through 4) were simulated with FLEOM, each with different sets of assumptions relating to changes in socio-economic conditions:

1. "Business-as-usual"
2. "Commodity market liberalization"
3. "Ecological commodity market liberalization"
4. "Dry year"

In all four scenarios, the four crops cotton, wheat, rice and maize entered the simulations, and livestock production was also considered. For the sake of simplicity, identical endowments with respect to agricultural equipment and irrigation water for all modeled farms located inside the test WUA were assigned. However, due to the different sizes of the farms, the distinct geographic location within the WUA and differences in soils, the farms differed considerably. The basic framework conditions of the scenarios are summarized in Table 17.1.

Scenario 1 describes the situation under "business-as-usual" and is the base scenario. In 2009, the state-set raw cotton price of 370 UZS (=28 US cents at an exchange rate of 1 US\$ = 1,319 UZS) per kilogram was slightly below the world market price. By the order of the state, at least 55% of the area of each individual farm specializing in cotton production had to be allocated to cotton. Additionally, 1 ton of wheat and rice had to be sold at a price prescribed by the government. Total available water for irrigation was defined by multiplying the farm size by 18,000 m^3 ha^{-1}, which is the average volume of irrigation water in a normal year (Müller 2006).

Scenario 2 describes a situation where the commodity market has been partly liberalized under the assumption that the state order for cotton as well as for the mandatory delivery of a certain amount of wheat and rice to state agencies has been abolished. In this scenario, the price for raw cotton was raised to that of the world market (440 UZS kg^{-1} in 2007). Also, prices for cotton oil cake, wheat grain, cotton seed hulls, and cotton and wheat seed were increased by the same rate (19%). Since fertilizers under scenario 1 were subsidized by the state, under scenario 2, real market prices for fertilizers were used. According to the Commodity Exchange

Table 17.1 Costs, resource availability and assets of the modeled farms for four scenarios; 1 US$ = 1,319 UZS

Parameter	Scenario 1 Business-as-usual	Scenario 2 Market liberalization	Scenario 3 Eco-liberalization	Scenario 4 Dry year
Product prices (UZS kg^{-1})				
Raw cotton	370	440	440	370
Cottonseed cake	210	250	250	210
Wheat, grains	400	476	476	400
Wheat straw	70			
Rice, grains	900			
Rice straw	75			
Maize, grains	350			
Maize stem	80			
Cotton seed hulls	180	214	214	180
Fertilizer prices (UZS kg^{-1})				
Ammonium nitrate	200	230	230	200
Ammonium phosphate	405	527	527	405
Potassium chloride	290	377	377	290
Ammonium sulfate	180	207	207	180
Diesel price (UZS litre^{-1})	730			
Seed price (UZS kg^{-1})				
Cotton	970	1 154	1 154	970
Wheat	450	535	535	450
Rice	900			
Maize	350			
Inter-farm canal efficiency (0–1)[a]	0			
Field-canal efficiency (0–1)[a]	0			
Max. availability of irrigation water (m^3 ha^{-1})	18,000	18,000	18,000	5,400
Interest rate for credit (%)	25			
Water price (UZS m^{-3}) for cotton, wheat, rice, maize	0	0	20	0
Max. leaching of nitrate (kg N ha^{-1})	Unconstrained			
Max. seasonal increase in soil salinity (dS m^{-1})	Unconstrained			
Share of farm land allocated to cotton according to state procurement task (%)	55	0	0	55
Wheat to be sold according to state procurement task (tons)	1	0	0	1

(continued)

17 Economic-Ecological Optimization Model of Land and Resource Use...

Table 17.1 (continued)

Parameter	Scenario 1 Business-as-usual	Scenario 2 Market liberalization	Scenario 3 Eco-liberalization	Scenario 4 Dry year
Rice to be sold according to state procurement task (tons)	1	0	0	0
Money in farm (mln UZS)	100			
Credit available for farm (mln UZS)	200			
(Family) working hours available (h)	12,000			
Number of cows and calves	20			
Number of bulls	15			
Number of poultry	40			

[a]0 = unchanged, i.e., as current, 1 = lined, lossless canal

Table 17.2 Size of modeled farms in Pakhlavan-Makhmud Water User Association

Farm	Size (ha)
1	90
2	121
3	135
4	161
5	83
6	84
7	147

Office in Urgench, the market prices for ammonium nitrate, ammonium phosphate, potassium chloride, and ammonium sulfate were approximately 15%, 30%, 30%, and15%, respectively, above the state price.

Scenario 3 follows scenario 2, but is guided by the recurrently postulated view that unconstrained use of irrigation water that is free of charge causes an irrational use of irrigation water, thus a water price of 20 UZS (=1.5 US cents) per cubic meter of water is introduced. This price was repeatedly suggested during a workshop in Tashkent in 2007 by representatives of the Ministry of Agriculture and Water Resources (MAWR) of Uzbekistan.

Scenario 4 assesses potentially optimal production under conditions of water scarcity. Similar to the two pronounced dry years 2000/2001, actual water availability was reduced to 30% of the water available in a normal year such as in 2007.

The Pakhlavan-Makhmud Water User Association (PM-WUA) located in south-eastern Khorezm served as the test area for the simulations of various changes in farm environments (Table 17.2). A detailed biophysical classification of the predominating soils, groundwater levels, as well as field boundaries and irrigation network was available in the field survey of Akramkhanov (2005). In the survey year

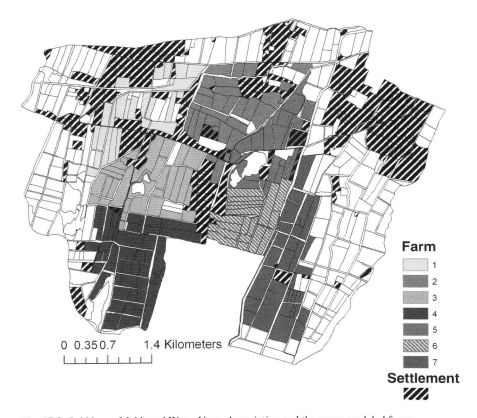

Fig. 17.3 Pakhlavan-Makhmud Water Users Association and the seven modeled farms

2005, the PM-WUA consisted of around 100 individual private farms. Recent reforms imposed by the Uzbek state led to a re-organization of landholdings and to a dissolution and subsequent aggregation of these small farms (cf. Djanibekov et al. 2011b; Veldwisch et al. 2011). Given the currently insecure land ownership (lease), but also to keep the FLEOM test applications simple and highlight the generic sub-regional land-use planning-tool character of FLEOM, the area of the PM-WUA (821 ha) was arbitrarily divided into seven modeled farms with 83–161 ha arable land (Table 17.2; Fig. 17.3).

17.3.2 Management Scenario Results

17.3.2.1 Scenario 1 "Business-as-Usual"

Given the state procurement task to produce cotton on 55% of the land in scenario 1, land allocation to cotton dominated in the PM-WUA in this simulation (Fig. 17.4a). However, cotton under these defined production conditions could not compete with

17 Economic-Ecological Optimization Model of Land and Resource Use... 275

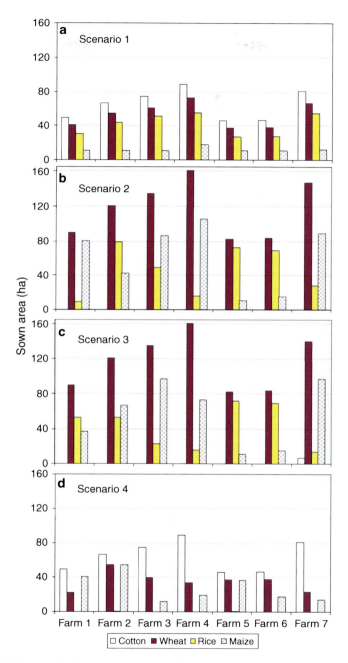

Fig. 17.4 Farm crop allocation in the four scenarios

wheat-rice or wheat-maize (double) cropping, and thus cotton was not grown above the level defined by the state procurement task. On the other hand, on all farms, all available land was cultivated – no field was set aside – and with 145%[1] the land-use rate was at a maximum. Rice, as the most profitable crop, occupied between 32% and 37% of the land of the model farms, and dominated as a summer crop in this scenario. Maize was produced for fodder for own livestock production and occupied 8–13% of the land.

Even though rice ranked third after cotton and wheat in area coverage in this scenario, total water use for this crop was highest (Fig. 17.5a). This was due to the fact that rice fields are kept continuously flooded during the cropping season until about one week before rice maturity, and the per-hectare water use is thus several times higher than for cotton, wheat or maize. An average irrigation water amount between 27,325 m^3 ha^{-1} (farm 5) and 39,212 m^3 ha^{-1} (farm 4) was applied to produce a rice yield of between 5.0 and 6.6 t ha^{-1}. Total irrigation water use for the other three crops followed overall crop abundance, with cotton production consuming more water than wheat or maize (Fig. 17.5a).

Fixed farm costs, which accounted for land tax, salaries of permanent labor, transportation expenditures, and costs for diesel (for field operations), were the type of expenditures that ranked highest in this scenario. These costs were followed by expenses for fertilizers and the payments to the bank for repaying credits and interest. The latter costs ranged between 5,222 US$ for the smallest farm 5 and 33,189 US$ for the largest farm 4.

Under the assumptions in scenario 1, total farm profits ranged between 94,729 US$ and 151,306 US$ (Fig. 17.6). Larger farms tended to make more profits than smaller farms. However, also the location, i.e., the differences in the natural resource base and distance to the water intake, influenced this trend. In fact, on a per-hectare basis, farms below 100 ha in size had the highest profit. This occurred as we had assumed that the considered input parameters (Table 17.1) 'money in farm', 'credit available for farm' and 'family working hours available' were the same for all farms. Smaller farms on an area basis thus had comparably more money and free labor available. As expected, the profits of 940 and 995 US$ ha^{-1} were lowest for farms 4 and 7, respectively. These were located in the south of the PM-WUA where the comparably less-productive sandy soils were most abundant (Fig. 17.2).

17.3.2.2 Scenario 2 "Commodity Market Liberalization"

Given its higher profitability, double cropping (wheat-rice or wheat-maize) as compared to growing cotton only dominated land use when state procurement was substituted by free marketing of the farm products, as was realized in scenario 2.

---

[1] The maximum land use rate is 145%, because 45% of the land (=100% – 55% for cotton) can be double cropped.

17 Economic-Ecological Optimization Model of Land and Resource Use... 277

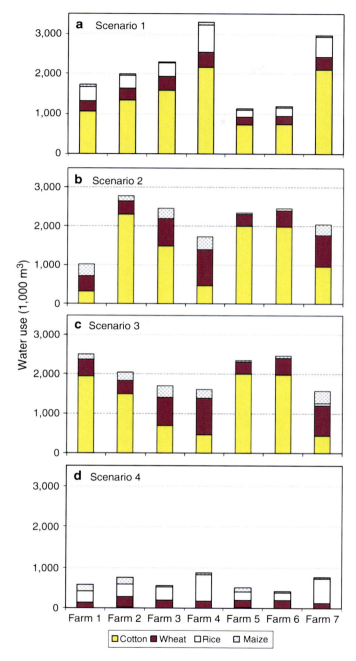

Fig. 17.5 Total water use by crops in the four scenarios

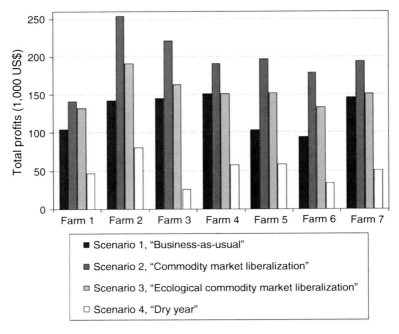

Fig. 17.6 Total farm profits in the four scenarios

Furthermore, under such conditions, cotton completely disappeared in the PM-WUA (Fig. 17.4b), and land-use rate was at, or close to, the maximum of 200%.

Even though rice was economically most profitable, its share dropped as compared to scenario 1 on some farms in favor of maize. This was due to the fact that the upper water consumption limit of 18,000 m^3 ha^{-1} had been reached when considering the PM-WUA as a whole. To meet this constraint, the optimization procedure reduced rice production on the least productive sandy fields.

Except for the farms where maize was the main summer crop, total water use increased under scenario 2 as compared to the business-as-usual scenario (Fig. 17.5b). The higher water use was related to the increased land use (double cropping) rather than the increased use of water per hectare and higher per hectare production. Rather the contrary was the case, i.e., the average maize yield of 4.9 t ha^{-1} had slightly decreased as compared to scenario 1 (5.1 t ha^{-1}). This indicates that spreading rather than concentrating crop production inputs was a more viable option under the assumptions of scenario 2.

Total profits were markedly higher on all farms under scenario 2 as compared to scenario 1 (Fig. 17.6). This meant that – in case water was provided free of charge – procurement tasks for cotton reduced farm profitability even when fertilizers were subsidized and provided below the market price. Farms 2, 3, 5 and 6 with comparatively more area under rice were able to achieve the highest increase in profits. Some farms had the highest profits per hectare of farmland. In this regard, differences between farms related to the productivity of land, i.e., less productive sandy soils in

17 Economic-Ecological Optimization Model of Land and Resource Use... 279

the south of the PM-WUA, and farm size were more marked than in scenario 1. For example, the farm profit per hectare of farmland of 2,385 US\$ ha^{-1} of farm 5 (smallest farm) was almost double that of farm 4 (largest farm located in the south) that could only achieve 1,189 US\$ ha^{-1}.

17.3.2.3 Scenario 3 "Ecological Commodity Market Liberalization"

If the Uzbek government were to abolish the state procurement tasks for cotton (as simulated in scenario 2), and thus waive the claim on the state income from cotton exports, it was assumed that the state would consequently have no financial means for the operation and maintenance of the irrigation and drainage network. Under these assumptions, these costs would have to be completely shouldered by the farmers. Alternatively, a potential state income from water pricing could account for this. Water pricing is often suggested as a possible steering instrument to promote the rational and more eco-friendly use of irrigation water (OECD 2011).

With a water price incurred on top of the settings of scenario 2, the land-use rate of farms 3, 4 and 7 decreased by 11%, 20% and 4%, respectively, whilst farms 1, 2, 5 and 6 maintained the maximum (200%) land-use rate. As an effect of the water price, the area under rice decreased on farms 2, 3 and 7. Relative crop allocation did not change notably on farms 4, 5 and 6. On farm 1, however, rice cultivation increased from 10 (scenario 2) to 53 ha. Here, the optimization rule applied that FLEOM optimizes total WUA profits using the common resource water as one constraining parameter. Reducing rice production on farms 2, 3 and 7 while increasing it on farm 1 is a particular result of this built-in optimization scheme. The optimal solution under scenario 3 also included the growth of cotton on 7 ha of land on farm 7.

Compared to scenario 2, the total water consumption under scenario 3 was reduced on farms 2, 3, 4 and 7 (Fig. 17.5). As more rice was cropped on farm 1, total water consumption increased. In total, for the whole PM-WUA, a water price of 1.5 US cents m^{-3} only triggered water saving to a maximum of 4% as compared to scenario 2.

Total farm profits decreased on all farms (Fig. 17.6), which had been expected, since additional production costs were imposed. However, profits were at least equal to or higher than under scenario 1, indicating that the increase in profits due to the abolishment of the state procurement task for cotton more than counterbalanced the increase in costs due to the introduction of a water price and higher fertilizer prices.

In total, 14,200,000 m^3 of irrigation water was used by the PM-WUA in scenario 3. Thus, the state income from payments for irrigation water (1.5 US cents per m^3) amounted to 213,000 US\$ for the PM-WUA with its 812 ha arable land.

17.3.2.4 Scenario 4 "Dry Year"

The land-use rate decreased considerably under conditions of water scarcity, as can be assumed to occur in a dry year, and this was simulated in scenario 4 (Fig. 17.4d). As expected, water scarcity mostly affected rice cultivation, but in total all double-cropping activities were reduced as compared to scenario 1. Rice was only produced in negligible quantities by farm 2 on 1 ha and by farm 5 on 0.35 ha. Similar to scenario 1, the area allocated to cotton did not exceed the area-percentage assigned to fulfill the state procurement task. The findings show that when keeping to this production target in dry years with an assumed reduction in water availability of 30% compared to a normal year, farm profits dropped substantially (Fig. 17.6). Compared to scenario 1, total water consumption was reduced by 55–76% (Fig. 17.5d).

17.3.3 Discussion

Double cropping of rice and maize as summer crops after wheat was the most profitable cropping strategy in all four scenarios. Double cropping was constrained mainly by the availability of water and/or by the obligation to fulfill the state procurement production of cotton. This is not surprising considering the higher market price for rice in Khorezm. Further simulations with varying market prices for rice, cotton and wheat and with varying prices for irrigation water demonstrated under which conditions cotton production would become lucrative for farmers.

A commodity market liberalization as simulated in scenario 2 has the potential to increase farmers' profits even when prices of fertilizers and seeds were increased to reflect unsubsidized market conditions. Average water use reached the specified maximum amount of 18,000 m^3 ha^{-1}, which would have been even higher if this constraint had been eliminated. The simulations of this scenario revealed that cotton production had no comparative advantages and that it would completely disappear in this scenario, and thus the state income from cotton exports. This indicates on the one hand that the state procurement of cotton seems to indirectly mitigate excessive use of irrigation water, and that without the state procurement system, scarcity and conflicts over irrigation water even in normal years are likely to occur. On the other hand, the findings of scenario 2 also demonstrate that, if the state remains responsible for the maintenance of the irrigation system as is currently the case, alternative state incomes would have to be generated. This demand could be satisfied by introducing a water price as is recurrently suggested (Rogers et al. 2002; OECD 2011).

In response to the introduction of a price for irrigation water (scenario 3), water consumption *per se* did not drop as remarkably as had been expected. This is very likely due to the low water price of 1.5 US cents m^{-3}, which is a too low incentive for a substantial reduction in irrigation water consumption as recently argued. Djanibekov (2008) estimated that an irrigation water price of at least 2.5 US cents m^{-3} could lead to a significant reduction of irrigation water use in Khorezm (see also cf. Djanibekov et al. 2011a). Further simulations that also take into account various application alternatives are necessary to demonstrate the dependency between a price for water and the impact on its (rational) use.

During the on-going debates about the effectiveness of water pricing, it has repeatedly been argued that a water price alone will not lead to the desired outcome of more rational water use (Dinar et al. 1997; Tsur et al. 2004), as the relationship between the two issues, water price and rational use of water, is more complex than can be expressed by a simple, linear water price – water use function. In this regard, simulations with FLEOM are limited if the constraining (potentially non-linear) framework conditions, e.g., a resilience-to-change behavior, are not clearly understood and introduced into the model. However, the FLEOM scenario results show clearly that under favorable settings – in our case an abolishment of state procurement tasks – farm grain-crop production levels and profits could be maintained at similar levels (compare scenario 1 and 3). This means that a water price does not necessarily threaten rural livelihoods as could be argued. Nevertheless, this instrument needs to be carefully designed, as under the present socio-economic farming conditions it may not lead to a strong reduction in water use.

The state income generated by introducing a water price is significant. For the PM-WUA alone, 21,000 US$ were generated in 1 year for an area of 821 ha arable land. Scaling up this income to the whole Khorezm region (assuming ca. 200,000 ha), annually more than 50,000,000 US$ could be set aside for the urgently needed repair and maintenance of the irrigation and drainage infrastructure. These investments would eventually trigger a boost in crop production by improved provision and more efficient use of irrigation water.

There are a few lessons to be learned from the results of the dry-year simulation in scenario 4. Maintaining the state procurement tasks for producing cotton on 55% of the land during a dry year should be re-assessed. Lowering this target would be an instrument to sustain farmers' incomes and livelihoods in such dry years. At the same time, this could be an efficient way of economizing labor and machinery, as taking land out of production using alternative crops with a lower water demand, and concentrating irrigation on less land during a dry year is more profitable, as would also be the case for harvesting more, i.e., better-irrigated cotton on less land.

As compared to yield ranges observed in reality (Shi et al. 2007), yields produced by FLEOM simulations did not show such large variations. For instance, raw-cotton yields were never below 1.95 t ha^{-1}, whereas in reality, especially in a dry year, much lower yields can be expected. At first sight, this seems to be a flaw of the model. However, FLEOM does not mimic real cropping patterns and yields, but rather optimizes land use in an economic-ecological way in response to the given constraints. It is therefore not surprising that yield variation under optimized conditions is comparatively low. The same characteristic is detectable when, for instance, temporal and spatial distribution of grain yields of developing countries are compared with those of developed countries where production conditions are close to optimal.

17.4 Scope for FLEOM as a Discussion and Decision Instrument

The presented evaluation of selected scenarios demonstrates that in response to making exogenous changes in the FLEOM specification, the model produces consistent and plausible outputs, and that it can thus already be used for quite complex scenario simulations. Increasing complexities in turn creates new challenges. Our first experience with the model showed that simulation results were sometimes difficult to comprehend at first sight, but that multi-disciplinary teams were able to understand causalities.

Great care was given to the detailed setup and description of the bio-physical settings for the multiple CropSyst simulation that built the basis of the input-output table. Also, bio-physical model calibration and validation received special attention. The importance of these underlying data sets cannot be overestimated, as all further integrated simulations rely on these.

The developed version of the model has a flexible and user-friendly interface and allows for one-click scenario formulation. However, up-to-datedness of integrated simulation models is a constant concern, especially when development of such a tool takes several years, as in our case. This is even more so true for the current situation of agriculture and farm enterprises in Uzbekistan, which underwent considerable institutional change in the past and most likely, will experience further changes in the near future (cf. Djanibekov et al. 2011b). Generally speaking, the current setup of FLEOM allows a multitude of changes via the user interface, and thus convenient updating of the agricultural system under study. Only a few rather 'hard coded' cases (e.g., maximum allowance of water use for a particular crop) might require changes in the GAMS code, and thus some basic understanding of this economic software.

Furthermore, equally important is FLEOM's in-built generic character that allows simulating any farming environment (single farm, farmers' association, WUA) as long as bio-physical data and maps are available. However, putting together all necessary data for a single WUA requires considerable time.

Apart from the total availability of irrigation water, FLEOM also allows setting limits to other environmental hazards, such as the pollution of groundwater by excessive nitrate leaching or the deterioration of land by secondary soil salinization. The four scenarios presented therefore only provide a first insight into the potentials of FLEOM for assessing the production systems of farms in a single WUA in Khorezm.

Another aspect is the educational potential of FLEOM, i.e., for gaining new theoretical insights into the interdependencies within the simulated agricultural system. For such purposes, the exact quantities of endogenous variables in a specific situation are not compulsory.

In conclusion, FLEOM has the potential to serve as a discussion- and decision-support instrument for medium-level, regional stakeholders with basic computer skills; computer novices on the other hand might be overwhelmed by the complexity of the software (above all the GIS environment). In any case, some training is advisable to (a) familiarize the users with the basic concept of FLEOM and to highlight its limitations, (b) to provide a steep learning curve and to maintain the enthusiasm, and (c) to be able to discuss results in a peer group. We believe that it is this group of users that may benefit directly the most from a "FLEOM experience", but they may also be in the position to "translate" FLEOM simulation results into more easily (non-computerized) recommendations for farmers.

References

Akramkhanov A (2005) The spatial distribution of soil salinity: detection and prediction. PhD dissertation, Bonn University, Bonn

Dinar A, Rosegrant MW, Meinzen-Dick R (1997) Water allocation mechanisms: principles and examples. World Bank: policy research working paper 1779, World Bank, Washington, DC

Djanibekov N (2008) A micro-economic analysis of farm restructuring in Khorezm region, Uzbekistan. PhD dissertation, Bonn University, Bonn

Djanibekov N, Bobojonov I, Djanibekov U (2011a) Prospects of agricultural water service fees in the irrigated drylands, downstream of Amudarya. In: Martius C, Rudenko I, Lamers JPA, Vlek PLG (eds) Cotton, water, salts and Soums – economic and ecological restructuring in Khorezm, Uzbekistan. Springer, Dordrecht/New York/Heidelberg/Berlin

Djanibekov N, Bobojonov I, Lamers JPA (2011b) Farm reform in Uzbekistan. In: Martius C, Rudenko I, Lamers JPA, Vlek PLG (eds) Cotton, water, salts and Soums – economic and ecological restructuring in Khorezm, Uzbekistan. Springer, Dordrecht/New York/Heidelberg/Berlin

Micklin P (2007) The Aral Sea disaster. Annu Rev Earth Planet Sci 35:47–72

Müller M (2006) A general equilibrium approach to modeling water and land use reforms in Uzbekistan. PhD dissertation, Rheinischen Friedrich-Wilhelms-Universitat Bonn, Bonn

OECD (2011) Water – the right price can encourage efficiency and investment. http://www.oecd.org/document/31/0,3746,en_2649_34285_45799583_1_1_1_1,00.html. Last accessed Jan 2011

Rogers P, de Silva R, Bhatia R (2002) Water is an economic good: how to use prices to promote equity, efficiency, and sustainability. Water Policy 4:1–17

Roll G, Alexeeva N, Aladin N, Plotnikov I, Sokolov V, Sarsembekov T, Micklin P (2005) Aral Sea: experiences and lessons learned brief. Lake basin management initiative. www.ilec.or.jp/eg/lbmi/reports/01_Aral_Sea_27February2006.pdf

Shi Z, Ruecker GR, Mueller M, Conrad C, Ibragimov N, Lamers JPA, Martius C, Strunz G, Dech S, Vlek PLG (2007) Modeling of cotton yields in the Amu Darya river floodplains of Uzbekistan integrating multi-temporal remote sensing and minimum field data. Int J Agron 99:1317–1326

Sommer R, Djanibekov N, Salaev O (2010) Model-based optimization of land and resource use at farm-aggregated level with the integrated model FLEOM – model description and first application. ZEF - discussion papers on development policy no. 139

Stöckle CO, Donatelli M, Nelson R (2003) CropSyst, a cropping systems simulation model. Eur J Agron 18:289–307

Tischbein B, Awan UK, Abdullaev I, Bobojonov I, Conrad C, Forkutsa I, Ibrakhimov M, Poluasheva G (2011) Water management in Khorezm: current situation and options for improvement (hydrological perspective). In: Martius C, Rudenko I, Lamers JPA, Vlek PLG (eds) Cotton, water, salts and Soums – economic and ecological restructuring in Khorezm, Uzbekistan. Springer, Dordrecht/New York/Heidelberg/Berlin

Tsur Y, Roe T, Doukkali R, Dinar A (2004) Pricing irrigation water: principles and cases from developing countries. Resources for the future, Washington, DC

UNESCO (2000) Water-related vision for the Aral Sea basin for the year 2025, 2000th edn. UNESCO Division of Water Science, Paris, p 238

Veldwisch GJ, Mollinga P, Zavgorodnyaya D, Yalchin R (2011) Politics of agricultural water management in Khorezm, Uzbekistan. In: Martius C, Rudenko I, Lamers JPA, Vlek PLG (eds) Cotton, water, salts and Soums – economic and ecological restructuring in Khorezm, Uzbekistan. Springer, Dordrecht/New York/Heidelberg/Berlin

Chapter 18
Water Patterns in the Landscape of Khorezm, Uzbekistan: A GIS Approach to Socio-Physical Research

Lisa Oberkircher, Anna Haubold, Christopher Martius, and Tillmann K. Buttschardt

Abstract A method is presented that uses a Geographic Information System (GIS) for integrating social science with physiographic data. Basis for the method is an empirical study on water use in irrigated agriculture in the Khorezm region. The boundaries of agricultural fields are used to link physiographic land characteristics such as proximity to the Amudarya river, land elevation etc. to survey answers of farmers, thereby explaining patterns of water access and water-saving practices. The results are based on (1) a statistical analysis across three water user associations (WUAs) and (2) visual observations within WUAs. The statistical analysis shows that proximity to the river influences water access and water-saving practices. One WUA thereby represents a specific tail-end situation within Khorezm's irrigation system with an exceptional prevalence of water-saving practices. In the visual analysis, the location of farms within the WUA and the land elevation are used to explain patterns of the survey answers. Despite data and method constraints and despite sometimes contrasting results from the statistical and visual analyses, it is concluded that the method has three advantages as compared to disciplinary, non-spatial approaches: (1) interdisciplinary data management is possible, including the identification of data gaps and consistency checks, (2) immediate data visualization facilitates quick monitoring of research processes and delivers starting points for disciplinary in-depth analyses, and (3) interdisciplinary data integration allows for

---

L. Oberkircher (✉) • T.K. Buttschardt
Working Group Applied Landscape Ecology/Ecological Planning,
Institute of Landscape Ecology, Robert-Koch-Str. 28, 48149 Münster, Germany
e-mail: lisa.oberkircher@uni-bonn.de; tillmann.buttschardt@uni-muenster.de

A. Haubold • C. Martius
Center for Development Research (ZEF), Walter-Flex-Str. 3, 53113 Bonn, Germany
e-mail: ahaubold@uni-bonn.de; gcmartius@gmail.com

C. Martius et al. (eds.), *Cotton, Water, Salts and Soums: Economic and Ecological Restructuring in Khorezm, Uzbekistan*, DOI 10.1007/978-94-007-1963-7_18,
© Springer Science+Business Media B.V. 2012

truly socio-physical analyses. The analytical depth reached in the empirical study is thereby primarily a result of the quality and quantity of the collected data. Considerations concerning the protection of informants determine which kind of analyses can be conducted and published.

Keywords Interdisciplinary research • Geographic information system (GIS) • Socio-physical analysis • Water access • Water-saving

18.1 Introduction

In landscape studies, humans are often conceptualized as external disturbances, and the rationales of their practices are not included in the analyses. Social science, on the other hand, is typically not undertaken in spatially explicit ways, which is a shortcoming when social aspects of land and water management need to be monitored and assessed, e.g., in the context of combating land degradation (Buenemann et al. 2011), which is the case in coupled human-ecological systems (Reynolds et al. 2007). This chapter presents a method[1] that uses a Geographic Information System (GIS) for the integration of social science with physiographic data, and thus uses GIS as a tool for interdisciplinary landscape research. The method uses the boundaries of agricultural fields to link physiographic land characteristics to survey answers of farmers.

The basis is an empirical study on irrigation and water-saving practices that was conducted under the ZEF/UNESCO project on land and water use in the Khorezm region in Uzbekistan. Khorezm is located approximately 350 km south of the present boundaries of the drying Aral Sea. Its irrigation system receives water from one of the (now intermittent) tributaries of the Aral Sea, the Amudarya river. Agriculture demands over 90% of the water resources in Khorezm (cf. Tischbein et al. 2011) with water use being technically very inefficient.[2] Table 18.1 gives an overview of the basic characteristics of agriculture in Khorezm.

Technical studies show that water-saving practices could increase the irrigation water use efficiency in Uzbekistan significantly (e.g. Horst et al. 2005; Paluasheva 2005; Forkutsa et al. 2009; Hornidge et al. 2011). However, very few of these practices have been adopted by farmers in Khorezm. The empirical study presented

---

[1] We recognize that the use of this term may raise some controversies, as natural scientists would tend to see this as an application of social-science data within the general method of analysing spatial data with Geographical Information Systems, rather than a method in itself. However, due to the interdisciplinary approach of this paper, the argument is partly rooted in social science methodology and thinking and we hence see it as seminal to enrich the methodology discussion in social research.

[2] An overall technical efficiency of approx. 30% and field application efficiency in the range of 45% were measured in a sub-unit of the irrigation system in Khorezm (Hornidge et al. 2011).

18 Water Patterns in the Landscape of Khorezm, Uzbekistan…

Table 18.1 Basic characteristics of agriculture in Khorezm (Based on Hornidge et al. 2011)

Hydrology	Precipitation approx. 90 mm, annual average potential evapotranspiration approx. 1,500 mm, shallow groundwater levels averaging 1.61 ± 0.51 m over the region and the period 1990–2006 (cf. Tischbein et al. 2011)
Soils	Medium and heavy loam soil textures prevailing, clayey light loam and sandy loam soils less widespread, approx. 50% of soils moderately or strongly saline (Ibrakhimov et al. 2007)
Crops	Main crops in summer are cotton and rice (the latter either sown directly or transplanted onto fields previously cropped with wheat); furthermore fruits, vegetables and fodder crops (e.g., maize) are grown.
Irrigation system	Canal system length of 16,233 km (Conrad 2006) reaching from the Amudarya to the borders of Turkmenistan and Karakalpakstan, irrigation water pumped or diverted by gravity from canals; drainage system of 9,255 km length (open ditches and collectors) (Ibrakhimov 2004)
Farm units	After independence in 1991, state farms (*sovkhozes*) were turned into collective farms (*kolkhozes*), then into joint-stock companies (*shirkats*, literally associations), by 2006 largely dismantled and divided into private farms (Veldwisch 2008), in 2008 consolidated to usually >80 ha for cotton and wheat producing farms and 5 ha for horticultural farms. Farms consist of land owned by and leased from the state and are headed by an individual farmer who – usually together with other household members – runs the farm
Planned economy	Area- and production-based yield quota for the state-ordered crops cotton and wheat; compulsory sale to the state at fixed prices for part of the harvest; agricultural norms to regulate cropping patterns and agricultural practices, norm compliance monitored and enforced (Müller 2006; Wehrheim and Martius 2008)

hereafter aims to shed light on the underlying rationales of water saving in Khorezm, i.e., on who adopts practices, which practices, and why. The focus is on the interrelation of physiography, perceptions and behavior.

18.2 Research Approach

18.2.1 GIS as Boundary Object

In this study, we conceptualize the arena of the water use activities of farmers as a landscape. The concept landscape emphasizes three characteristics that apply at any given moment: (1) a spatial pattern, (2) a combination of interlinked physiographic and human/societal elements, and (3) the simultaneous existence of a bio-physical reality and a social construct based on human perception processes. A landscape view hence has the advantage of facilitating both a spatial analysis and an integration of natural and social science perspectives.

288

L. Oberkircher et al.

Table 18.2 Natural and social science perspectives and GIS-supported boundary crossing

	Natural sciences	Social sciences
Physiographic landscape elements	Physiographic elements are analyzed with regard to their spatial pattern, their temporal dynamics and their interaction	Physiographic elements are considered the material setting around human interaction, the source of resources or the carrier of symbolic meaning. Spatial pattern is rarely considered in the analysis
Human/societal landscape elements	Human/societal elements are considered external to the analysis. Human actors are discussed as disturbance of the physiography and as recipients of the normative results of the analysis	Human/societal elements are analyzed with regard to cognitive patterns, values, practices and interactions
Added value of a GIS-supported boundary crossing	Spatial observations remain possible while human landscape elements become part of the system of analysis. Associations between physiographic and human landscape elements can be investigated. The analysis takes into account human/societal factors when deriving normative results	The material setting becomes part of the system of analysis, i.e., context becomes analytical content. Physiography as the source of resources or carrier of symbolic meaning can be analyzed in a more differentiated way and with regard to its spatial pattern. The impact of human practices on the physiography can be investigated

In the sense of Löwy (1992) and Mollinga (2008), landscape functions as a boundary concept: "Boundary concepts are words that function as concepts in different disciplines or perspectives, refer to the same object, phenomenon, process or quality of these, but carry (sometimes very) different meanings in those different disciplines or perspectives. In other words, they are different abstractions from the same 'thing'" (Mollinga 2008:24). While boundary concepts make communication between disciplines possible, so-called boundary objects permit the practical integration of data. They are the actual instruments for joint analyses. In this study, we use a GIS as such a boundary object, thereby translating the landscape into a set of GIS layers containing both physiographic as well as social science data.

Table 18.2 shows the departure points of the natural and the social science perspectives, respectively, with regard to the physiographic and human landscape elements. It furthermore presents the added value that we envision by a GIS-supported crossing of disciplinary boundaries.

18.2.2 Previous Interdisciplinary GIS Studies

While the relationship between human activities and environmental systems remains little understood, global challenges of a socio-physical character (e.g. climate change) have stimulated new approaches in the last decade, particularly in the field of natural resource management. In this process, GIS has increasingly found its way into interdisciplinary studies that aim at incorporating the aspect of space into analyses. Besides technological progress, this also reflects the rising engagement of social theory with the concept of space (Soja 2003), which has brought social science closer to traditionally spatial disciplines such as environmental science.

Wahid et al. (2008) argue that the spatial interplay of environmental and socioeconomic predictors determines land degradation in the Lam Phra Phloeng watershed of Thailand and therefore included village surveys into their study. Their approach, however, applies regression models while GIS is not used, and space thus is only implicitly incorporated. Likewise, the study of Hoffman and Todd (2000) on land degradation in South Africa is interdisciplinary with regard to the considered data, but does not provide an explicit spatial analysis. Le Blanc and Perez (2008) explore the regional differences in population density and rainfall variation in Sub-Saharan Africa in order to assess future water stress. The study combines GIS data on population density and rainfall with climate change scenarios, but does not integrate survey data in the sense of individual interviews. Also on a macro-level, Huby et al. (2007) integrated socio-economic and environmental data to characterize different rural environments in England.

Messina and Walsh (2005) use socioeconomic and demographic survey data, GIS coverage of resource endowments and geographic accessibility, remote sensing data, and data derived from expert interviews to explore the causes and consequences of land-use and land-cover change in the Amazon region of Ecuador. The study does not include interviews of individual land users. Wright et al. (2009) conducted research on GIS approaches of natural resource management associated with public participation in the USA. Their study provides an interesting insight into the power structures created or decomposed by the use of GIS but does not deal with the integration of social and physical data.

Hence, previous approaches either (1) do not make space explicit in a way that allows visual analyses, (2) remain on a macro level without emphasis on individual natural resource users, or (3) do not truly integrate data in an interdisciplinary way. Hence, the here presented methodology is innovative due to the fact that it allows the interdisciplinary study of local interrelations between physiography and individual natural resource users.

18.2.3 Objectives of the Study and Hypotheses

Method development. The primary objective is the method development. The process of interdisciplinary data integration, the data requirements for a successful joint analysis, and the strengths and weaknesses of the proposed method are to be investigated through its application in an empirical study. Irrigation water use in Khorezm was hereby chosen as case study because the interaction between the physical and the social sphere are extremely relevant, differing data reliability allows for the detection of data requirements, and qualitative data previously collected by the first author in the study area could be used as background knowledge for the analysis.

Empirical study. The empirical study itself has a separate objective. Previous studies in Khorezm have concluded that to develop adequate water-saving strategies, it is important to take into account the heterogeneity of farmers with regard to the physiography of their land, because the physiographic differences influence both perceptions and water use practices of farmers. How and to which extent this influence has an effect, has not been investigated in detail. The objective of the empirical study is to shed light on this question

Based on previous studies (Veldwisch 2008; Oberkircher 2011; Oberkircher and Hornidge 2011) the following hypotheses are formulated:

1. *WUA proximity to the river*: Farmers in WUAs closer to the river have better water access and apply fewer water-saving practices.
2. *Location of farms within WUAs.* Since irrigation water arrives through a centralized canal network, farmers with fields closer to the head of the canal have better water access and apply fewer water-saving practices.
3. *Land elevation.* While low land is less fertile, it is much easier to get water there (Veldwisch 2008). Water can often be received by gravity and pumps are not necessary or pumping costs are at least lower than for high land. Farmers with lower land have better water access and apply fewer water-saving practices.
4. *Orthogonal distance from the main canal*: Instead of using the already constructed field canals, farmers often dig small canals and divert water to their fields through temporary constructions particularly when water is scarce (Oberkircher and Hornidge 2011). This is only possible when land is located close to the main canals. Hence, farmers with land closer to the main canal have better water access and apply less water-saving practices.

18.3 Methods

18.3.1 Empirical Data Collection

The empirical study uses the landscapes of three water user associations as cases (WUAs, i.e., non-governmental, local level water management organizations): WUA 3 is located in the Kushkupir district at the tail end of the irrigation system at the border to Turkmenistan, WUA 1 in Gurlen district directly on the western bank of the Amudarya, and WUA 2 in medium proximity to the river (Khiva district).[3] The three WUAs are comparable with regard to overall agricultural system, regional culture, and selection of crops that is available. However, the proximity of each WUA to the river differs, which also causes differences with regard to water availability and living standards. Regarding the latter two characteristics, WUA 1 is most advantaged, followed by WUA 2 and the disadvantaged WUA 3.

---

[3] WUA (and farmer) names are not mentioned in the article due to anonymity concerns.

Table 18.3 Physiographic data and their sources

Data	GIS Format	Source and processing
Canals	Polyline	Secondary data from the state water management organizations BUIS and UPRADIK, verified and corrected with the help of Spot 5 satellite images
Field boundaries	Polygon	Secondary data from the Land Cadastre branch of Uzgiprozem,[4] verified and corrected with the help of Spot 5 satellite images
Farm boundaries	Polygon	Primary data collected with the help of the local land surveyor (*zemlemer*) who drew the farm boundaries on the existing field boundary map
Elevation	Raster	SRTM (Shuttle Radar Topography Mission) remote sensing data

Primary data collection in the WUAs took place between March and December 2009. Only farmers (Uzbek *fermers*) with a state-plan arrangement for cotton (main summer crop in Khorezm) participated in the study. Livestock, orchard, fishery, and silkworm fermers were not considered, since they use comparatively little water; peasants (Uzbek *dehqons*) were also not included because a representative survey of their perceptions and practices was beyond the scope of this study.

18.3.2 *Physiographic Data Collection*

The physiographic data collected for this study include the location of canals, field and farm boundaries as well as information on land elevation (Table 18.3).

18.3.3 *Social Science Data Collection*

The social science data were collected through a farmer survey covering topics such as irrigation satisfaction, water availability perception and farmer practices (Table 18.4). The expression 'drainage blocking' in one of the survey questions refers to the practice of raising groundwater levels by closing drains completely or partly with earth or weirs that prevent drainage water run-off.

---

[4]Uzbekskiy Institut po Proektirovaniyu Zemleustroystva (State Institute for Land Management Planning).

Table 18.4 Survey questions and their rationale

Topic	Rationale	Survey questions
Irrigation satisfaction	Irrigation satisfaction represents the actual water access that farmers had in 2009, i.e., the result of the combined use of all the assets they had at their disposal to get water	Did you irrigate as much as you wanted this year (2009) (Yes/No)
Water availability perception	2008 was the most water scarce year in Khorezm since (at least) independence[5] and water became abundant again only in late spring/early summer of 2009 with water availability then rising above average throughout 2009. How these differences in water availability were perceived by farmers is covered by questions on the water situation in spring and summer 2009. The difference between these questions and irrigation satisfaction is an emphasis on perception as opposed to actual water access	*Spring 2009 assessment*: How would you characterize the water situation this spring (2009)? (More than usual/Usual/Less than usual but enough/Less than usual and not enough) *Summer 2009 assessment*: How would you characterize the water situation this summer (2009)? (More than usual/Usual/Less than usual but enough/Less than usual and not enough)
Farmer practices	Oberkircher (2011) studied the response of farmers to the water scarcity prevailing in 2008. The survey questions reflect the practices she observed	*Rice cropping 2009*: Did you grow rice in 2009? (Yes/No) *Drainage blocking 2009*: Did you block the drainage in 2009? (Yes/No) *Drainage water use 2009*: Did you use drainage water in 2009? (Yes/No) *Changed cropping pattern*: When you cannot get water, do you change the crops you grow? (Yes/No) *Reduced cropping area*: When you cannot get water, do you reduce the size of the cropped area? (Yes/No)

'Drainage water use' refers to the practice of re-using drainage water for irrigation (usually by pumping it from the drain back into the canal system).

In WUA 2, five farmers were excluded from the survey because information on the location of their land was not available; the remaining 23 farmers were interviewed. All farmers of WUA 1 and 3 were interviewed (21 each). In sum, the sample consists of 65 cases.

---

[5] Based on overall physical water availability, the drought years 2000/2001 could be considered even water scarcer. However, in the perception of the Khorezmian farmers (as stated in interviews) 2008 was clearly the most water scarce year.

18   Water Patterns in the Landscape of Khorezm, Uzbekistan...

Table 18.5 Physiographic GIS layers

Layer	Method of creation
WUA proximity to the river	As mentioned in Sect. 18.2.3, proximity to the river is the characteristic that shapes overall differences between WUAs (proximity to the river here meaning proximity to the closest irrigation water intake along the river). All farmers were categorized based on the WUA they belong to. Farmers in WUA 1 were given the rank 1 for proximity to the river, likewise all farmers of WUA 2 and 3 the rank 2 and 3 respectively
Location along the canal within WUA	This raster represents the head- and tail-end gradient within each WUA based on the actual distance from the entrance of the canal into the WUA. Distance is here not measured in the sense of the straight-line distance but is instead based on the distance the water has to travel through the canal network to every point within the WUA. A point shape was created for each WUA with the distance from the water source assigned as attribute to each point. A kriging interpolation was applied to derive a raster that covers all points within the WUA
Land elevation	Elevation was represented by a SRTM digital elevation model dataset
Orthogonal distance from the main canal(s)	Buffers were created around the main canals in intervals of 50 m, then clipped to the boundaries of the WUAs. Fifty meter was chosen as most suitable interval based on run-times of the calculation and desired precision

18.3.4   Setting up the GIS

Physiographic layers. Table 18.5 gives an overview of the physiographic information captured in four layers of the GIS and the method of their creation.

Farm unit. The central unit of analysis is the polygons that represent the farm boundaries. Through this unit it was possible to reference all physiographic information to a farmer as a person. Figure 18.1a–c shows the farm boundaries within the three WUAs. Each letter-number combination symbolizes one farmer (who may also crop land that is not connected but consists of separate farm parts, i.e., fields, and therefore polygons); land represented by one shade/symbol belongs to the same farm.

The three continuous physiography rasters were converted into the farm unit by a zonal statistics tool and a raster-to-polygon conversion to attach a single value to the farm unit for each physiographic attribute, i.e., the mean of the input layer area covered by the polygon. The result was a polygon shape file that consisted of the farm boundaries and one attribute column each for location along the canal, orthogonal distance from the canal and land elevation. In addition to this, the rank of proximity to the river was incorporated.

Survey answers. The resulting shape file then contained all information on the independent variables of analysis. The dependent variables, i.e., the survey answers with regard to irrigation satisfaction, water availability perception and farmer practices, were inserted into the same layer, which then stored all data in one shape file.

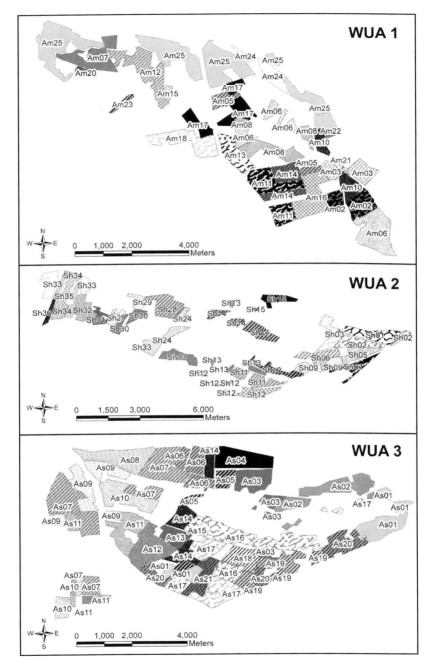

Fig. 18.1 Farm boundaries within the three case study WUAs

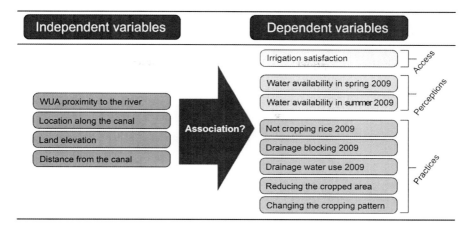

Fig. 18.2 Summary of the resulting variables incorporated into the statistical analysis

18.3.5 Analysis

All dependent and independent variables (Fig. 18.2) were exported from GIS, and a logistic regression analysis was conducted across WUAs. The following constraints of the statistical analysis have to be kept in mind when interpreting the results:

- *Small sample size.* For a logistic regression, the sample size n should be around 300 (Long and Freese 2006) to detect small to medium effects. With our sample of 65 farmers, we might consequently fail to reach significance or, equivalently, get estimates where even the direction (positive or negative) might not be clearly identified. Technically, the small sample limits the number of independent variables that can be incorporated into the model. As a minimum ratio, ten cases per variable should be reached (Rothman et al. 2008:422), i.e., six variables would be the maximum to use in a generalized linear model for this study. The logistic model was therefore simplified as much as possible. Four independent variables were used and even the dependent variables on water availability perception that would have required a multinomial regression were reformulated into binary variables.
- *Confounders.* Since this study is not based on an experiment but on real-life situations, the influence of unknown confounders can never be fully estimated.
- *Differences between WUAs.* Since the three WUAs were analyzed jointly, the variables of analysis had to be comparable between WUAs. The physiographic variables were therefore normalized by recoding them into respective values within a [0;1] interval for each WUA. The underlying rationale is that, for instance, in the case of land elevation not the absolute value is relevant but the relative value within the WUA with reference to the height of the canal entry point to the WUA.

296   L. Oberkircher et al.

Figure 18.2 summarizes the resulting variables that were incorporated into the statistical analysis.

In addition to the statistical analysis, maps were exported from the GIS that represent different attributes of layers. These were used for a visual analysis of local phenomena within each WUA.

18.4 Results from the Empirical Study and Discussion

18.4.1 Statistical Analysis of Phenomena Across WUAs

Table 18.6 shows variables that were identified as significant (and relevant) in the statistical analysis. As expected, in several regressions the variable 'WUA proximity to the river' showed that the further away from the river, the worse the water situation was perceived and the more practices in response to water scarcity were observed.

In one regression, the variable 'location within the WUA' had a significant effect. However, the direction of the effect was contrary to what could have been expected and what was observed during the field research. Namely, according to the statistical analysis, the further the farms were from the head of the canal, the smaller was the number of farmers reducing their cropping area in the case of water scarcity.

Table 18.6 Results of logistic regressions

	WUA proximity to the river	Location along the canal within WUA	Land elevation	Orthogonal distance from the main canal
Irrigation satisfaction				
Water availability in spring 2009	Significant at the 5% level ($p=0.000$)			
Water availability in summer 2009				
Rice cropping	Significant at the 5% level ($p=0.000$)			
Drainage blocking 2009				
Drainage water use 2009	Significant at the 5% level ($p=0.034$)			
Reducing the cropping area	Significant at the 5% level ($p=0.001$)	Significant at the 5% level ($p=0.025$)		
Changing the cropping pattern				

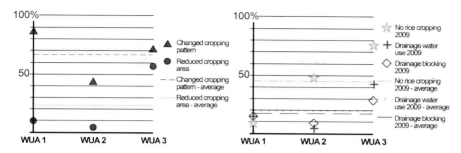

Fig. 18.3 Percentage of farmers who change their cropping pattern/reduce their cropping area in response to water scarcity and percentage of farmers who did not crop rice/used drainage water/blocked the drainage in 2009

There are three possible explanations for this: (1) the result is a random peculiarity of the farmers of the three WUAs revealed in the analysis because the population is too small to eliminate the phenomenon, (2) there is an unknown confounder that influences both the dependent variables and the location within the WUAs, or (3) farms at the tail end of a WUA in fact have better water access. To be able to clarify this, the first two options should be eliminated by increasing the sample size and incorporating further variables into the analysis in a follow-up study.

Farmer practices. Since the findings with the limited number of parameters show that the WUA proximity to the river plays an important role with regard to farmer practices, these were analyzed again regarding their occurrence in each WUA (Fig. 18.3).

Rice growing. Overall, 55% of the sample farmers grew rice in 2009. However, there is a strong gradient between the WUAs. WUA 1 is a rice growing area (91%) while rice is more of an exception in WUA 3 (24%).

Drainage blocking and drainage water use. Drainage blocking and drainage water use are practiced by 20% and 17%, respectively, of the farmers overall, but the majority of these are from WUA 3 (29% and 43%, respectively). Previous surveys show that both practices are (correctly) considered bad for the soil by farmers in Khorezm and only applied when water is scarce or its supply unreliable,[6] which is often the case at the tail end of the irrigation system represented by WUA 3.

Changed cropping pattern and reduced cropping area. 66% of the farmers overall change their cropping pattern when they expect not to get sufficient water at the right time. However, this is more so the case in WUAs 1 (86%) and 3 (71%). The former is a rice growing area with rice often being grown on land exempted from the state quota system. On this so-called free area, farmers themselves choose the crops they want to grow and can therefore react to changes in water availability, which they do particularly for rice that bears the risk of drying out quickly with respect to the way it is presently cultivated in Khorezm (as paddy rice).

---

[6]This statement is based on previous qualitative research conducted in the study area.

In WUA 3, little rice is grown, but farmers are used to having to adapt to water scarcity, which occurs regularly in the downstream regions of Khorezm. In 2008, people were observed to grow maize and sorghum on their free areas, since these crops need less water than rice and many vegetables. The location of WUA 3 at the very tail end of the irrigation system is also acknowledged as such by the authorities, and even land under the state plan may sometimes be dealt with in a more flexible manner than elsewhere. During the water scarce year of 2008, farmers had to leave land fallow, even though some of it was under the state quota. Previous findings indicate that farmers invited the authorities to judge the situation of water scarcity in their WUA, thereby aiming to have their practices tolerated because of the extreme tail-end situation.[7] This is also reflected in the survey answer on reducing the cropping area, for which WUA 3 scored very high (57%) compared to WUA 1 (10%) and WUA 2 (4%).

In general, all practices that function as a response to water scarcity (including growing less rice) were observed most in WUA 3 and less so in the other WUAs (apart from changing the cropping pattern in WUA 1, which is explained above as a particularity of rice cropping). Farmers of WUA 2 were found to apply the practices in response to water scarcity least frequently – however only slightly less frequently than those of WUA 1. The association between proximity to the river and water-saving practices is thus not a linear one. Instead, there appears to be a boundary somewhere between WUA 2 and WUA 3 where a particular tail-end situation starts and practices in response to water scarcity are more frequent. Other WUAs at the very tail end as well as between WUA 2 and WUA 3 should be investigated to determine where exactly this change takes place, and whether there are specific reasons for this (e.g., a hydraulic structure or a WUA that frequently exceeds its official share of water). Conceptually, this would also allow a more precise definition of what characterizes a tail-end situation in Khorezm.

18.4.2 Visual Analysis of Phenomena Within WUAs

Irrigation satisfaction. As expected, farmers in WUA 1 near the river show a generally high irrigation satisfaction of 86% without a physiographic pattern within the WUA. In WUA 2, there is no clearly visible pattern of irrigation satisfaction within the WUA either, and satisfaction is overall lower (70%).

In WUA 3 at the tail end of the canal system, the question on irrigation satisfaction yielded a distinguishable pattern of answers, visible when presented in the form of a map (Fig. 18.4a). The majority of farmers (67%) were satisfied with their water access. However, two groups of farmers were not, i.e., one at the entrance and another in the south-west of the WUA. It seems surprising at first sight that farms located at the very head of the canal should have had problems with water access. However, the land elevation map (Fig. 18.4b) shows that the farms in this part of the

---

[7]This statement is based on previous qualitative research conducted in the study area.

18 Water Patterns in the Landscape of Khorezm, Uzbekistan...

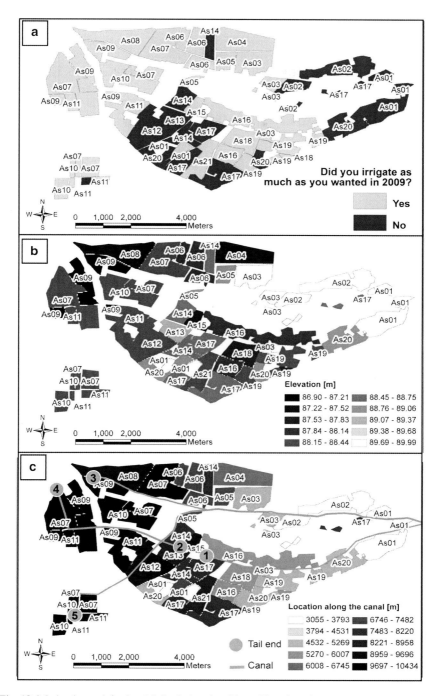

Fig. 18.4 Irrigation satisfaction (**a**), land elevation (**b**), and location along the canal (**c**) in WUA 3

WUA are relatively high in comparison with the rest of the WUA. Farmers therefore rely on pumping for irrigation. It was frequently observed during the field research in WUA 3 that the electricity necessary for pumping was deliberately turned off in response to requests of tail-end farmers who wanted the water to reach their fields. Farmers in the middle reaches of the canal, who were able to irrigate their (lower-lying) land by gravity, hence had better water access than those at the head of the canal, who could only irrigate when electricity was provided. The survey answers reflect this situation.

The second group of dissatisfied farmers is located around two of the five local tail ends of the WUA (Fig. 18.4). This area is not further away from the head of the canal than the other three tail ends. However, qualitative observations revealed that there is a difference with regard to the socio-political capital that farmers can make use of. It is likely that particularly one farmer, who holds a very large land lease of the WUA and used to be in an influential position before decollectivization, was able to use his influence to get water to the three tail ends where he holds land. Hence, the two other tail ends with little socio-political capital were disadvantaged. Previous qualitative studies indicate that socio-political capital is an influential factor in water management besides land physiography. The example of WUA 3 shows that it would be useful for further analyses to quantify socio-political capital in subsequent studies.

Water availability perception. It has already been mentioned that water availability in Khorezm was scarce in spring 2009 but abundant in summer 2009. At the tail end of the irrigation system, in WUA 3, this overall trend is reflected in the survey answers, while there are little local differences. Regarding the spring season, everybody stated that there was less water than usual, and only 10% of the farmers thought there was nevertheless enough water. In summer, in contrast, everybody perceived water to be enough, and the majority (67%) stated that there was more water than usual.

In WUA 1, the difference between the spring and summer assessment is similarly pronounced. However, for summer 2009 (Fig. 18.5a), WUA 1 also shows a physiographic pattern of answers, with land elevation (Fig. 18.5b) influencing the perception of the situation. While this may be related to similar pump and electricity issues as in WUA 3, high land elevation also leads to relatively lower groundwater levels. Since groundwater contributes significantly to crop development in Khorezm (Forkutsa et al. 2009), high land requires more water, and water availability may hence have been perceived as usual rather than particularly abundant, because groundwater levels may have lagged behind with regard to water abundance, thus reflecting the drought situation slightly longer than the canal water supply. This shows that it is important to include separate questions on actual water availability and the perception of it into surveys because perceptions are influenced by many more aspects than mere factual water availability.

In the case of WUA 2, there is no clear pattern of answers. However, there seem to be two rather distinct regions within the WUA where people tend to have better access to water (which is revealed in the answers on irrigation satisfaction, water availability spring 2009 and water availability summer 2009) (Fig. 18.6). One of

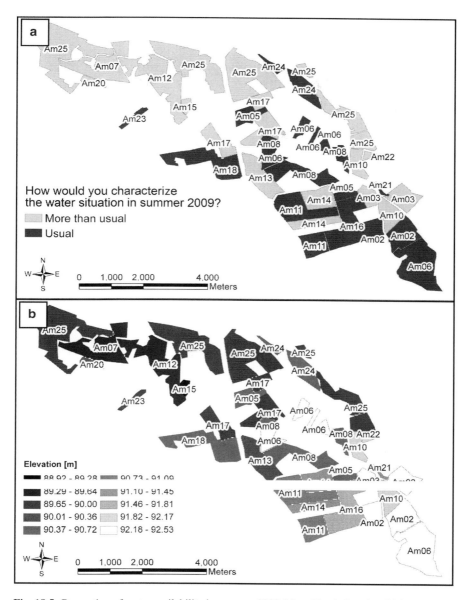

Fig. 18.5 Perception of water availability in summer 2009 (**a**) and land elevation (**b**) in WUA 1

these regions is located at the entrance of the WUA and one at the tail end of the canal, where land elevation is particularly low. However, there are also other areas in the WUA with these characteristics that do not reflect the same water access in the survey answers. This cannot be explained by the data at hand, but further details could be clarified with, for instance, qualitative research.

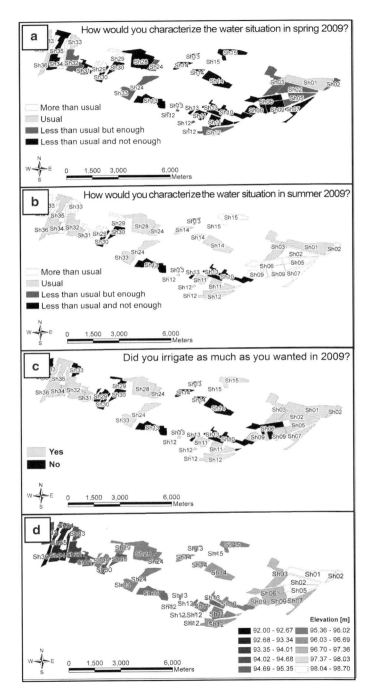

Fig. 18.6 Perception of water availability in spring (**a**) and summer 2009 (**b**), irrigation satisfaction 2009 (**c**), and land elevation (**d**) in WUA 2

18 Water Patterns in the Landscape of Khorezm, Uzbekistan...

Orthogonal distance from the canal. Orthogonal distance from the canal was not found to influence perceptions or practices of farmers – neither in the statistical analysis across WUAs nor in the visual analysis within WUAs. That Oberkircher and Hornidge (2011) found this variable to influence practices may be due to the fact that *dehqon* farmers were included in their analysis. For the much smaller peasant plots, self-made canals are probably much more relevant than for the larger land of farmers that this study deals with.

Farmer practices. The statistical analysis of farmer practices across WUAs revealed the influence of the WUA proximity to the river on farmer practices. The visual analysis on the local level did not yield any further results. This is most likely due to the influence of further physiographic variables that were not included in this analysis. Particularly with regard to drainage water and drainage blocking, it is likely that the proximity to drains and the soil texture influence farmers' practices. In WUA 1, for instance, only three farmers used drainage water in 2009. For one of them we know that his land is located directly next to a central drain of the WUA. Collecting reliable data on soil texture and the location of drains was beyond the scope of this study, but should be considered for further studies applying the here presented methods.

18.4.3 Data and Method Limitations

Farm unit. Although the farm unit is a fundamentally important part of the analysis, its characteristics pose certain methodological constraints. Several farmers hold land in different locations within one WUA, but the characteristics of their land as well as the survey answers were incorporated into the analysis in the form of a mean. Leaving room for differentiated answers of survey questions based on different parts of the farm would improve the depth of the analysis. A more fundamental question is whether a farmer with his/her personal biography can be well represented by his/her farm unit. In the case of farmers who have held their land for a long time, their perceptions and practices will certainly be connected to experiences with the land. However, farmers who may have worked on different land in the past may still perceive and act on these earlier experiences, which is difficult to capture with the method presented here unless biographic questions are included into the survey.

Data reliability. Figure 18.7 shows scatter plots of the area of farms based on the GIS boundary calculations and the area stated as total irrigated area by farmers in the survey (excluding farmer Am25 from WUA 1 and As07 from WUA 3 since they hold very large areas, which would skew the figure and hamper its readability). We had expected to find the GIS area to be larger than the total irrigated area, since part of the land within the GIS farm boundaries is left fallow every year, particularly in WUA 3 where land is only put under the state plan when water availability is sufficient (hence a rather large difference between the GIS and survey area in the case of WUA 3). Unexpected was an underestimation of the irrigated area in the GIS as is the case for several farmers in WUA 1 (and one outlier in WUA 2).

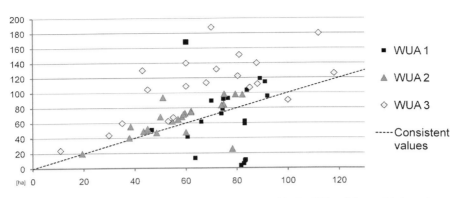

Fig. 18.7 Differences between the farm area as represented in the GIS and the total irrigated area as stated by farmers in the survey

To determine the reason for this, we compared the official boundary of WUA 1 with satellite images and investigated the fields in the eastern part of the WUA further. We found that the official data on farm boundaries as well as the WUA boundary were not correct, and therefore several fields in the east of the WUA close to the river were missing. It is highly likely that these fields are cropped by the five farmers that are underestimated in the GIS area, which may have disguised local phenomena within WUA 1 that could have otherwise been detected in the visual analysis. For the statistical analysis across WUAs, which mainly showed differences due to the variable proximity to the river, this error is unlikely to have caused incorrect results.

Anonymity and protection of informants. One extremely important consideration when applying the method presented here is the protection of the informants who participate in the survey. As soon as the visual analysis is made transparent by publishing maps, the answer of each informant in the survey can be immediately traced to him/her based on the location and shape of his/her land, even if codes are used instead of names. This has general implications for the kind of analyses that can be conducted and is even more problematic in a politically sensitive context such as agriculture in Uzbekistan. In this study, for instance, maps that show who cropped rice were not displayed due to the sensitivity of this topic.

18.5 Conclusions

The method presented here is able to integrate survey data and physiographic data for a joint spatial analysis. For the case study, it provided insights into the influence of physiography on perceptions and practices and yielded the following results with regard to the hypotheses formulated at the onset of the study:

1. *WUA proximity to the river*: Farmers in WUAs closer to the river indeed apply more water-saving practices, albeit there is not a linear association for all

practices. Instead, WUA 3 seems to represent a specific tail-end situation that increases the prevalence of water-saving practices. The perception of water availability in spring 2009 (a water-scarce time) was also significantly associated with proximity to the river.

2. *Location of farms within WUAs.* The hypothesis was not confirmed in the statistical analysis. To investigate this further, the sample size needs to be increased and confounders such as, for instance, socio-political capital need to be incorporated into the analysis. It may also be necessary to disaggregate these analyzes to field instead of farm level. The visual analysis suggested that farm location (in the sense of mean of the field locations) within the WUA can have an influence on water access (cf. WUA 3).

3. *Land elevation.* The visual analysis resulted in three examples that confirm the hypothesis. The statistical analysis did not identify land elevation as significant variable, which could have been caused by the limited sample size. Previous research in the study region indicated that the elevation of cropland determines water availability and in turn water use (Ibrakhimov 2004).

4. *Orthogonal distance from the main canal*: Neither the statistical analysis nor the visual analysis confirmed the hypothesis.

Since the results of the visual analysis make sense for the respective cases but were not found to consistently hold true in all three WUAs, further research should be conducted to gain a better understanding of the presented phenomena. It is recommended to include further physiographic data into the analysis such as soil texture, soil salinity, and information on the drainage system. Developing ways to include social aspects such as socio-political capital would also be very relevant albeit challenging with regard to quantification.

The GIS approach to socio-physical landscape research presented here has the following advantages as compared to non-spatial and disciplinary approaches:

Data management. Both physiographic and survey data can be stored jointly with a spatial reference. Data can be easily retraced and data gaps identified. Data from different sources can be compared with regard to consistency (e.g., as shown for the farm area above).

Description and exploration. The possibility of creating maps with survey data allows an immediate visualization of the collected data. This makes descriptions based on visual observations possible across data formats and delivers starting points for further research by highlighting patterns among the investigated farmer sample and the physiographic characteristics of the land.

Analysis. Both visual analysis and the statistical analysis based on an export of GIS data to statistics software allow a joint analysis of social and physiographic phenomena and particularly of their interrelations. In this sense, GIS can function as a boundary object for interdisciplinary research. The analytical depth that can be reached depends primarily on the quality of the collected data and the sample size of the survey. The protection of informants determines which analyses can be conducted (and published).

References

Buenemann M, Martius C, Jones JW, Herrmann SM, Klein D, Mulligan M, Reed MS, Winslow M, Washington-Allen RA, Lal R, Ojima D (2011) Integrative geospatial approaches for the comprehensive monitoring and assessment of land management sustainability: rationale, potentials, and characteristics. Land Degrad Dev 21:1–14

Conrad C (2006) Remote sensing based modeling and hydrological measurements to assess the agricultural water use in the Khorezm region, Uzbekistan. PhD dissertation, University of Wuerzburg, Wuerzburg

Forkutsa I, Sommer R, Shirokova YI, Lamers JPA, Kienzler K, Tischbein B, Martius C, Vlek PLG (2009) Modelling irrigated cotton with shallow groundwater in the Aral Sea Basin of Uzbekistan: I. Water dynamics. Irrig Sci 27(4):331–346

Hoffman M, Todd TS (2000) A national review of land degradation in South Africa. The influence of biophysical and socio-economic factors. J South Afr Stud Spec Issue: Afr Environ: Past Present 26(4):743–758

Hornidge A, Oberkircher L, Tischbein B, Schorcht G, Bhaduri A, Manschadi AM (2011) Reconceptualizing water management in Khorezm, Uzbekistan. Nat Resour Forum 35(4)

Horst MG, Shamutalov SS, Pereira LS, Gonçalves JM (2005) Field assessment of the water saving potential with furrow irrigation in Fergana, Aral Sea Basin. Agric Water Manage 77:210–231

Huby M, Owen A, Cinderby S (2007) Reconciling socio-economic and environmental data in a GIS context: an example from rural England. Appl Geogr 27(1):1–13

Ibrakhimov M (2004) Spatial and temporal dynamics of groundwater table and salinity in Khorezm (Aral Sea Basin), Uzbekistan. PhD dissertation, Bonn University, Bonn

Ibrakhimov M, Khamzina A, Forkutsa I, Paluasheva G, Lamers JPA, Tischbein B, Vlek PLG, Martius C (2007) Groundwater table and salinity. Spatial and temporal distribution and influence on soil salinization in Khorezm region (Uzbekistan, Aral Sea Basin). Irrig Drain Syst 21(3–4):219–236

Le Blanc D, Perez R (2008) The relationship between rainfall and human density and its implications for future water stress in Sub-Saharan Africa. Ecol Econ 66(2–3):319–336

Löwy I (1992) The strength of loose concepts - boundary concepts, federative experimental strategies and disciplinary growth: the case of immunology. Hist Sci 30(90):371–396

Long JS, Freese J (2006) Regression models for categorical dependent variables using stata, 2nd edn. Stata Press, College Station

Messina JP, Walsh SJ (2005) Dynamic spatial simulation modeling of the population – environment matrix in the ecuadorian Amazon. Environ Plan B: Plan Des 32(6):835–856

Mollinga PP (2008) The rational organisation of dissent. Boundary concepts, boundary objects and boundary settings in the interdisciplinary study of natural resources management. ZEF working paper series no. 33. Center for Development Reserch, Bonn

Müller M (2006) A general equilibrium approach to modeling water and land use reforms in Uzbekistan. PhD dissertation, Rheinischen Friedrich-Wilhelms-Universitat Bonn, Bonn

Oberkircher L (2011) On pumps and paradigms. Water scarcity and technology adoption in Uzbekistan. Soc Nat Resour 24(12)

Oberkircher L, Hornidge AK (2011) "Water is Life" - Farmer rationales and Water-Saving in Khorezm, Uzbekistan - A lifeworld analysis. Rural Socio 76(3)

Paluasheva G (2005) Dynamics of soil saline regime depending on irrigation technology in conditions of Khorezm oasis. Scientific support as a factor for Sustainable Development of Water Management, Taraz

Reynolds JF, Stafford-Smith DM, Walker B, Lambin EF, Turner BL, Mortimore M, Batterbury SPJ, Downing TE, Dowlatabadi H, Fernandez RJ, Herrick JE, Huber-Sannwald E, Jiang H, Leemans R, Lynam T, Maestre FT, Ayarza M (2007) Global desertification: building a science for dryland development. Science 316:847–851

Rothman KJ, Greenland S, Lash TL (2008) Modern epidemiology, 3rd edn. Lippincott-Williams-Wilkins, Philadelphia

Soja EW (2003) Postmodern geographies: the reassertion of space in critical social theory. Verso Press, London

Tischbein B, Awan UK, Abdullaev I, Bobojonov I, Conrad C, Forkutsa I, Ibrakhimov M, Poluasheva G (2011) Water management in Khorezm: current situation and options for improvement (hydrological perspective). In: Martius C, Rudenko I, Lamers JPA, Vlek PLG (eds) Cotton, water, salts and Soums – economic and ecological restructuring in Khorezm, Uzbekistan. Springer, Dordrecht/New York/Berlin/Heidelberg

Veldwisch GJ (2008) Cotton, rice & water: transformation of agrarian relations, irrigation technology and water distribution in Khorezm, Uzbekistan. PhD dissertation, Bonn University, Bonn

Wahid SM, Babel MS, Gupta AD, Routray JK, Clemente RS (2008) Degradation-environment-society spiral: a spatial auto-logistic model in Thailand. Nat Resour Forum 32:290–304

Wehrheim P, Martius C (2008) Farmers, cotton, water and models: introduction and overview. In: Wehrheim P, Schoeller-Schletter A, Martius C (eds) Continuity and change Land and water use reforms in rural Uzbekistan. Socio-economic and legal analyses for the region Khorezm, Studies on the agricultural and food sector in Central and Eastern Europe. 43. Leibniz Institute of Agricultural Development in Central and Eastern Europe (IAMO), Halle, pp 1–16

Wright DJ, Duncan SL, Lach D (2009) Social power and GIS technology: a review and assessment of approaches for natural resource management. Ann Assoc Am Geogr 99(2):254–272

Chapter 19
Modeling Irrigation Scheduling Under Shallow Groundwater Conditions as a Tool for an Integrated Management of Surface and Groundwater Resources

Usman Khalid Awan, Bernhard Tischbein, Pulatbay Kamalov, Christopher Martius, and Mohsin Hafeez

Abstract To restructure land- and irrigation-water use in Khorezm towards sustainability and economical feasibility, the current water use demands improvement. This requires increasing water use efficiency as much as possible, while at the same time minimizing negative impacts on the production system. These objectives can be reached with an integrated management of the irrigation and drainage system. To develop optimal management strategies, models describing the water distribution (irrigation scheduling model) and analyzing the impact on the groundwater (groundwater models) will be very helpful. In the Water Users Association (WUA) Shomakhulum, located in the southwest of Khorezm and with an irrigated area of approximately 2,000 ha, current irrigation strategies were monitored. Overall irrigation efficiency of the sub-unit representing the WUA is rather low (33%). Besides the poor state of

---

U.K. Awan (✉)
Center for Development Research (ZEF), Walter-Flex-Str. 3, Bonn 53113, Germany

University of Agriculture, Faisalabad, Pakistan
e-mail: ukawan@uni-bonn.de

B. Tischbein • C. Martius
Center for Development Research (ZEF), Walter-Flex-Str. 3, Bonn 53113, Germany
e-mail: tischbein@uni-bonn.de; gcmartius@gmail.com

P. Kamalov
Urgench State University, Khamid Olimjan str., 14, Urgench 220100, Uzbekistan
e-mail: kamolpulat@mail.ru

M. Hafeez
International Centre of Water for Food Security, Charles Sturt University, Locked Bag 588, Building 24, Wagga, NSW 2678, Australia
e-mail: mhafeez@csu.edu.au

C. Martius et al. (eds.), *Cotton, Water, Salts and Soums: Economic and Ecological Restructuring in Khorezm, Uzbekistan*, DOI 10.1007/978-94-007-1963-7_19,
© Springer Science+Business Media B.V. 2012

the irrigation infrastructure, major reasons for the low efficiency are on the one hand a lack of detailed and up-to-date information on the system and its temporal behavior, and on the other hand missing options to consider detailed information in the procedures to establish water distribution plans. To tackle these issues, the irrigation scheduling model FAO CROPWAT was applied as an alternative to the current rather rigid water distribution planning. Feeding the model with detailed information on the irrigation system and its behavior (application efficiency by field-water balancing, network efficiency based on ponding experiments) provided a powerful tool to improve water use. As the groundwater in Shomakhulum is shallow, the model was further developed in order to assess the importance of the capillary rise. Therefore, the soil-water model Hydrus-1D was applied. The results of the study show that capillary rise is an important factor in water balancing and can contribute a maximum of 28% of crop-specific evapotranspiration in cotton, 12% in vegetables and 9% in winter wheat. In-practice irrigation scheduling, when simulated and assessed with the CROPWAT model, showed a 7–42% reduction in cotton yield. If the overall irrigation efficiency is improved to 56%, water saving of 41% can be achieved. Introducing alternative crops to cotton can result in 6% water saving. About 15–20% of the water can be saved by leaving marginal lands, i.e., land of low quality, out of the production.

Keywords Irrigation scheduling • CROPWAT • HYDRUS • Water saving

19.1 Introduction

During the past few decades, world-wide competition for water among different users has increased manifold and is expected to increase further. The development of new resources is not economically and environmentally viable. Therefore, raising the efficiency of the use of the existing water resources will play a key role to meet the increasing demands of water.

The majority of irrigation schemes around the world operate with a low overall efficiency of about 30% (Sarma and Rao 1997). Often, inappropriate system design causes low overall efficiency, but even with proper design, adequate management is essential for the effective operation and maintenance of irrigation water delivery systems. Worldwide evidence shows that significant improvements are to be gained through irrigation scheduling (e.g. Malano et al. 1999). Although yield and seasonal evapotranspiration (ET) relationships have been widely used for the management of water resources, the effect of timing of water application is also of key importance (Hanks 1983; Vaux and Pruitt 1983; Howell 1990). Irrigation scheduling should define when and how much to irrigate a crop. Given the complexity, several computerized simulation models to support and improve irrigation scheduling were developed (Kincaid and Heermann 1974; Smith 1992; Mateos et al. 2002). For the amount and timing of water application, these models use soil water accounting over the root zone, whereas crop water production functions, which relate crop yield to the amount of water applied, are often used to optimize on-farm water application.

Unfortunately, most of the irrigation scheduling models cannot compute/incorporate the groundwater contribution, which is an important parameter for water budgeting under shallow groundwater conditions, particularly in arid and semi-arid regions. Ayars and Schoneman (1986) concluded that groundwater can contribute up to 37% of the total evapotranspiration (ET) of cotton under a water table depth of 1.7–2.1 m, and Wallender et al. (1979) reported that cotton can extract even 60% of its ET from a saline (electrical conductivity (EC) = 6.0 dS m^{-1}) shallow water table. Pratharpar and Qureshi (1998) observed that in areas with shallow water tables, the irrigation requirements can be reduced to 80% of the total crop ET without reducing crop yield.

The recent developments in modeling enable simulation of the capillary rise under different agro-climatic conditions. These models simulate the water flow in the unsaturated zone (Simunek et al. 1998; Hammel et al. 2000; Costantini et al. 2002; Mandal et al. 2002; Vrugt and Bouten 2002; Srinivasulu et al. 2004; Vanderborght et al. 2005). However, the selection of the model depends on the purpose for which the model is elaborated (Bastiaanssen et al. 2004), whilst the accuracy of the results depends on the assumptions and simplifications made in the model and their relation to site-specific conditions, and also on the accuracy of the input data (Zavattaro and Grignani 2001).

The present research was conducted in the Khorezm region of Uzbekistan, which is located at the downstream reach of the Amudarya river, and is seen as a model region for the irrigated Amudarya Lowlands (Martius et al. 2009). Virtually all the soils in the region are saline to a different degree, and at least 48% were moderately and highly saline as of 2006 (Ibrakhimov et al. 2007). Soil salinization in Khorezm is primarily a consequence of the shallow saline groundwater, since the regional average groundwater tables range from 1.0 to 1.2 m below the surface during leaching and irrigation events (Ibrakhimov et al. 2007). It is postulated that groundwater tables become shallow due to substantial losses from the irrigation network, which is enhanced by deficits of the drainage system and by the regional flatness and subsequent absence of a regional lateral groundwater flow. Despite the fact that groundwater levels are shallow, water insufficiency is a pressing issue in the downstream areas of the region, for which all water is tapped from the Amudarya river. During years with average and low discharge in the Amudarya river, the middle and tail ends of the irrigation network usually suffer from insufficient amounts and timing of the water supply.

Present water distribution guidelines originate from the 1960s following the widespread development of areas for irrigated agriculture in the Aral Sea Basin (Sadikov 1979; Rakhimbaev et al. 1992). In these guidelines, water requirements of major crops such as cotton were elaborated for different agro-hydrological, climatic and ecological areas, but the contribution of shallow groundwater to soil moisture enhancement is not accounted for in detail. This is a one of the reasons for the excessive use of irrigation water.

The current irrigation strategies based on the conventional norms were therefore monitored and compared with the results obtained from an integrated water management (IWM) tool. The capillary rise obtained from the HYDRUS-1D model (Simunek et al. 1998) was introduced to the irrigation scheduling model. The FAO CROPWAT program (Clarke et al. 1998) was selected as a baseline model for this study. Procedures for calculating crop water demands and hence irrigation requirements are based on methodologies presented in various FAO Irrigation and Drainage Papers (Doorenbos and Pruitt 1977; Doorenbos and Kassam 1979; Allen et al. 1998). Advantages of the model are (1) the flexibility to react to changing situations, (2) an option for developing site-specific solutions, and (3) the ability to compile scenarios. The HYDRUS-1D model was selected to calculate the capillary rise. The model includes sophisticated and useful tools to simulate soil water content. Moreover, the water fluxes (on which then the soil water contents are based) in HYDRUS are based on the Richard's equation. HYDRUS has been frequently applied in the recent past (Ventrella et al. 2000; Hernandez 2001; Bitterlich et al. 2004; Meiwirth and Mermoud 2004; Sommer et al. 2004; van der Grift et al. 2004; Vanderborght et al. 2005) but not intensively in the irrigated areas of the Aral Sea Basin.

19.2 Materials and Methods

19.2.1 Study Site

The Khorezm region, located in the northwest of Uzbekistan, is a lowland with elevations ranging from 112 to 138 m a.s.l.. It is part of the desert region of Central Asia and has an arid continental climate (Suslov 1961). Monthly-averaged daily maximum temperatures exceed 33°C and low relative humidity (48.1%) occurs between May and August. The annual long-term average precipitation of 90–100 mm does not have a significant influence on surface water or groundwater. The potential evapotranspiration (annually ca. 1,150 mm) rises from April until a peak in July.

The Shomakhulum WUA, located in the southwest of Khorezm, was chosen to assess the water use practices, since its environmental conditions are representative for the irrigated areas in Khorezm with respect to groundwater table and salinity, soil characteristics, cropping patterns and climate. A further reason for this selection was the fact that the WUA had been chosen by the agricultural management authorities of Uzbekistan in 2005 for testing the introduction of water pricing, and excellent collaboration could be established with the WUA management, which facilitated the field research. In addition, the Shomakhulum WUA has distinct hydrological boundaries, which facilitate water monitoring and accounting. Soils are predominantly loamy to sandy loamy according to the USDA classification (Knight 1937).

19.2.2 Irrigation Scheduling Model (CROPWAT)

The FAO CROPWAT program was used to model optimal irrigation schedules and to analyze options for water saving and needs in water deficit situations. The optimal use of water for irrigation requires an accurate estimate of crop water requirements against the available water supply from irrigation network, rainfall and capillary rise.

19.2.2.1 Crop Water Use

Crop water use is commonly expressed in terms of crop evapotranspiration (ETc), which was computed here using the FAO CROPWAT model. The required weather data to calculate the reference evapotranspiration (ETo) were collected with an automatic weather station installed in the WUA. ETo was converted to crop ETc using crop physiological parameters (development stages and coefficients, rooting depths, depletion fractions), which were obtained from the Central Asian Scientific Research Institute of Irrigation (SANIIRI), Uzbekistan.

19.2.2.2 Capillary Rise (Hydrus-1D)

Capillary rise is dynamic in nature due to its dependence upon various physical parameters, such as crops, soils, groundwater levels, and groundwater salinity. To allow the spatial dynamics of groundwater levels and the soil characteristics to be represented – factors that determine capillary rise and in turn water balance and then recharge – the WUA in this study was divided into smaller spatial units for which these characteristics were uniform (hydrological response units, HRUs). For the formation of HRUs, the soil texture data from five different profiles at different depths (0–30, 30–60, 60–90, 90–120 and 120–150 cm) was collated from secondary sources (ZEF/UNESCO GIS center in Urgench). Out of these five soil profiles, four had a dominantly silt loam soil texture, whereas the fifth had a silty clay loam. Thiessen polygons were drawn using a GIS to consider spatial distribution of the soil texture attributes (Fig. 19.1). The advantage of using Thiessen polygons was twofold: (1) Interpolation techniques such as Kriging (Cressie 1992) and Inverse Distance Weighted (Nalder and Wein 1998) create the representative value of the parameter considered whereas Thiessen polygons creates the polygon, which was the prerequisite for creating the HRUs, and (2) The Thiessen polygon is the only interpolation technique which works with sparse data sets – a limitation in the study region due to sparse data sets on soil texture and groundwater levels.

Groundwater level data for the 15 groundwater observation wells in the WUA were collated from *TEZIM* (the irrigation system authority). Using the Thiessen polygon method in a GIS environment, the WUA was divided into polygons on the basis of 15 observation wells (Fig. 19.1). Polygons were drawn using the 15 wells

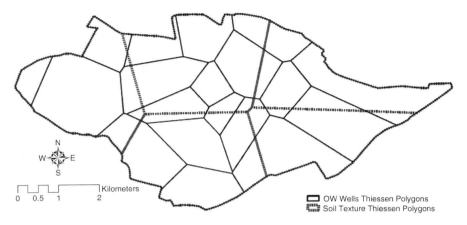

Fig. 19.1 Thiessen polygons for soil texture and groundwater levels in the WUA Shomakhulum

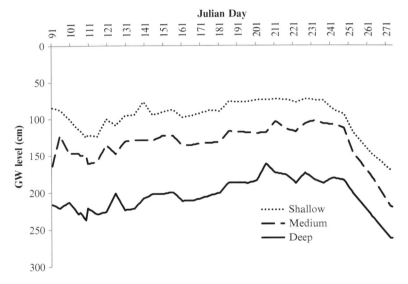

Fig. 19.2 Average daily values of groundwater level (*GW*) for the shallow, medium and deep groundwater level zones in the Shomakhulum WUA for year 2007

as the center point for each polygon, and the groundwater level measured at the well is considered representative for the area of land included in the respective polygon.

The groundwater levels in the polygons were classified as shallow (0–100 cm), medium (100–150 cm) and deep (more than 150 cm) during the peak irrigation season (June–August) (Fig. 19.2). The temporal behavior of the groundwater level in the classes defined above follows the same characteristic pattern.

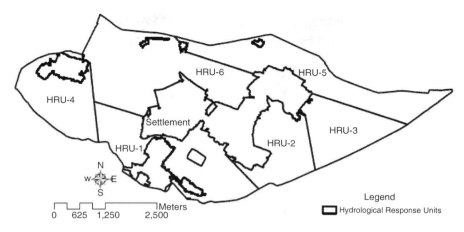

Fig. 19.3 Hydrological response units in the WUA Shomakhulum

Table 19.1 Characteristics of hydrological response units in the WUA Shomakhulum

Hydrological response unit	Soil texture	Groundwater level	Resulting HRU type
HRU-1	Silt loam	Shallow	S-SL
HRU-2	Silt loam	Shallow	S-SL
HRU-3	Silt loam	Shallow	S-SL
HRU-4	Silty clay loam	Deep	D-SCL
HRU-5	Silt loam	Medium	M-SL
HRU-6	Silt loam	Medium	M-SL

Thiessen polygons around the groundwater observation wells were overlaid on the soil texture polygons in a GIS environment to formulate the HRUs (Fig. 19.3). Each HRU represents the same class of groundwater level and soil texture value (Table 19.1). The category of the HRUs with shallow groundwater levels and silt loam soils is named S-SL, that of the HRUs with medium groundwater levels and silt loam soil texture is named M-SL, and that of the HRUs with deep groundwater levels and silty clay loam soil texture is named D-SCL. The cropping pattern in each HRU is described in the following section.

19.2.2.3 Initial and Boundary Conditions

Water flow parameters were taken from a database provided by HYDRUS; parameter selection from the database was guided by the information on the soil conditions in the WUA (silt loam, silt clay loam). The capillary rise was observed from a node equal to the maximum rooting depth for that specific crop restricted by the groundwater condition.

For the bottom boundary conditions, the groundwater data collated from the WUA were used. Potential evaporation (E) and transpiration (T) for dominant crops (cotton, wheat and vegetables; represented by their first letter in the matrix below) in each polygon were calculated by the FAO dual crop coefficient approach (Allen et al. 1998). ETc (and irrigation and rainfall) was then used as the upper boundary condition. Subsequently, the capillary rise was calculated using the following matrix:

$$
\begin{array}{cc}
\text{Crop} & \text{HRU} \\
\left\{ \begin{array}{c} C \\ W \\ V \end{array} \right. & \left. \begin{array}{c} S - SL \\ M - SL \\ \end{array} \right\}
\end{array}
\qquad
\begin{array}{cc}
\text{Crop} & \text{HRU} \\
\left\{ \begin{array}{c} C \\ W \\ V \end{array} \right. & \left. \begin{array}{c} D - SCL \\ \\ \end{array} \right\}
\end{array}
$$

$$\text{(a)} \qquad\qquad\qquad \text{(b)}$$

19.2.3 Scenarios by Integrated Modeling Approach

The main interest in using models results from their ability to simulate alternative irrigation scenarios relative to different levels of allowed crop water stress and to various constraints in water availability (Pereira et al. 2009). Different scenarios were analyzed in this study and are presented below.

Optimal irrigation schedule and the irrigation schedule in practice: Improvement in irrigation efficiency under the typical Khorezmian conditions can lead to lower groundwater recharge and hence lower groundwater table; this leads to lower capillary rise and in turn to changed (higher) net irrigation depths (quota). The optimal irrigation schedule investigated in this study not only refers to the hydrological response units (HRUs) but also to improvements of the field application (FAR) and conveyance (CR) ratios to reach the target values (Table 19.2). Target values of FAR are taken from Bos and Nugteren (1974), and those of CR are taken from Jurriens et al. (2001). Overall irrigation efficiency by definition is the product of CR and FAR. The reference situation is based on the product of the current values of FAR (0.43) and CR (0.76). This scenario is the business-as-usual (BAU) scenario, and represents low irrigation efficiency. In Scenario B, the target value of CR (0.84) was multiplied by the current value of FAR (0.43). In this scenario, the improvement in the irrigation efficiency is caused by increasing the CR. However, this improvement is not so marked, as the CR is already close to the target value. In scenario C, the target value of FAR (0.67) was multiplied by the current value of CR (0.76). This scenario shows that an improvement in FAR can significantly improve the irrigation efficiency. Scenario D is the product of the target values of FAR and CR. This scenario represents the maximum irrigation efficiency, which can be achieved by improving both the FAR and CR.

Strategies to cope with situation of low water availability: In years of low water availability, fulfillment of crop water requirements might not be possible, and deficit irrigation cannot be avoided. Deficit irrigation strategies need to be conceived such that they lead to the lowest possible impact of non-avoidable water stress on yield.

19 Modeling Irrigation Scheduling Under Shallow Groundwater...

Table 19.2 Improved irrigation efficiency scenarios

Scenario	Application ratio	Conveyance ratio[a]	Irrigation efficiency ratio
S-A (BAU)	0.43	0.76	0.33
S-B	0.43	0.84	0.36
S-C	0.67	0.76	0.51
S-D	0.67	0.84	0.56

[a]Conveyance ratio considers percolation/seepage (and evaporation), but not the operational losses, which can hardly be measured

As a prerequisite, information on the impact of water stress on yield considering the site-specific situation is needed. For that purpose, the impact of deficit irrigation was analyzed assuming a 25% and 50% reduction of available water resources.

Water saving scenarios (WSS): In the water saving scenarios (WSS), different water saving options were introduced. In WSS-1, high water demand crops (e.g., cotton and rice) were replaced with low water demand crops (vegetables). The share of cotton was reduced from 50% to 30% and 20%, respectively, and replaced with vegetables. Rice is grown on only 2% of the irrigated area in the WUA, therefore it was totally eliminated from the crop portfolio to see the impact on water saving.

In contrast, for the years with expected high water availability, when 20% and 30% of the present cotton area is replaced with rice, there will be an expected rise in water use; this was quantified in this scenario.

In WSS-2, marginal land (soils having low productivity either due to soil salinity or to a coarse soil texture (sand)) was eliminated from the irrigation plan. Martius et al. (2004) cited that 15–20% of the Khorezm area (based on the soil bonitet (soil fertility indicator) values) consists of marginal land areas, which vary however in size and are scattered over the region.

In WSS-3, irrigation efficiency at system level was evaluated. At present, 63–53% of the water is lost by farmers during field irrigation, whereas 24% is lost during conveyance in the irrigation network, and that together results in a low irrigation efficiency of 33% (Awan 2010). Assuming the target values of Bos et al. (2005), the current efficiency of the irrigation system in the WUA can be increased to 56%. The amount of water saved by increasing the irrigation efficiency was quantified in this scenario. The amount of water saved by improving the irrigation efficiency and the reduction in capillary rise contribution due to improved efficiency are also discussed.

19.3 Results

19.3.1 Optimal Irrigation Schedule and the Irrigation Schedule in Practice

The optimal irrigation schedules for all the improved irrigation efficiency scenarios for cotton (scenarios S-A, S-B, S-C and S-D) show that the capillary rise

Table 19.3 Simulation results of irrigation scheduling (optimized) for cotton under S-SL, M-SL and D-SCL characteristics of the HRUs (Scenario A)

S-SL		M-SL		D-SCL	
Irrigation interval (days)	Irrigation quota (mm)	Irrigation interval (days)	Irrigation quota (mm)	Irrigation interval (days)	Irrigation quota (mm)
51	55	63	61	28	42
16	64	16	69	33	58
14	71	11	71	13	65
11	72	11	68	11	71
13	72	11	74	11	68
13	70	13	72	11	69
17	79	20	86	12	70
				16	78
Total	483		501		521

Table 19.4 Simulation results of irrigation scheduling (optimized) for cotton under S-SL, M-SL and D-SCL characteristics of the HRUs (Scenario B)

S-SL		M-SL		D-SCL	
Irrigation interval (days)	Irrigation quota (mm)	Irrigation interval (days)	Irrigation quota (mm)	Irrigation interval (days)	Irrigation quota (mm)
60	58	63	61	44	50
13	64	13	69	20	60
11	73	11	72	12	67
10	67	11	71	10	69
11	71	12	70	10	68
12	72	12	72	11	70
15	77	17	82	12	71
				16	79
Total	482		497		534

Table 19.5 Simulation results of irrigation scheduling (optimized) for cotton under S-SL, M-SL and D-SCL characteristics of the HRUs (Scenario C)

S-SL		M-SL		D-SCL	
Irrigation interval (days)	Irrigation quota (mm)	Irrigation interval (days)	Irrigation quota (mm)	Irrigation interval (days)	Irrigation quota (mm)
26	41	34	46	27	54
33	58	25	59	16	62
13	64	14	64	11	66
11	73	11	71	10	69
10	68	11	71	11	73
11	72	12	73	11	68
12	72	13	71	12	68
14	74	18	79	19	83
Total	522		534		543

19 Modeling Irrigation Scheduling Under Shallow Groundwater...

Table 19.6 Simulation results of irrigation scheduling (optimized) for cotton under S-SL, M-SL and D-SCL characteristics of the HRUs (Scenario D)

S-SL		M-SL		D-SCL	
Irrigation interval (days)	Irrigation quota (mm)	Irrigation interval (days)	Irrigation quota (mm)	Irrigation interval (days)	Irrigation quota (mm)
33	45	33	41	24	40
25	59	25	53	33	57
14	64	14	60	14	65
11	70	11	71	11	72
11	72	11	68	10	69
11	67	11	68	11	73
13	72	13	71	12	73
17	77	17	70	13	71
			80	28	95
Total	526		582		615

contribution has a significant impact on the irrigation schedule (Tables 19.3–19.6). In all scenarios, the irrigation quota for the HRUs increases in the order S-SL < M-SL < D-SCL. The quota is influenced by the depth of the groundwater levels; it is shallowest in S-SL and deepest in D-SCL. In the S-SL HRU, the lower irrigation quota is due to a greater capillary rise contribution, whereas the higher irrigation quota in D-SCL is due to lower capillary rise resulting from less shallow groundwater levels.

The irrigation quotas in the different irrigation efficiency scenarios do not differ much in the S-A and S-B HRUs for all characteristics. The small difference between the irrigation quotas of S-A and S-B is due to the small difference in capillary rise contribution for these scenarios (Tables 19.3 and 19.4). However, irrigation quota for S-C and S-D differs substantially from that of S-A and S-B. The increase in S-C and S-D compared to S-A and S-B is again due to the difference in capillary rise contribution. The low capillary rise in S-C and S-D increased the irrigation quota in these scenarios.

The irrigation schedule officially recommended for cotton by the irrigation planners for the WUA is presented in Table 19.7. This schedule not only leads to higher irrigation quotas, but also to yield losses. The high yield losses are due to inflexible irrigation timing (norm-based), which cannot completely satisfy the time-depending water requirements. Compared to the optimal irrigation schedule of cotton simulated by CROPWAT, the irrigation schedule implemented by the government officials shows not only yields losses (7%) but also water wastage (9%) through deep percolation.

The comparison of the official schedule and the schedules practiced by the farmers reveals that the farmers do not necessarily follow the recommended irrigation schedules, but often apply untimely and large amounts of irrigation water. When the irrigation scheduling of cotton typically used by farmers (Table 19.7) was fed into

Table 19.7 Irrigation scheduling recommended by WUA officials and currently practiced by farmers

Official recommendation		Farm-1		Farm-2	
Irrigation interval (days)	Irrigation quota (mm)	Irrigation interval (days)	Irrigation quota (mm)	Irrigation interval (days)	Irrigation quota (mm)
40	11	70	355	40	117
10	17	6	183	51	130
10	58			14	32
10	64				
10	68				
10	68				
10	77				
10	68				
10	58				
10	47				
10	17				
Total	553		538		279

the CROPWAT program, the results were not only yield reductions but also water losses. At Farm-1, the farmer applied two irrigations, which resulted in 354 mm of irrigation water lost by deep percolation and a 42% reduction in cotton yield. The farmer applied more water than necessary in the first irrigation and just 6 days after the first irrigation, he started to apply the second irrigation. At Farm-2, the farmer applied three irrigations, which resulted in a 33% yield loss and 84 mm irrigation water loss due to deep percolation during the whole cotton vegetation season. Questioned about this, the farmer replied that he used irrigation water when he had the opportunity, as the water might not be available in the future, and he did not want to miss the opportunity. This in fact happened, and the farmer could not match the next scheduled irrigation with that of other farmers, i.e., he could not irrigate his fields when this was needed at the later occasion.

It is important to mention that irrigation schedule data for the farmers is based on only two farm samples. The irrigation schedule practiced by the farmers depends upon several factors such as water availability, operational problems with pumps due to electricity cut-offs or pump mal-functioning, and on agreement between different farmers about who will apply water on which day.

19.3.2 Optimizing the Irrigation Scheduling with Reduced Water Supply

The effects of the deficit irrigation on crop yield are presented in Table 19.8. If 25% of the surface water supply is proportionally deducted from each irrigation quota of the optimal irrigation schedule (Table 19.3), yield losses vary from

19 Modeling Irrigation Scheduling Under Shallow Groundwater... 321

Table 19.8 Effect of reduced water supply on crop yields (% of the potential yield lost)

HRUs	Proportionally reduced irrigations		Reduced number of irrigations	
	YR (%) with 25% RWS	YR (%) with 50% RWS	YR (%) with 25% RWS	YR (%) with 50% RWS
S-SL	10	22	16	24
M-SL	11	25	13	26
D-SCL	18	30	20	29

YR yield reduction, *RWS* reduced water supply

10% to 18% of the potential yield, and for a 50% water reduction, the yield losses are 22–30%.

The second option to manage the reduced surface water supply is to reduce the number of irrigation events. This would result in yield reductions of 24–29% at 50% deficit irrigation for cotton, wheat and vegetables, whereas with 25% deficit irrigation, yield losses would be 13–20% (Table 19.8). This smaller difference in yield is due to the selection of an irrigation time when the impacts of water stress on yield are minimal.

19.3.3 WSS 1 – Change in Cropping Pattern by Introducing Low Water Demand Crops

Different types of vegetables are the most promising crops in the study area due to their low net irrigation requirements of about 340 mm (against 601 mm for cotton). Reducing the cotton area from currently 50–30% and replacing it with vegetables would save about 6% water. If 30% of the cotton area is replaced with vegetables, 9% of water could be saved while concurrently increasing farmers' incomes (Bobojonov et al. 2008). If rice, which currently covers 2% of the cropped area, is eliminated from the system, water use could be reduced by up to 1%. On the other hand, if 20–30% of the present cotton area is replaced with rice, the rise expected in net irrigation water requirements could amount to 7–10%.

19.3.4 WSS 2 – Not Cropping Marginal Areas

Eliminating marginal areas from the irrigation plan is another option to save 15–20% surface water. This would be extremely beneficial in years of water shortage (Djanibekov 2008), as 15–20% water saving could contribute strongly to lowering the impact of water stress of the routine crops such as cotton. The water saved by keeping the marginal area out of the production plans in normal years of water availability could also be used for cultivating crops with higher income potential such as rice. Under the existing official cropping pattern, only a smaller share of rice can be cultivated due to higher water requirements (the local practice in Khorezm is to provide permanent flooding to the rice basins) of rice against the limited water supply. Alternatively, the saved water could be used to improve the ecological situation of the landscape by planting trees or creating fish ponds as recently advocated (Rudenko 2008).

19.3.5 WSS 3 – Improving the Irrigation Efficiency

By increasing the current average application efficiency from 43% to 67%, a total of 36% water can be saved. Improving the application and irrigation network efficiency to the target values, 41% of water can be saved. By improving the irrigation network efficiency from 76% to 84% without changing the current application efficiency, 8% of water can be saved. It is feasible to achieve an overall irrigation efficiency of at least 56% in the WUA, which would result in water savings of about 41%.

19.4 Discussion

Norms and recommendations currently in use for water resource use in Central Asia were developed during the Soviet era of collective farming systems. Despite the reforms initiated after independence, the former practices still dominate the current practices. The share of the areas with medium and heavy salinization and shallow groundwater levels is constantly growing, indicating the inefficiency of those practices (Ibrakhimov et al. 2007). Shallow groundwater levels are dynamic in nature and play an important role in root-zone water balancing. Therefore, to update the norms for the existing conditions, introduction of flexible schedules based on model approaches is a prerequisite for sustainable water resources management in the area.

Water balance models are used to develop the optimum irrigation schedule (Brown et al. 1978; Odhiambo and Murty 1996; Mishra 1999). Pereira (1989) developed the IRRICEP and ISAREG water balance models, and tested their performance against field data. Paulo et al. (1995) validated the IRRICEP model for selected sectors of the Sorraia irrigation system in Brazil. Khepar et al. (2000) developed a water balance model to predict deep percolation loss during wet and dry periods. However, these models are applied under conditions where groundwater is deep and where groundwater does not affect the root zone water balance. George et al. (2000) reported that the problem of irrigation scheduling is complicated by a number of factors (weather, response of crop to irrigation, spatial and temporal variability of infiltration characteristics, soil water availability, etc.), and therefore a user-friendly irrigation-scheduling model is required.

Due to shallow groundwater conditions in the WUA Shomakhulum, the capillary rise contribution to crop water needs cannot be neglected (Forkutsa et al. 2009). None of the existing irrigation scheduling models estimates the capillary rise contribution, which is an important parameter in the context of Khorezm, and for similar regions in Central Asia and elsewhere. Flat-terrain, flood-irrigated wheat and cotton production areas are abundant throughout Central Asia and the Caucasus (Martius unpublished observations).

The results of the irrigation schedule for cotton under the business-as-usual scenario show that 9% of the irrigation water can be saved, compared to the norms, by following the optimum irrigation schedule. It is only possible to model this water saving due to the introduction of the capillary rise contribution in the CROPWAT model. The capillary rise contribution is within the range of the studies conducted in the region by Forkutsa et al. (2009) and studies conducted in regions with similar climatic conditions, e.g., Pakistan (Kahlown et al. 2005). Capillary rise has a significant impact on the surface water requirements and needs to be addressed when designing the irrigation norms.

The results of the scenarios show that by improving the irrigation efficiency to the target value (from scenario S-A to scenario S-D), the irrigation quota (net irrigation depth) for cotton would increase by 8–15% compared to the business-as-usual scenario. This increase in surface water demand is due to the decline of the groundwater levels. Through the capillary rise contribution, 19% of the average surface water demand for all the crops can be met.

Instead of improving the irrigation scheduling, a different option would be to improve the overall irrigation efficiency in the WUA. As the major losses occur during the field application, these could be reduced if the farmers were to implement water saving approaches such as improving the furrow irrigation (double sided irrigation or surge flow; cf. Tischbein et al. 2011), by installing equipment for discharge control at field level (e.g., siphons), and by introducing modern irrigation techniques (drip irrigation). Application efficiency would also be improved by using laser leveling, especially on large fields (Masharipova 2009). However, farmers may need financial support for implementing some of these modern, costly techniques. An intensified maintenance of canals could already improve conveyance efficiency at much lower costs. General lining is not recommended due to the shallow and fluctuating groundwater tables, and also for cost reasons. A better coordination of operational activities could also help to avoid or reduce the overflow of water from the channels into the drains.

By improving the irrigation efficiency to the target value, 41% of surface water can be saved. However, this would lead to a decline in the groundwater levels and cause an 11% reduction of the capillary rise contribution to crop water requirements, which would need to be complemented by higher irrigation water amounts.

Currently, the decline in groundwater levels is not a feasible option for the farmers nor for water management institutions, as groundwater provides the necessary soil moisture to the roots (at WUA level, 19% of the crop water requirements) at the existing level. Farmers use it as a precaution against unreliable water supplies, e.g., by blocking drainage outlets (cf. Tischbein et al. 2011). Forkutsa et al. (2009)

reported that farmers grow cotton under these shallow groundwater levels even if they do not have a surface water supply.

Therefore, the impact of efficiency improvement on the current irrigation and drainage system/situation assessed in this study clearly shows that the groundwater contribution will be reduced. Therefore, the compensation for the safety-net function of the groundwater needs to be considered by institutional strengthening to make the water supply at farm level more reliable.

19.5   Summary and Conclusions

The importance of groundwater has not yet been considered adequately by the water managers in the development of water management strategies in Khorezm. The developed norms do not consider the contribution of groundwater, while farmers intervene on an individual basis for a beneficial use of groundwater by blocking the drains. An integrated water resources management tool was developed that considers the impact of groundwater on surface water requirements in a detailed way. The tool can help the farmers and WUA managers to improve the currently practiced scheduling by integrating surface water and groundwater. The tool highlights the importance of groundwater in the WUA Shomakhulum. Although 41% of the surface water can be saved by improving the irrigation efficiencies near to the target values, this would result in a decline in the groundwater levels. Such a decline would increase the uncertainty of water availability at farm level, as shallow groundwater levels provide 19% of the crop water requirement for the whole WUA, and farmers use groundwater as a safety net because of the unreliable water supplies. An option to improve the irrigation efficiency while at the same time keeping the groundwater as a safety net is to improve the infrastructure to control drainage outflows, and to operate this infrastructure in coordination with surface water management (controlled drainage) taking into account spatio-temporal water availability. Institutionalizing controlled drainage has the potential to make better use of groundwater than the current practice of the farmers, who utilize the groundwater on an individual basis.

References

Allen RG, Pereira LS, Raes D, Smith M (1998) Crop evapotranspiration: guidelines for computing crop requirements. Irrigation and drainage paper 56. FAO, Rome

Awan UK (2010) Coupling hydrological and irrigation schedule models for the management of surface and groundwater resources in Khorezm, Uzbekistan. Ecology and development series no. 73. ZEF/University of Bonn, Bonn

Ayars JE, Schoneman RR (1986) Use of saline water from a shallow water table by cotton. Trans ASAE 29:1674–1678

19 Modeling Irrigation Scheduling Under Shallow Groundwater... 325

Bastiaanssen WGM, Allen RG, Droogers P, D'Urso G, Steduto P (2004) Inserting man's irrigation and drainage wisdom into soil water flow models and bringing it back out: how far have we progressed? In: Feddes RA, de Rooij GH, van Dam JC (eds) Unsaturatedzone modeling, UR frontis series. Kluwer Academic, Wageningen, pp 263–299

Bitterlich S, Durner W, Iden SC, Knabner P (2004) Inverse estimation of the unsaturated soil hydraulic properties from column outflow experiments using free-form parameterizations. Vadose Zone J 3(3):971–981

Bobojonov I, Lamers J, Martius C, Berg E (2008) A decision support tool for ecological improvement and income generation for smallholder farms in irrigated dry lands, Uzbekistan. In: Environmental problems of Central Asia and their economic, social and security impacts, NATO Advanced Research Workshop, Tashkent, 01–05 Oct 2007

Bos MG, Nugteren J (1974) On irrigation efficiencies. International Institute for Land Reclamation and Improvement (ILRI), Wageningen, 19:138

Bos MG, Burton MA, Molden DJ (2005) Irrigation and drainage performance assessment: practical guidelines. CABI Publishing, Trowbridge, p pp. 155

Brown KW, Turner FT, Thomas JC, Deuel LE, Keener ME (1978) Water balance of flooded rice paddies. Agric Water Manag 1:277–291

Clarke D, Smith M, El-Askari K (1998) Cropwat for Windows: user guide. University of Southampton, Southampton

Costantini EAC, Castelli F, Raimondi S, Lorenzoni P (2002) Assessing soil moisture regimes with traditional and new methods. Soil Sci Soc Am J 66:1889–1896

Cressie N (1992) Statistics for spatial data. Wiley, New York

Djanibekov N (2008) A micro-economic analysis of farm restructuring in Khorezm region, Uzbekistan. PhD dissertation, Bonn University, Bonn

Doorenbos J, Kassam AH (1979) Yield response to water. Irrigation and drainage paper 33: 193. FAO, Rome

Doorenbos J, Pruitt WO (1977) Crop water requirements. Irrigation and drainage paper 24: 144. FAO, Rome

Forkutsa I, Sommer R, Shirokova YI, Lamers JPA, Kienzler K, Tischbein B, Martius C, Vlek PLG (2009) Modeling irrigated cotton with shallow groundwater in the Aral Sea Basin of Uzbekistan: I. Water dynamics. Irrig Sci 27(4):331–346

George BA, Shende SA, Raghuwanshi NS (2000) Development and testing of an irrigation scheduling model. Agric Water Manag 46(2):121–136

Hammel K, Weller U, Stahr K (2000) In: Graef F et al (eds) Soil water balance in Southern Benin - characteristics and conclusions. Adapted farming in West Africa: issues, potentials and perspectives. Verlag Ulrich e.Grauer, Stuttgart, pp 321–330

Hanks RJ (1983) Yield and water-use relationships: an overview. In: Taylor HM, Jordan WR, Sinclair TR (eds) Limitations to efficient water use in crop production. ASA, CSSA, and SSSA, Madison, pp 393–411

Hernandez TX (2001) Rainfall-runoff modeling in humid shallow water table environments. MSc dissertation, University of South Florida, Tampa

Howell TA (1990) Relationships between crop production and transpiration, evapotranspiration, and irrigation. In: Steward BA, Nielson DR (eds) Irrigation of agricultural crops, agronomy monograph 30. ASA, CSSA, and SSSA, Madison, pp 391–434

Ibrakhimov M, Khamzina A, Forkutsa I, Paluasheva G, Lamers JPA, Tischbein B, Vlek PLG, Martius C (2007) Groundwater table and salinity: spatial and temporal distribution and influence on soil salinization in Khorezm region (Uzbekistan, Aral Sea Basin). Irrig Drain Syst 21(3/4):219–236

Jurriens M, Zerihun D, Boonstra J (2001) SURDEV: surface irrigation software. International Institute for Land Reclamation and Improvement, Wageningen

Kahlown MA, Ashraf M, Zia-ul-Haq (2005) Effect of shallow groundwater table on crop water requirements and crop yields. Agric Water Manag 76(1):24

Khepar SD, Yadav AK, Sondhi SK, Siag M (2000) Water balance model for paddy fields under intermittent irrigation practices. Irrig Sci 19(4):199–208

Kincaid DC, Heermann DF (1974) Scheduling irrigations using a programmable calculator. Agricultural Research Service-NC-12, p 55

Knight HG (1937) New size limits for silt and clay. Soil Sci Soc Am Proc 2:592

Malano HM, Chien NV, Turral HN (1999) Asset management for irrigation and drainage infrastructure: principles and case study. Irrig Drain Syst 13(2):109–129

Mandal UK, Sarma KSS, Victor US (2002) Profile water balance model under irrigated and rain fed systems. Agron J 94:1204–1211

Martius C, Lamers JPA, Wehrheim P, Schoeller-Schletter A, Eshchanov R, Tupitsa A, Khamzina A, Akramkhanov A, Vlek PLG (2004) Developing sustainable land and water management for the Aral Sea Basin through an interdisciplinary research. In: Seng V, Craswell E, Fukai S, Fischer K (eds) Water in Agriculture, ACIAR Proceedings, Canberra 116:45–60

Martius C, Froebrich J, Nuppenau EA (2009) Water resource management for improving environmental security and rural livelihoods in the irrigated Amudarya lowlands. In: Brauch HG, Spring UO, Grin J, Mesjasz C, Kameri-Mbote P, Behera NC, Chourou B, Krummenacher H (eds) Facing global environmental change: environmental, human, energy, food, health and water security concepts, Hexagon series on human and environmental security and peace, 4. Springer, Berlin/Heidelberg/New York, pp 749–762

Masharipova H (2009) Effect of laser leveled field to irrigation application efficiency. MSc Dissertation, Urgench State University, Urgench

Mateos L, Lopez-Cortijio I, Sagardoy JA (2002) SIMIS: the FAO decision support system for irrigation scheme management. Agric Water Manag 56:193–206

Meiwirth K, Mermoud A (2004) Simulation of herbicide transport in an alluvial plain. In: Pahl-Wostl C, Schmidt S, Rizzoli AE, Jakeman AJ (eds) iEMSs 2004 international congress: complexity and integrated resources management. iEMSs, Switzerland, pp 951–955

Mishra A (1999) Irrigation and drainage needs of transplanted rice in diked rice fields of rain fed medium lands. Irrig Sci 19:47–56

Nalder IA, Wein RW (1998) Spatial interpolation of climatic normals: test of a new methods in the Canadian boreal forest. Agric For Meteorol 92:211–225

Odhiambo LO, Murty VVN (1996) Modeling water balance components in relation to field layout in lowlandpaddy fields. II. Model application. Agric Water Manag 30(2):201–216

Paulo AM, Pereira LA, Teixeira JL, Pereira LS (1995) Modelling paddy irrigation. In: Pereira LS, vanden Broek BJ, Kabat P, Allen RG (eds) Crop water simulation models in practice. Wageningen Press, Wageningen

Pereira LA (1989) Rice water management. PhD dissertation, Technical University, Lisbon

Pereira LS, Paredes P, Cholpankulov ED (2009) Irrigation scheduling strategies for cotton to cope with water scarcity in the Fergana Valley, Central Asia. Agric Water Manag 96(5):723–735

Pratharpar SA, Qureshi AS (1998) Modeling the efficacy of deficit irrigation on soil salinity, depth of water table transpiration in semiarid zones with monsoonal rains. Water Resour Dev 15:141–159

Rakhimbaev FM, Bezpalov NF, Khamidov MK, Isabaev KT, Alieva D (1992) Peculiarities of crop irrigation in lower Amu Darya river areas. Fan, Tashkent

Rudenko I (2008) Value chains for rural and regional development: the case of cotton, wheat, fruit, and vegetable value chains in the lower reaches of the Amu Darya River, Uzbekistan. PhD dissertation, Hannover University, Hannover

Sadikov AS (1979) Irrigation of Uzbekistan: contemporary state and perspectives of irrigation development in the Amu Darya River Basin. Fan, Tashkent, 3:255

Sarma PBS, Rao VV (1997) Evaluation of an irrigation water management scheme - a case study. Agric Water Manag 32(2):181

Simunek J, Sejna M, van Genuchten MT (1998) The HYDRUS-1D software package for simulating the one-dimensional movement of water, heat and multiple solutes in variably-saturated media, Version 2.0. US salinity laboratory. Agricultural research service, US department of agriculture, Riverside, pp 1–177

Smith M (1992) CROPWAT, a computer program for irrigation planning and management. Irrigation and drainage paper 46. FAO, Rome

Sommer R, Vlek PLG, Deane de Abreu T, Vielhauer K, de Fátima Rodrigues Coelho R, Fölster H (2004) Nutrient balance of shifting cultivation by burning or mulching in the Eastern Amazon - evidence for subsoil nutrient accumulation. Nutr Cycl Agroecosyst 68:257–271

Srinivasulu A, Rao CS, Lakshmi GV, Satyanarayana TV, Boonstra J (2004) Model studies on salt and water balances at Konanki pilot area, Andhra Pradesh, India. Irrig Drain Syst 18:1–17

Suslov SP (1961) Physical geography of Asiatic Russia. W. H Freeman and Company, San Francisco/London

Tischbein B, Awan UK, Abdullaev I, Bobojonov I, Conrad C, Forkutsa I, Ibrakhimov M, Poluasheva G (2011) Water management in Khorezm: current situation and options for improvement (hydrological perspective). In: Martius C, Rudenko I, Lamers JPA, Vlek PLG (eds) Cotton, Water, Salts and Soums – Economic and Ecological Restructuring in Khorezm, Uzbekistan. Springer, Dordrecht

van der Grift B, Passier H, Rozemeijer J, Griffioen J (2004) Integrated modeling of cadmium and zinc contamination in groundwater and surface water of the Kempen Region, The Netherlands. IEMSs 2004 international congress: complexity and integrated resources management, Switzerland, Jan 2011, pp 1235–1240

Vanderborght J, Kasteela R, Herbst M, Javaux M, Thiéry D, Vanclooster M, Mouvet C, Vereecken H (2005) A set of analytical benchmarks to test numerical models of flow and transport in soils. Vadose Zone J 4:206–221

Vaux HJJ, Pruitt WO (1983) Crop-water production functions. In: Hillel D (ed) Advances in irrigation 2. Academic, New York, pp 61–97

Ventrella D, Mohanty BP, Simunek J, Losavio N, van Genuchten MT (2000) Water and chloride transport in a fine-textured soil: field experiments and modeling. Soil Sci SocAm J 165(8):624–631

Vrugt JA, Bouten W (2002) Validity of first-order approximations to describe parameter uncertainty in soil hydrologic models. Soil Sci Soc Am J 66:1740–1751

Wallender WW, Grimes DW, Henderson DW, Stromberg LK (1979) Estimating the contribution of a perched water table to the seasonal evapotranspiration of cotton. Agron J 71:1056–1060

Zavattaro L, Grignani C (2001) Deriving hydrological parameters for modelling water flow under field conditions. Soil Sci Soc Am J 65:655–667

Chapter 20
Estimation of Spatial and Temporal Variability of Crop Water Productivity with Incomplete Data

Maksud Bekchanov, John P.A. Lamers, Aziz Karimov, and Marc Müller

Abstract Crop water productivity (WP) in irrigated agriculture is a key for food and environmental security, in particular when water becomes scarce, as has been predicted for downstream regions in the Aral Sea basin. The assessment of WP for each field crop is hampered when farmers cultivate several crops at the same time and on the same fields and cannot record water allocation for each crop. Since results from the commonly used Ordinary Least Squares (OLS) approach turned out to be unreliable and in some cases even had negative values, we combined a behavioral approach with mixed estimation methods to estimate water allocation for each crop over larger areas and over various years with limited data availability. Unobserved crop specific input data was derived from aggregated data using the mixed estimation method for the case study region Khorezm, located in northwestern Uzbekistan. On the basis of actual water usage, spatial (for different administrative districts in Khorezm) and temporal (for the years 2004–2007) distributions of WP for cotton, wheat, rice, vegetables (including melons), and fruits were estimated and visualized through contour diagrams. All crops, except forage, used much more irrigation water than the recommended amount, with cotton and rice being the highest water consumers. For example, cotton was almost 64% over-irrigated compared to the recommended amounts. Even though rice was cropped on a relatively small share of the total land in Khorezm (less than 10% of the total arable land), about 30% of the total crop irrigation water was applied on rice. WP depends on district or farms' location declining proportionally to the distance from the water source, due to high conveyance losses and low yields monitored at the tail ends of the irrigation

---

M. Bekchanov (✉) • A. Karimov • M. Müller
Center for Development Research (ZEF), Walter-Flex-Str. 3, 53113 Bonn, Germany
e-mail: maksud@uni-bonn.de; akarimov@uni-bonn.de; molinero@uni-bonn.de

J.P.A. Lamers
ZEF/UNESCO Khorezm Project, Khamid Olimjan Str., 14, 220100 Urgench, Uzbekistan
e-mail: j.lamers@uni-bonn.de

C. Martius et al. (eds.), *Cotton, Water, Salts and Soums: Economic and Ecological Restructuring in Khorezm, Uzbekistan*, DOI 10.1007/978-94-007-1963-7_20,
© Springer Science+Business Media B.V. 2012

329

system. Extremely high water losses on both conveyance and field level revealed much scope for water saving by implementing different water-wise options such as lining canals, introducing best water use practices and producing less water consuming crops. This would, in turn, allow increasing crop yields and WP particularly in downstream districts. Furthermore, the comparison of WP among different crops and different districts allowed determining the potential suitability of certain districts for certain crops, which suggests that a more regionally differentiated cropping portfolio, according to water availability, soil quality and similar parameters, would improve overall system WP, and hence sustainability of the agricultural production.

Keywords Crop specific water allocation • Mixed estimation method • Irrigated agriculture • Water saving • Khorezm

20.1 Introduction

Irrigated agriculture plays a dominant role in the economy of Uzbekistan, as well as in the Khorezm region, consuming 92% of all water used in the country (Dukhovny and Sokolov 2002). Vast irrigation and drainage systems have been developed since the 1960s to provide croplands with irrigation water from the Amudarya and Syrdarya rivers. In 1999, the immense irrigation systems used 83% of the average water resources available in all rivers in the Aral Sea Basin (Martius et al. 2004). However, the efficiency of the irrigation network in the Khorezm region does not exceed 55%, owing to, in particular, the deterioration of the infrastructure and poor management of irrigation water. Main irrigation methods for crop cultivation are furrow and basin irrigation with field application efficiencies of less than 50% (Awan 2010; Bekchanov et al. 2010b) indicating substantial water losses. In contrast, the probability that farmers receive sufficient water in this downstream region has been decreasing every year since the early 1990s (Müller 2006). The present combination of physical water scarcity, as shown by reduced natural water flow and increased water demand in the upstream countries in conjunction with global climate change effects, and economic water scarcity caused by the deteriorated irrigation network and poor water resources management (Martius et al. 2004), have provoked an increased frequency of water shortages. This makes water a critical resource for the maintenance of regional livelihoods.

Despite these changes over time, crop water application recommendations developed during the Soviet Union period are still in use. They have been formulated to obtain highest yields since water at that time was not considered a limiting factor. However, with water becoming a limiting factor as experienced today, WP, which is the physical or monetary output obtained per unit of water, should gain at least equal priority as land productivity (Oweis and Hachum 2006). However, a major obstacle in estimating crop specific WP is the lack of available data on crop specific water use. Statistical sources for agricultural production systems in Uzbekistan used to report data on production volumes and land areas of different crops while presenting total water use according to administrative districts only. Thus, bridging the information gap on water inputs used by various crops is a crucial step for quantitatively assessing the efficiency of agricultural production systems.

Although detailed experimental measurements at the field level are the best source for crop-specific data on irrigation water use, such measurements are too resource-demanding for the local water user associations (WUAs) and farmers. Since this dilemma frequently occurs in the context of transition economies, several approaches have been considered to obtain the necessary data using available, incomplete data sets (e.g. Just et al. 1990; Golan et al. 1996; Lence and Miller 1998). This study applies the methods proposed by Just et al. (1990), who suggested two approaches for estimating variable input allocations among activities. The first approach is based on behavioral rules whereby input/land ratios are assumed to follow accepted practices. The second approach is based on profit maximization, where input use depends on the changes in input and output prices. We used here the first approach considering the availability of data and improved replication of real data compared to the second approach. The Mixed Estimation Method (MEM), as proposed by Theil and Goldberger (1961), was employed to assess crop specific water allocation across the administrative districts in the case study region Khorezm and across years, relying on aggregate water use and average per hectare water requirements of main crops. While allowing the inclusion of additional information on the parameters to be estimated, this approach methodologically improves the Ordinary Least Squares (OLS) model. On the basis of the empirical and estimated data, spatial and temporal variation of WP was estimated for a better understanding of how effectively water is used in irrigated agriculture.

The study presented here complements the crop-specific water allocation analysis of Müller (2008) by incorporating temporal and spatial WP variation analysis. The overall objectives were to (i) identify the actual water distribution across districts and crops, (ii) determine the reasons of crop WP variation, and (iii) estimate the suitability and comparative advantages of the administrative districts to cultivate certain crops in terms of WP.

20.2 Data and Computation Methods

20.2.1 Data Sources

Data on water intake during the irrigation period in the Khorezm region was obtained from regional water resources management departments (e.g. OblSelVodKhoz 2008) while harvested area, produced quantities, and yields of different crops across the districts were based on the data published by the Regional Statistical Department (e.g. OblStat 2008). The recommended norms of crop water demand per hectare were taken from regional water resources management department guidebooks (HydroModRay 2002).

20.2.2 Crop Specific Water Allocation

The estimation of crop specific water allocation followed the method by Just et al. (1990) and Müller (2006). The behavioral approach describes water allocation in a region with a group of r districts (here $r = 1, 2 \ldots 10$) producing k crops (here $k = 1, 2 \ldots 7$) over a period of t years (here $t = 2004, 2005 \ldots 2007$). The statistical analysis consisted of estimating the allocation of water among crops. The two types of information used for these estimations included the area allocated to the production of crop k in district r in year t (A_{krt}) and the aggregate water use in each district r in year t (W_{rt}). Then,

$$W_{rt} = \sum_{k=1}^{7} W_{krt}, \tag{20.1}$$

where W_{krt} is the (unobserved) volume of water allocated to produce crop k in district r in year t. Information on W_{rt} existed for each district of Khorezm over the analyzed years, whereas the actual allocation of water for the different crops within fields is rarely recorded or not measured at all. In contrast, annual data on cropland allocation on field and district level is recorded by farmers and the statistical department.

The estimated relationship of the allocation of water for various crops is based on assumptions about farmers' behavior. Production functions of farmers are assumed to have constant returns to scale, e.g., their decisions consist of water/land ratios and land allocations (Just et al. 1990).

Let $w_{krt} = W_{krt} / A_{krt}$ be the amount of water use per hectare to produce crop k in district r in year t. Then the decomposition of the systematic element of w_{krt} denoted by $\overset{\cdot}{w}_{krt}$ will be:

$$\overset{\cdot}{w}_{krt} = \alpha_k + \beta_r + \gamma_t, \tag{20.2}$$

where α_k is a coefficient for crop effects, β_r is for the district effects, and γ_t is for the time effects.

When combining (20.1) and (20.2), the following equation is derived:

$$W_{rt} = \sum_{k=1}^{7} (\alpha_k + \beta_r + \gamma_t) \cdot A_{krt} + \varepsilon_{rt}, \tag{20.3}$$

where ε_{rt} is a random error which is assumed to be normally and independently distributed for empirical purposes. The coefficients of Eq. 20.3 are estimated by regressing total water use on the areas allocated to each of the crops crossed with dummy variables, which correspond to the district and time effect. The results were normalized for the year 2004 and the district Gurlen in each case to avoid singularity of the matrix of explaining variables. The behavioral estimate of the allocation of water to crop k in district r at time t was then derived from the multiplication of the estimated per hectare water use by land allocated to the crop.

20 Estimation of Spatial and Temporal Variability of Crop Water Productivity... 333

The Eq. 20.3 with OLS turned out to be unreliable, sometimes even resulting in negative values, which was previously observed as well (Müller 2006). Since this was most likely caused by a comparatively limited number of available observations, the MEM approach was used as described by Theil and Goldberger (1961) with the inclusion of additional (prior) information on the parameters to be estimated (see Müller 2008 for more details).

Determination coefficient (R^2) for the regression was estimated according to Greene (2003):

$$R^2 = \frac{SSR}{SST} = 1 - \left[\frac{e'e}{y'M^0 y}\right],$$

(20.4)

with: e: $n \times 1$ matrix for error terms of the regression model (n is number of observations)

y: $n \times 1$ matrix for total water use ($1,000\ \mathrm{m}^3$)

M^0: $n \times n$ idempotent matrix whose diagonal elements are all $(1 - 1/n)$ and off-diagonal elements are $(-1/n)$

20.2.3 Water Productivity

Following the estimation of water use per cropped area, water productivity (WP_{krt}) was estimated as:

$$WP_{krt} = \frac{Q_{krt}}{W_{krt}} = \frac{y_{krt} \cdot A_{krt}}{w_{krt} \cdot A_{krt}} = \frac{y_{krt}}{w_{krt}},$$

(20.5)

Where Q_{krt} is total production amount of crop k in district r at time t, and y_{krt} is the corresponding yield.

Considering its very small share in total water consumption (Hydromet 2007), drainage water re-usage for irrigation and groundwater use for satisfying rural household needs were not included in the estimations of water productivity. The results may therefore in some cases have been slightly overestimated, but due to this systematic approach, the relative WP values are likely to remain unchanged.

In order to show how WP values change across the districts and years, contour charts were designed for each crop. Contour charts are analogous to contour maps which show how altitude changes with longitude and latitude and allows visualizing classification groups of the districts according to WP levels through the observed years. The borders of each WP level group (called clusters here) were determined by applying an elementary classification, e.g., based on obtaining four groups with equal ranges considering maximum and minimum WP values.

20.3 Results and Discussion

The water use estimates based on the Mixed Estimation Method (MEM) show that the estimated crop effect coefficients substantially differ from the prior information for crops such as cotton, fodder crops, rice, and vegetables (Table 20.1).

This indicates in turn that the values recommended by HydroModRay (2002) did not match the actual water application for these crops on the fields during the years examined. The highest relative deviation of 64% (9,200 m^3 ha^{-1} (estimated) vs 5,620 (recommended) m^3 ha^{-1}) was found for cotton. Even though the high area share for cotton most likely increased the explanatory power of the related parameter in the estimation model, it nevertheless shows that water application losses on field level were comparatively high. The recommended amount of irrigation water for cotton production is calculated based on about 50% field application efficiency in cotton cultivation (cf. Tischbein et al. 2011). Using this value, the estimated 64% relative deviation can be interpreted as an irrigation water use efficiency of about 30% (crop water use/total water applied $= (100\% - 50\%)/(100\% + 64\%)$). This derived irrigation water use efficiency is well in line with the efficiency measured during field experiments by Awan (2010) in Kushkupir district in Khorezm. Our derived estimation of cotton water use per hectare, which amounted to 9,200 m^3 ha^{-1}, is also comparable with the remote-sensing based findings of Conrad (2006), who estimated an average of 8,310 m^3 ha^{-1} of actual evapotranspiration for cotton fields in 2005 over entire Khorezm. These indirect comparisons show the validity of the MEM approach when used for estimating gaps in incomplete datasets.

Rice water use was estimated to be 33,000 m^3 ha^{-1} (3,300 mm), and rice was therefore the crop with the highest water consumption. Moreover, despite its comparatively low area share (about 10%), rice production consumed almost 30% of the total crop water use, and its per-hectare water use was four to six times higher than that of other crops. This indicates the scope to conserve water when decreasing rice production in the region. Furthermore, reducing water use in rice production may have a strong social impact if the unused water can be directed to areas with high water tensions. On the other hand, since rice is a high-cash crop that permits large profits compared to all other crops, attempts at establishing administrative rules to restrict the rice area may face huge challenges in Khorezm. Khorezm's population consists to almost 70% of rural people, of which 27% live below the poverty line of 1 US\$ day^{-1}, with high and rising unemployment rates (Müller 2006). Introducing less water demanding rice varieties or high cash crops may therefore at present be options with the highest potential to economize the scarce resources as recently suggested (Bekchanov et al. 2010b; Devkota 2011).

The district effects conveyed virtually the same conclusions as those derived from the prior data (Table 20.1). Water usage per hectare in the districts not bordering the Amudarya River, such as Kushkupir, Khiva, and Yangiarik, was higher than in the districts along this river. This seems counter-intuitive but makes sense when considering the water losses as previously postulated (Müller 2008). In the downstream districts, more water per hectare needs to be transported via the irrigation channels

20 Estimation of Spatial and Temporal Variability of Crop Water Productivity... 335

Table 20.1 OLS based regression and MEM estimation results and prior information on crop water use and regional and time effects (1,000 m^3 ha^{-1})

		Ordinary Least Squares (OLS)		Prior information		Mixed Estimation Method (MEM)		
		B	σ_b	β_0	$\sqrt{\Sigma_0}$	β_{bay}	σ_{bay}	t_{bay}
Crop effects	Cotton	20.57	3.79	5.62	1.87	9.20	1.10	8.34
	Fodder	4.98	4.50	8.42	2.81	5.68	2.19	2.60
	Fruits	37.58	32.36	6.29	2.10	6.19	2.08	2.97
	Grains	−2.52	8.35	4.49	1.50	4.37	1.40	3.13
	Other	21.26	52.40	6.72	2.24	6.67	2.24	2.98
	Rice	10.22	5.47	26.2	8.73	32.86	2.68	12.25
	Vegetables	5.22	13.95	6.29	2.10	7.34	2.03	3.61
District effects	Bagat	2.15	1.84	6.52	1.84	5.58	0.46	12.18
	Kushkupir	2.16	1.04	7.16	1.04	6.12	0.47	13.11
	Urgench	0.23	1.14	4.56	1.14	3.47	0.38	9.05
	Khazarasp	−0.63	1.07	3.14	1.07	1.35	0.34	3.96
	Khiva	2.66	1.97	7.43	1.97	6.66	0.54	12.32
	Khanka	0.86	1.08	4.16	1.08	2.96	0.37	7.97
	Shavat	−0.22	1.43	4.72	1.43	3.64	0.42	8.57
	Yangiarik	2.61	1.28	6.79	1.28	5.52	0.47	11.84
	Yangibazar	−0.70	1.16	5.06	1.16	3.83	0.47	8.12
Annual effects	2005	0.16	0.30	0.15	0.30	0.12	0.19	0.65
	2006	−0.48	0.50	−0.56	0.50	−0.45	0.22	−2.10
	2007	−1.57	0.62	−1.84	0.62	−1.12	0.23	−4.87

Notes: B is the regression coefficient and σ_b is its standard deviation according to OLS estimations; β_0 and $\sqrt{\Sigma_0}$ are prior information about the regression coefficient and its standard deviation; β_{bay}, σ_{bay}, and t_{bay} are the regression coefficient, standard deviation and t-statistics value according to MEM estimations

to compensate for water losses that occur during the transport of water from the river node to these districts. Since irrigation water is transported to the fields in open and unlined channels, percolation, seepage and evaporation losses as well as overflow to the drainage system are common in the entire Khorezm region (Manschadi et al. 2010). According to Veldwisch (2008), in 2005, a year with sufficient water supply, almost 50% of all incoming water ended up in the drainage systems flowing directly to desert sinks. This often has technical reasons, e.g., non-adjustable pump discharge rates (Awan 2010; Awan et al. 2011). On the other hand, when high seepage losses occur during conveyance, the water may end up in the groundwater and thus becomes indirectly available to satisfy crop water demand in those districts through which the channels run (Forkutsa et al. 2009). This may therefore lead to overestimating WP of the crops in such districts. But although irrigation management organizations presently still add about 30% surplus water during irrigation events to compensate for seepage and conveyance losses, recent results showed that seepage losses in the main distribution channel during the season were only 2%, which is merely a fraction of the initially estimated seepage losses (Awan 2010). This in turn indicates that

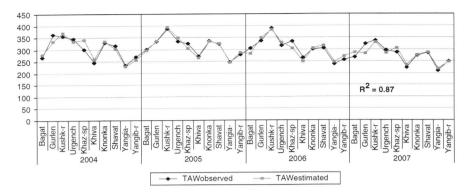

Fig. 20.1 Observed and estimated total agricultural water use (TAW, 1,000,000 m³) in the Khorezm region, 2004–2007

the estimated WP for crops in districts through which channels run are less biased by the contribution of the groundwater as previously postulated. Despite the close vicinity of Bagat to the river, high water consumption was estimated for this district, consistent with high water losses during conveyance, because most fields in this district are as far from the river as fields in downstream districts.

The average annual effects did not deviate significantly but showed an annual decrease in water usage during the last 2 years examined.

The MEM estimation for total water consumption across the districts of Khorezm replicated the sample information with an R^2 of 0.87, which indicates a sufficiently strong fitness. This was also evidenced visually by a match of the observed and estimated values (Fig. 20.1). Since these estimates are in line with previous results and the model used replicated the empirical data sufficiently accurate, the results allowed for further calculations of aggregated water use and crop specific water productivity.

The estimated values of crop specific water use per hectare allowed estimating total field level water consumption across districts and crops (Fig. 20.2). The share of cotton water consumption varied substantially across districts from 34% to 56%, whilst the share of grain water consumption ranged from 5% to 14%. The share of rice water usage was highest in Gurlen (50%) and Khazarasp (44%), indicating that almost half of the total water withdrawal was used for rice. Vegetables consumed up to 10% of the total water use in Khiva district, but their share did not exceed 2% in districts such as Khazarasp. The share of fodder in water use varied between 3% and 9% across districts.

Based on the derived water consumption amounts by crops and across districts for each year and considering the actual volume of crop production, temporal and spatial WP dynamics could be depicted. Variation analysis shows that for raw cotton, WP fluctuated in the range of 140–320 kg m^{-3} over the years 2004–2007 and across districts (Fig. 20.3). Cluster 1 comprises the districts Khiva (171–182 kg m^{-3}), Kushkupir (154–166 kg m^{-3}), and Yangiarik (176–180 kg m^{-3}). The low cotton

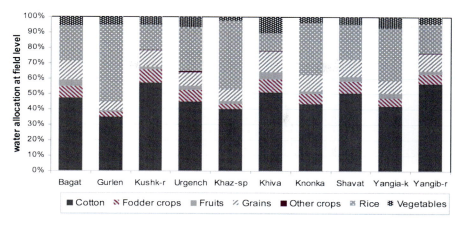

Fig. 20.2 Water allocation by crops and across districts (average of 2004–2007)

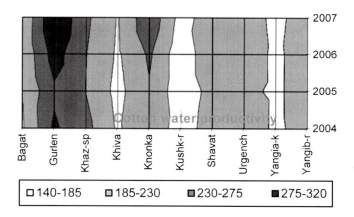

Fig. 20.3 Temporal-spatial variation of cotton water productivity (kg m^{-3}) in 2004–2007

WP in these districts was caused mainly by higher water usage for cotton production compared to other districts. In these districts, 14,000–15,000 m^3 ha^{-1} (1,400–1,500 mm) water were used while in other districts this amounted to only 9,000–13,000 m^3 ha^{-1} (900–1,300 mm). However, water use per hectare here also included canal conveyance losses due to seepage, percolation and evaporation and not only net water usage on field level. The results here may therefore deviate from findings obtained from direct field measurements (Forkutsa et al. 2009).

In the initial year of the study period, Bagat was grouped in Cluster 1 but then moved to Cluster 2 in the other 3 years. Cluster 3 comprised Gurlen in 2005, Khazarasp in 2006, and Khanka in 2006 and 2007. All other districts were grouped in Cluster 2 during all years, with WP values ranging between 185 and 230 kg m^{-3}. The increase in cotton WP over the districts in 2007 was caused by a decreased

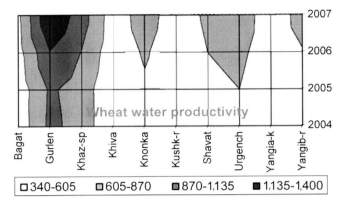

Fig. 20.4 Temporal-spatial variation of wheat water productivity (kg m^{-3}) in 2004–2007

water supply in this year (Table 20.1). The highest cotton WP in Gurlen is a result of a combination of relatively high cotton yields and low water consumption per hectare in this district.

A similar classification analysis showed different contour graph images for grain WP (Fig. 20.4). In this case, Cluster 1, which ranged between 340 and 605 kg m^{-3}, covered most of the area in Khorezm. Yet, as a result of comparatively low yields and the inconveniences associated with further distances to the water source, downstream districts such as Khiva, Kushkupir, and Yangiarik fell into Cluster 1 as well during all observed years. In spite of its location advantages and high wheat yields, Bagat also ended up in Cluster 1 because of high water losses during conveyance. Cluster 4 included only Gurlen, where the highest grain yield (4.5 t ha^{-1}) was obtained in 2007. Another reason was that this district is situated closer to the river and therefore experienced relatively lower water losses during conveyance compared to other districts.

Cluster 3 included Gurlen in the other years (2005–2007) and Khazarasp in 2007. Cluster 2 comprised Khanka, Shavat, and Urgench in 2006–2007, Khazarasp in 2004–2006, and Yangibazar in 2007. The main cause for the transition of Khazarasp, Shavat, and Urgench districts from Cluster 1 to Cluster 2 was the decrease in annual average water use on the one hand, and the yield increase (from 4.3 to 4.6 tons ha^{-1}, from 4.3 to 4.5 tons ha^{-1}, and from 4.6 to 4.8 tons ha^{-1}, respectively) on the other hand.

The highest rice WP (266 kg m^{-3}) was observed in Khiva in 2007 (Fig. 20.5). Cluster 3 comprised Gurlen and Khazarasp in 2007. Rice WP was lowest (Cluster 1) in Kushkupir, Khanka and Yangibazar districts through most years. The increasing rice WP in Yangiarik was caused by high yields of rice in this district in spite of low irrigation efficiency. Although Khazarasp had the highest water-use efficiency, it also showed the lowest WP (77 kg m^{-3}) in 2004, which was due to the lowest rice yields (2.6 tons ha^{-1}) in this year. Highest WP values were observed in Khiva, as rice yields were substantially high in spite of high water consumption per hectare.

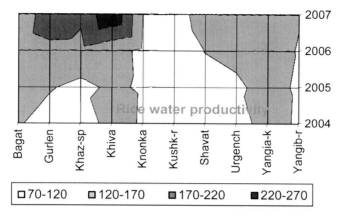

Fig. 20.5 Temporal-spatial variation of rice water productivity (kg m^{-3}) in 2004–2007

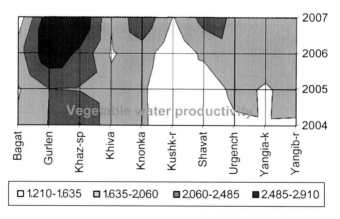

Fig. 20.6 Temporal-spatial variation of vegetable water productivity (kg m^{-3}) in 2004–2007

Vegetable WP analyses allowed identifying the potentially most attractive zones for vegetable production (Fig. 20.6). Gurlen and Khazarasp can be considered the most promising districts to cultivate vegetables due to the high WP. Kushkupir again fell into Cluster 1, with WPs that ranged from 1,008 to 1,614 kg m^{-3} in all study years. Cluster 1 also included Shavat and Yangiarik in 2004 and 2005, Urgench and Yangibazar in 2004, Bagat in 2005 and 2007, and Khiva in 2006. The main reason for the low WP was again caused by the long distance of these districts from the water source and the consequent high losses during the entire process of irrigation water delivery.

Fruit WP across districts and over years varied in the range of 570–1,790 kg m^{-3} (Fig. 20.7). Cluster 4 (1,485 and 1,790 kg m^{-3}) included Khazarasp in 2004 and 2007 and Gurlen in 2006 and 2007. These two districts in the other years were

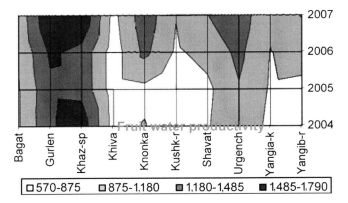

Fig. 20.7 Temporal-spatial variation of fruit water productivity (kg m^{-3}) in 2004–2007

included in Cluster 3, which additionally comprised Khanka and Urgench in 2006 and 2007. Khiva fell into Cluster 1 while Kushkupir and Yangiarik in years 2004–2006 were grouped in Cluster 1 but transferred to Cluster 2 in 2007.

The findings revealed that, in general, WP of all crop groups except rice in the downstream districts was lower than that in the upstream districts, which was very likely due to a combination of high distribution and conveyance losses and relatively lower yields. Shi et al. (2007) also found an inverse relationship between the distance of fields from main irrigation channels and cotton yields, caused by generally lower water availability in the more remote areas. This was also postulated for the entire Khorezm region based on region-wide remote sensing analyses (Conrad et al. 2011) and on the analyses of a single WUA in the Khorezm region (Awan 2010).

The previous annual analysis of aggregated WP in economic terms across the districts in Khorezm (Bekchanov et al. 2010a) covering the years 2000–2007 revealed a significant correlation between temporal-spatial water availability and agricultural revenue generation. Despite higher WP in drier years as shown here, yields are lower in general and even less irrigation water is available to downstream water users (Bekchanov et al. 2010a).

Increasing crop WP without compromising on sustainable yields, and consequently maintaining the comparative advantages of downstream districts, requires improving the presently low efficiency of water distribution in the conveyance system (Manschadi et al. 2010). The rehabilitation of the irrigation networks by, for example, lining would contribute to increasing efficiency in water transportation. Awan (2010), who assessed irrigation efficiency in the Khorezm region, also postulated the huge scope for water saving within the irrigation system to efficiently cope with current and future water shortages. However, costs and benefits of rehabilitation measures should be investigated further to determine the economic efficiency of canal lining, which requires huge investments. Although these estimates are still due for the Khorezm region, some indications can be gained from previous estimates for the entire Aral Sea Basin (Micklin 1988). An estimated 10–22 km^3 of water could be saved every year in the basin by rehabilitating and renovating the irrigation systems and concrete lining of irrigation canals (Micklin 1988). However, the costs for the entire basin would be enormous and were estimated to be around 16,000,000,000 US$ (Micklin 2002). On the other hand, recent ponding trials showed that due to the shallow groundwater levels during the growing season, seepage losses with a major distribution channel did not exceed 2%, which is thus much lower than previously estimated (Awan 2010). Hence the present policy of adding at least 30% irrigation water to compensate for seepage and conveyance losses in the channels needs to be reviewed. Cheaper but less durable options and immediate actions could be covering the canal bottoms with special polyethylene films, which is particularly relevant in sandy soil zones of the irrigation canals. Also improved low cost water use practices such as alternate dry furrow, short furrow, and double side irrigation (Abdullaev et al. 2009) or advanced water-conservation technologies such as drip irrigation and laser guided land leveling on field level (Bekchanov et al. 2010b) may help farmers to increase water use efficiency.

Considering that the production of state order crops – cotton and winter wheat – occupies more than 70% of the irrigated area (cf. Djanibekov et al. 2011), diversification of the cropping pattern also could contribute improving the irrigation water use efficiency and environmental sustainability in the region (Bobojonov et al. 2011). For example, less water consumptive alternative crops, such as sorghum, mungbean and sweet maize, could provide very high yields and enhance crop rotations and are very attractive especially in downstream locations with very low probability of sufficient irrigation water supply for cotton or rice production (Müller 2006). Moreover, the cultivation of alternative crops such as indigo and potato could improve farm income with less water (Bobojonov et al. 2011). However, since WP of crops differentiates across districts, in addition to water availability, soil productivity and groundwater level and salinity are main indicators for the suitability of cultivating a particular crop in a district or field. Crops diversification, in turn, could be supported by easing the presence of the state order, lack of access to seed and planting materials, and an underdeveloped market infrastructure which includes the processing sector and storage and trade channels (Bobojonov and Lamers 2008).

20.4   Summary and Conclusions

Increased demand for agricultural commodities and decreased rates of irrigation water availability over the years make water a critical resource for providing sustainable livelihoods in the Aral Sea Basin. When water becomes a limiting factor, improving water productivity (WP) should become as important as improving land productivity, which dominates the present thinking of local experts. However, calculating crop specific WP requires crop specific water use, which is currently not recorded. Mixed estimation methods were seen to be an effective tool for estimating detailed crop specific water allocation, which in turn allowed accounting water allocation and loss as well as WP over the study years and districts in the Khorezm region.

The model results show that cotton production, which occupies more than half of the total irrigated area in the region, consumes about 64% more water than the recommended amount. The resulting low water use efficiency of cotton production indicates a huge scope for water saving. In spite of its low share (ca. 10% on average) in total cropped areas, rice cultivation consumed about 30% of total crop water usage. Therefore, replacing paddy rice varieties with less water consumptive rice varieties or replacing paddy rice and cotton with other less water consumptive cash crops would allow substantial water savings in the region. Technical options for this are currently being investigated (Devkota 2011).

Considering the combined seepage and evaporation losses during water conveyance in the whole region and that decreased water availability in downstream districts and WUAs is becoming even more severe in drier years, there is a dire need to increase conveyance efficiency. Such improvements may also contribute to increased WP and the comparative advantages of downstream farming zones with respect to the production of agricultural commodities. In addition, due to low water use efficiency, implementing field level water-wise options have much scope to maintain high WP and, therefore, to add to welfare increase particularly in downstream irrigated areas.

The findings on crop specific WP values allowed classifying districts according to their comparative advantages with respect to the cultivation of certain crops, and thus may guide future policy formulations aiming at optimal cropland allocation in the Khorezm region. Taking into account that cotton occupies more than half of the irrigated areas in each district crop diversification could allow improved water use efficiency, secure farm incomes, and effective crop rotations. However, since WP for each crop varies according to district as the results have shown, planting and watering crops should be planned according to the district's situation along the irrigation system, soil quality, and impact on the environmental sustainability, instead of following blanket recommendations for the whole district. The portfolio of crops grown needs to become more diversified, which would make it necessary, as a precondition, for the present constraints of the state order to be relaxed, and for the input-output markets as well as the market infrastructure to be improved.

References

Abdullaev I, De Fraiture C, Giordano M, Yakubov M, Rasulov A (2009) Agricultural water use and trade in Uzbekistan: situation and potential impacts of market liberalization. Int J Water Resour Dev 25(1):47–63

Awan UK (2010) Coupling hydrological and irrigation schedule models for the management of surface and groundwater resources in Khorezm, Uzbekistan. Ecology and development series no. 73, ZEF/University of Bonn, Bonn

Awan UK, Tischbein B, Martius C, Hafeez M, Kamalov P (2011) Modeling irrigation scheduling under shallow groundwater conditions as a tool for an integrated management of surface and groundwater resources. In: Martius C, Rudenko I, Lamers JPA, Vlek PLG (eds) Cotton, water, salts and *Soums*: economic and ecological restructuring in Khorezm, Uzbekistan. Springer, Dordrecht/Berlin/Heidelberg/New York

20 Estimation of Spatial and Temporal Variability of Crop Water Productivity...

Bekchanov M, Karimov A, Lamers JPA (2010a) Impact of water availability on land and water productivity: a temporal and spatial analysis of the case study region Khorezm, Uzbekistan. Water 2:668–684

Bekchanov M, Lamers JPA, Martius C (2010b) Pros and cons of adopting water-wise approaches in the lower reaches of the Amu Darya: a socio-economic view. Water 2:200–216

Bobojonov I, Lamers JPA (2008) Analysis of agricultural markets in Khorezm, Uzbekistan. In: Wehrheim P, Schoeller-Schletter A, Martius C (eds) Continuity and change: land and water use reforms in rural Uzbekistan. Socio-economic and legal analyses for the region Khorezm. IAMO, Halle/Saale, pp 165–182

Bobojonov I, Lamers JPA, Djanibekov N, Ibragimov N, Begdullaeva T, Ergashev A, Kienzler K, Eshchanov R, Rakhimov A, Ruzimov J, Martius C (2011) Crop diversification in support of sustainable agriculture in Khorezm. In: Martius C, Rudenko I, Lamers JPA, Vlek PLG (eds) Cotton, water, salts and *Soums*: economic and ecological restructuring in Khorezm, Uzbekistan. Springer, Dordrecht/Berlin/Heidelberg/New York

Conrad C (2006) Remote sensing based modeling and hydrological measurements to assess the agricultural water use in the Khorezm region, Uzbekistan. PhD dissertation, University of Wuerzburg, Wuerzburg

Conrad C, Dech SW, Hafeez M, Lamers JPA, Tischbein B (2011) Remote sensing and hydrological measurement based irrigation performance assessments in the upper Amu Darya Delta, Central Asia. Phys Chem Earth (in press)

Devkota KP (2011) Resource Utilization and Sustainability of Conservation Based Rice-Wheat Cropping Systems. PhD dissertation, Bonn University, Bonn

Djanibekov N, Bobojonov I, Lamers JPA (2011) Farm reform in Uzbekistan. In: Martius C, Rudenko I, Lamers JPA, Vlek PLG (eds) Cotton, water, salts and *Soums*: economic and ecological restructuring in Khorezm, Uzbekistan. Springer, Dordrecht/Berlin/Heidelberg/New York

Dukhovny VA, Sokolov VI (2002) Lessons on cooperation building to manage water conflicts in the Aral Sea Basin. Technical documents in hydrology: PC-CP Series: 11

Forkutsa I, Sommer R, Shirokova YI, Lamers JPA, Kienzler K, Tischbein B, Martius C, Vlek PLG (2009) Modeling irrigated cotton with shallow groundwater in the Aral Sea Basin of Uzbekistan: I. Water dynamics. Irrig Sci 27(4):331–346

Golan A, Judge G, Miller D (1996) Maximum entropy econometrics: robust estimation with limited data. Wiley, New York

Greene WH (2003) Econometric analysis, 5th edn. Prentice-Hall, New Jersey

Hydromet (2007) The state water cadastre: surface and ground water resources use and quality, Center of Hydrometeorological Service at the Cabinet of the Ministers of Uzbekistan Tashkent

HydroModRay (2002) Average requirements for water use for different crops in Khorezm, Urgench. Khorezm Department of Land and Water Resources of the Ministry of Agriculture and Water Resources

Just RE, Zilberman D, Hochman E, Bar-Shira Z (1990) Input allocation in multicrop systems. Am J Agric Econ 72(1):200–209

Lence SH, Miller DJ (1998) Estimation of multi-output production functions with incomplete data: a generalized maximum entropy approach. Eur Rev Agric Econ 25(2):188–209

Manschadi AM, Oberkircher L, Tischbein B, Conrad C, Hornidge A-K, Bhaduri A, Schorcht G, Lamers J, Vlek PLG (2010) White Gold' and Aral Sea disaster – towards more efficient use of water resources in the Khorezm region, Uzbekistan. Lohmann Info 45(1):34–47

Martius C, Lamers JPA, Wehrheim P, Schoeller-Schletter A, Eshchanov R, Tupitsa A, Khamzina A, Akramkhanov A, Vlek PLG (2004) Developing sustainable land and water management for the Aral Sea Basin through an interdisciplinary research. In: Seng V, Craswell E, Fukai S (eds) Water in agriculture. ACIAR proceedings no. 116, Canberra

Micklin P (1988) Dessiccation of the Aral Sea: a water management disaster in the soviet union. Science 241(4870):1170–1176

Micklin P (2002) Water in the Aral Sea Basin of Central Asia: cause of conflict or cooperation? Eur Geogr Econ 43(7):505–528

Müller M (2006) A general equilibrium approach to modeling water and land use reforms in Uzbekistan. PhD dissertation, Bonn University, Bonn

Müller M (2008) Where has all the water gone? In: Wehrheim P, Shoeller-Schletter A, Martius C (eds) Continuity and change: land and water use reforms in rural Uzbekistan. IAMO series 43:89–104

OblSelVodKhoz (2008) Data on norms, limits and usage of irrigation water from 2004 to 2007, Urgench. Khorezm Department of Land and Water Resources of the Ministry of Agriculture and Water Resources

OblStat (2008) Agricultural indicators for Khorezm oblast, 2000–2007. Khorezm Region Department of Statistics

Oweis T, Hachum A (2006) Water harvesting and supplemental irrigation for improved water productivity of dry land farming in West Asia and North Africa. Agric Water Manag 80:57–73

Shi Z, Ruecker GR, Mueller M, Conrad C, Ibragimov N, Lamers JPA, Martius C, Struntz G, Dech S, Vlek PLG (2007) Modeling of cotton yields in the Amu Darya river floodplains of Uzbekistan integrating multitemporal remote sensing and minimum field data. Agron J 99:1317–1326

Theil H, Goldberger AS (1961) On pure and mixed statistical estimation in economics. Int Econ Rev 2(1):65–78

Tischbein B, Awan UK, Abdullaev I, Bobojonov I, Conrad C, Forkutsa I, Ibrakhimov M, Poluasheva G (2011) Water management in Khorezm: current situation and options for improvement (hydrological perspective). In: Martius C, Rudenko I, Lamers JPA, Vlek PLG (eds) Cotton, water, salts and *Soums*: economic and ecological restructuring in Khorezm, Uzbekistan. Springer, Dordrecht/Berlin/Heidelberg/New York

Veldwisch GJ (2008) Cotton, rice & water: transformation of agrarian relations, irrigation technology and water distribution in Khorezm, Uzbekistan. PhD dissertation, Bonn University, Bonn

Part VI
Economic System Management Reform Options

Chapter 21
A Computable General Equilibrium Analysis of Agricultural Development Reforms: National and Regional Perspective

Maksud Bekchanov, Marc Müller, and John P.A. Lamers

Abstract Agriculture plays a pivotal role in the economy of both the Republic of Uzbekistan and the region Khorezm, as substantiated by a contribution of 25% of GDP and 40% of regional income. Since the agrarian sector is the engine of rural development and welfare, the impact of various agricultural policy scenarios on macroeconomic interrelationships and private and governmental earnings was examined using a computable general equilibrium (CGE) model developed for the national and the regional level. A total of seven scenarios include the current set-up as well as alternative cumulative and non-cumulative scenarios. Considering the substantial contribution of the agricultural sector to the GDP and regional income and the dominance of cotton production in the agricultural sector of Uzbekistan, non-cumulative scenarios include a partial liberalization of cotton market and an upgrade of the total factor productivities of livestock and main crop production sectors. The national and regional databases included production, final demand, and input–output relations for 20 sectors of the economy, of which seven belonged to the agrarian sector. The establishment of the CGE model for the economies of both the region and the country permitted the comparison of policies on both levels. The model findings suggest, among other findings, that the scenario of the liberalization of the present cotton production policy would not necessarily have an immediate and substantial impact on national and regional income. However, this policy change would substantially decrease government (state budget) revenues on the regional scale while significantly benefiting the private sector due to the enormous reliance

M. Bekchanov (✉) • M. Müller
Center for Development Research (ZEF), Walter-Flex-Str. 3, 53113 Bonn, Germany
e-mail: maksud@uni-bonn.de; molinero@uni-bonn.de

J.P.A. Lamers
ZEF/UNESCO Khorezm Project, Khamid Olimjan Str., 14, 220100 Urgench, Uzbekistan
e-mail: j.lamers@uni-bonn.de

C. Martius et al. (eds.), *Cotton, Water, Salts and Soums: Economic and Ecological Restructuring in Khorezm, Uzbekistan*, DOI 10.1007/978-94-007-1963-7_21,
© Springer Science+Business Media B.V. 2012

of the regional economy, particularly government earnings, on cotton and cotton-related industry sectors. The scenario of increased livestock productivity would yield a higher positive impact on national and regional income than cotton market liberalization and upgrading crop production efficiency. In spite of its negative impact on private revenues at the national scale, it would not only bear much room for increasing private incomes of the rural population, but also would allow a wider implementation of advanced water saving technologies, in particular in remote rural areas, and promote more crop biodiversity, at regional scale. In terms of government earnings, a livestock productivity increase is estimated to have higher impact than a crop-production efficiency increase on the national scale. However, on the regional level, it is the latter option that would produce higher governmental (state budget) earnings. It is argued that the effectiveness of regional development strategies in Uzbekistan would be enhanced by accounting for regional characteristics and the comparative advantages of each region, instead of a blanket approach to all regions in a uniform nationwide program.

Keywords Social Accounting Matrix • General equilibrium model • Cotton market liberalization • Livestock productivity • National and regional income • Governmental earnings • Private sector • Uzbekistan • The Khorezm region

21.1 Introduction

Uzbekistan is one of the five Central Asian countries that gained independence from the Soviet Union block in 1991. Between 1924 and 1991, the region was part of a political and economic system in which in particular irrigated cotton production was stimulated since the 1960s and the country was chosen to become a raw cotton provider for the entire Union. This was made possible due to an enormous extension of the irrigated areas from about 2,500,000 ha in 1960 (FAO AQUASTAT 1997) to about 4,300,000 ha in 2000 (FAO 2003). Cotton production was of importance during the Soviet era and keeps playing an important role in present-day Uzbekistan, which is the sixth largest cotton producer in the world and ranks second in the list of cotton fiber exporting countries (Rudenko et al. 2008). In particular, in the Khorezm region, located in the northwest of Uzbekistan on the lower reaches of the Amudarya river and subject to water release from up-stream regions, cotton is cultivated on more than half of 270,000 ha of irrigated land. Cotton export provides as high as 99% of the entire export revenues (Rudenko et al. 2008). However, despite the dominance of cotton production and its high contribution to agricultural revenues, "...*the cotton industry was blamed for political repression, economic stagnation, widespread poverty and environmental degradation in the region*" (International Crisis Group 2005).

At present, more than 60% of the total population of Uzbekistan lives in rural areas and generates their income directly or indirectly from agriculture (Müller 2006). Due to the dominant role of agriculture, political reforms addressing this sector may impact on the livelihood security of rural households in a variety of ways. Following its independence from the Soviet Union in 1991, Uzbekistan implemented different economic restructuring programs targeting all sectors of the economy, including agriculture. The restructuring of previously state-owned collective farms into private farms and the introduction of water user associations for managing water supply are among the most obvious outcomes of those reforms, although significant positive changes in farming practices or rural income generation did not occur (Abdullaev et al. 2009). Despite the gradual reforms in the centrally enforced raw cotton production and trade control system, and despite governmental support of liberalization in agriculture, particularly in cotton production and processing, technical progress and investment are taking place only at a slow pace.

Cotton processing and export contribute a high share of foreign currency revenues and thus support the state budget, which in turn is used to pay for employees in public organizations such as schools, kindergartens, and medical services, which account for about 23% of total employment (CEEP 2006). In addition, because of the cotton-processing infrastructure inherited from the Soviet Union and the lack of financial resources to develop new non-cotton industries, cotton production and processing is still prioritized by national economic strategies. However, the profitability of the cotton sector for either farmers or the governmental budget was debated in various studies (Müller 2008; Rudenko et al. 2008) due to fluctuations in world market cotton prices between 1998 and 2001. Technological improvements in the last 20 years have likely decreased cotton production costs in other developed and less-developed countries, while technological depreciation and lack of funds for restoration of the irrigation infrastructure were a common problem in many former Soviet republics.

Against this background, a number of research questions related to the Uzbek cotton sector had to be addressed: Are there opportunities to gain more benefits for all stakeholders through promoting the non-cotton sectors? What are the effects of changing policies on macroeconomic and welfare indicators, as well as on land and water allocation? Are policies that appear adequate at the national level also adequate on a regional level? This paper uses national and regional equilibrium models to analyze development opportunities at the national and regional level for Uzbekistan, using the Khorezm district as a case study for the regional level. The different scenarios that we analyzed are (1) a liberalization of the cotton market and an increase in total factor productivity in the (2) livestock and (3) main crop production sectors.

The Computable General Equilibrium (CGE) approach was chosen due to its economy-wide scope and the representation of relevant actors, such as government and producers, in spite of its limitations which are in an extensive data requirement – often not available for transition countries – and various assumptions underlying the model calibration. For instance, production function parameters in most CGE models are calibrated or borrowed from the literature rather than estimated econometrically. The standard IFPRI CGE model (Lofgren et al. 2001) was therefore adapted to the case of transition economies (Wehrheim 2003; Müller 2006), which in turn was calibrated to the datasets for national and regional economies of Uzbekistan for 2005.

These datasets were prepared in the form of Social Accounting Matrices (SAMs). A SAM represents all flows of economic transactions within an economy both at the national and regional levels and is the backbone of CGE models.

In the following, first the compilation of both the regional and national SAMs is described. The following section briefly discusses how national and regional CGE models are implemented based on SAMs. The main results obtained under various scenarios for the economies on regional and national scale are discussed next. The final section summarizes the main findings of the CGE analysis.

21.2 National and Regional Social Accounting Matrices

A Social Accounting Matrix (SAM) represents a double-accounting framework for monetary flows in entire economies for a given year. The construction of a SAM on national or regional level requires extended data sets, which often have to be collected from various sources. Using multiple sources to construct the national and regional SAM usually results in unbalanced accounts. In the present study, in order to obtain well-balanced micro-SAMs, the SAM accounts for 20 sectors were compiled in a step-by-step procedure, moving from highly aggregated datasets to more detailed representations. First, macro-SAMs, which are an aggregated balance of macroeconomic flows directly available from the System of National Accounts (SNA), were constructed. Then, the macro-SAMs were further disaggregated into SAMs with six productive sectors, which are conditionally called meso-SAMs. Finally, the desired micro-SAMs with 20 sectors were elaborated on the basis of the meso-SAMs. Unbalanced meso-SAMs and micro-SAMs obtained after the disaggregation were here balanced based on a maximum entropy approach (Robinson et al. 2001). The detailed steps were previously described in Müller (2006).

21.2.1 Macro-SAM

A macro-SAM is a highly aggregated circular flow matrix that captures all income-expenditure relationships in an economy for 1 year and is based on macro-economic totals, which can be taken directly from SNA (Müller 2006). A micro-SAM differs from the (underlying) macro-SAM since it reveals the micro-economic structure of respective macro-totals (e.g. intermediate or final demand) by sectors and other subsets of the national accounts (Müller 2006). For instance, households can be split based on different criteria (e.g. income groups, regions, etc. Müller 2006).

21.2.1.1 Macro-SAM for Uzbekistan

The macro-SAM for Uzbekistan (Table 21.1) was compiled from different sources such as Asian Development Bank (ADB) and Uzbek organizations such as the Centre for Efficient Economic Policy (CEEP) and Center for Economic Research (CER). While import and export data could be extracted from key indicators of the ADB, primary factor accounts were filled based on the calculations from datasets from CER. Other accounts of the macro-SAM were provided by CEEP, or estimated as the residue between totals of rows and columns of the SAM (Müller 2006).

21 A Computable General Equilibrium Analysis of Agricultural Development Reforms... 351

Table 21.1 Macro-SAM[a] for Uzbekistan in 2005 (in billion UZS, current)

		Expenditures										
		ACT	COM	LAB	CAP	HHO	ENT	GOV	TAX	S-I	ROW	SUM
Revenues	Activities (ACT)		13,717									13,717
	Commodities (COM)					7,931		2,225		3,499	6,382[b]	20,038
	Wages (LAB)	5,160										5,160
	Operating surplus (CAP)	9,070										9,070
	Households (HHO)			5,160			7,378	274			389	13,201
	Enterprises (ENT)				8,099							8,099
	Government (GOV)						228		2,608			2,836
	Taxes (TAX)	-513	1,493		700	435	493					2,608
	Savings-Investment (S-I)					4,835		337			-1,673	3,499
	Rest of the world (ROW)		4,828[b]		271							5,099
	Total (SUM)	13,717	20,038	5,160	9,070	13,201	8,099	2,836	2,608	3,499	5,099	

Source: [a]CEEP (2006) and calculation results, [b]ADB (2006)

21.2.1.2 Macro-SAM for Khorezm

For constructing the regional macro-SAM for Khorezm (Table 21.2), data on the regional economy were provided by the Regional Statistical Department (Oblstat 2006). Although this source provided most of the data needed, several assumptions had to be made concerning the volume of capital expenditures (operating surplus). This in the end was estimated as the difference between the total value added and the product of the average number of employees times the average wage. Due to the absence of reliable information on trade between Khorezm and the rest of the country, the account "Rest of Uzbekistan" was not included. Export and import flows between the region and foreign countries were included in the "Rest of the World" account.

21.2.2 Meso-SAM

The constructed macro-SAMs represent a consistent overview of the economies at country- (Uzbekistan) and regional (Khorezm) levels for 2005 and formed the basis for all further computations. For this purpose, both SAMs were disaggregated into a SAM with six productive sectors (agriculture, industry, construction, trade, transport and communication, and other services) subsequently referred to as "meso-SAMs". The accounts of the meso-SAM were chosen to allow for as much information as possible to be used (Müller 2006). For example, "construction" and "trade" sectors are represented as the sectors that do not produce any commodities for consumption. Any payments to the construction sector are regarded as investments while the trade sector received the trade mark-ups from all sectors.

21.2.2.1 Meso-SAM for Uzbekistan

For the establishment of the meso-SAM for Uzbekistan (Table 21.3), data from the macro-SAM of 2005 and from the input–output matrix for 2005 by CER (CER-IOM, UNDP 2006) was used. These data sources resulted in the compilation of the input–output table (IOT). Value-added by the six productive sectors were obtained from the ADB database (www.adb.org/Documents/Books/Key_Indicators/2006/pdf/UZB.pdf). Components of value-added (compensation of employees, operating surplus), investment, export and import were calculated based on the shares previously reported for 2005 (UNDP 2006). Trade margins (mark-ups) for each sector were estimated based on national averages. Indirect taxes were calculated according to sectoral shares for the 13 sectors input–output model of the CER (UNDP 2006) and the relative shares of export, import and domestic consumption in those sectors. Other accounts of the meso-SAM were derived directly from the developed macro-SAM. These compilation steps resulted in a complete, yet unbalanced meso-SAM. The balancing procedure was completed by the use of macro-SAM as control-totals, following the works on Maximum Entropy applications by Robinson et al. (2001)

Table 21.2 Macro-SAM for the Khorezm region in 2005 (in billion UZS, current)

		Expenditures										
		ACT	COM	LAB	CAP	HHO	ENT	GOV	TAX	S-I	ROW	SUM
Revenues	Activities	ACT		1,118									1,118
	Commodities	COM	584				357		69		56	99	1,164
	Wages	LAB	289										289
	Operating surplus	CAP	246										246
	Households	HHO			289			223	13			0	525
	Enterprises	ENT				238							238
	Government	GOV						7		60			66
	Taxes	TAX		28		8	16	8					60
	Savings–Investment	S-I					153		−16			−81	56
	Rest of the world	ROW		18									18
	Total	SUM	1,119	1,164	289	246	525	238	66	60	56	18	

Source: Authors' results on the basis of OblStat (2006)

Table 21.3 Meso-SAM for Uzbekistan in 2005 (in billion UZS)

Macro	Name	Meso	№	1	2	3	4	5	6	7	8	9	10	11	12
	Agriculture	AAGR6	1							3,518					
	Industry	AIND6	2								11,824				
	Construction	ACON6	3									2,085			
	Trade	ATRD6	4										2,000		
ACT	Transport and communication	ATCM6	5											3,191	
	Other services	AOTS6	6												3,645
	Agriculture	CAGR6	7	512	1,297		7		11						
	Industry	CIND6	8	1,343	6,091	957	159	763	736						
	Construction	CCON6	9												
COM	Trade	CTRD6	10												
	Transport and communication	CTCM6	11	119	488	333	214	606	344						
	Other services	COTS6	12	108	348	36	199	92	302						
	Trade	TRCD6	13							279	542	142		11	244
TRD	Transport and communication	TRCM6	14							4	328	1		11	8
	Other services	TRCE6	15							6	444			9	5
LAB	Labor	LAB6	16	933	672	517	459	1,087	1,493						
CAP	Operating surplus	CAP6	17	2,868	2,470	225	945	626	1,236						
HHO	Households	HHO6	18												
	Shirkats	SHR6	19												
ENT	Farmer	FER6	20												
	Dehqon	DKH6	21												
	Other enterprises	ENT6	22												
GOV	Government	GOV6	23												
	Indirect taxes	ITD6	24							4	236	86	32	474	70
Ti	Import taxes	ITM6	25								167			62	1
	Export taxes	ITE6	26								213			146	1
Td	Direct taxes	DTX6	27												
Tf	Factor taxes	RES6	28	9	603	18	18	18	35						
Ts	Subsidies	SUB6	29						−513						
S-I	Capital account	S-I6	30												
ROW	"Rest of the world"	ROW6	31							63	4,303	14		326	121
		Total	32	5,893	11,969	2,085	2,000	3,191	3,645	3,874	18,056	2,329	2,033	4,231	4,095

13	14	15	16	17	18	19	20	21	22	23	24	25	26	27	28	29	30	31	32
					2,375														5,893
					145														11,969
																			2,085
																			2,000
																			3,191
																			3,645
					1,696					200							60	89	3,874
					1,680												1,022	5,305	18,056
																	2,329		2,329
1,217	352	463																	2,033
					923					292								911	4,231
					1,112					1,732							89	77	4,095
																			1,217
																			352
																			463
																			5,160
																			8,370
			5,160			405	648	1,727	4,598	274								389	13,201
				439															439
				703															703
				1,727															1,727
				5,231															5,231
									228		904	231	359	928	700	−513			2,836
																			904
																			231
																			359
						435	34	54		404									928
																			700
																			−513
					4,835					337								−1,673	3,499
				271															5,099
1,217	352	463	5,160	8,370	13,201	439	703	1,727	5,231	2,836	904	231	359	928	700	−513		3,499	5,099

356 M. Bekchanov et al.

and Müller (2006). The model was implemented with the General Algebraic Modelling System (GAMS), using the numerical solver CONOPT3.

21.2.2.2 Meso-SAM for Khorezm

The meso-SAM for Khorezm (Table 21.4) was obtained by the same procedures as described for the elaboration of the national meso-SAM. Data were mainly provided by the Regional Statistics Department (Oblstat 2006). However, in the absence of data to construct the IOT, the input–output coefficients (IOCs) from the national meso-SAM were used for building the regional IOT. Household and commodity consumption expenditures by sectors were calculated according to sectoral shares for these accounts in the national meso-SAM. In calculating capital and labor expenditures, labor was estimated as the product of the average wage and the average number of employees while capital (operating surplus) was assessed as the difference between the values added and labor expenditures for each of the six sectors. Trade margins, resource and indirect tax accounts by each of the sectors were derived from the regional average rates. Import and export volumes by sectors were provided by regional statistics (Oblstat 2006).

As previously described for the national meso-SAM, the regional meso-SAM was also balanced applying the maximum entropy approach. The balancing procedure revealed various differences between the original or derived unbalanced and the final balanced datasets at the national and regional levels. This could not be avoided since some information was not available from any source and therefore could only be indirectly derived without any possibility to cross-check for consistency.

21.2.3 Micro-SAM

Given the intended focus on the agricultural and closely related sectors, the still highly aggregated meso-SAM accounts for agriculture and industry were disaggregated further into seven and nine sub-sectors, respectively. Similar to the use of macro-SAM as control-totals in the previous step, now the meso-SAMs formed the boundary conditions for the targeted micro-SAMs with 20 sectors.

21.2.3.1 Micro-SAM for Uzbekistan

The structure of micro-SAM for 2005 follows closely the micro-SAM for 2001 by Müller (2006), but differs from the CER-IOM for 2005 as data for only 13 sectors of the economy were made available in the latter. To disaggregate the agricultural sector, IOCs from the micro-SAM developed for 2001 (Müller 2006) were used. In contrast, for a more suitable disaggregating of the industrial sector, IOCs of the

CER-IOM for 2005 as reported by UNDP (2006) were used. The missing data in the CER-IOM for 2005, such as for light manufacturing and the cotton processing sub-sector, were derived from data and shares between sectors as developed previously for the micro-SAM 2001 (Müller 2006), to complete the disaggregation. In the third step, the capital account was disaggregated into land, water and other capital accounts according to the shares and factors derived from the micro-SAM 2001. Factor accounts were split into sub-sectors due to shares from the CER-IOM for 2005 and the micro-SAM for 2001. Values for "the Rest of the World", and "Household" accounts were calculated based on shares from CER-IOM for 2005 and the micro-SAM for 2001. The rest of the accounts were completed according to the balanced Meso-SAM.

21.2.3.2 Micro-SAM for Khorezm

To complete the micro-SAM of the Khorezm region, IOCs of the consumption expenditure from the national SAM were used. The calculations of the different disaggregated account values followed the same procedures as used during the completion of the national micro-SAM. Foreign trade volumes by sectors were obtained from OblStat (2006), while trade margins and indirect taxes were determined in accordance with average sector rates.

The thus obtained unbalanced micro-SAMs for both levels were balanced once more running a GAMS solver using the maximum entropy method. The final micro-SAMs were used for the elaboration of the base scenario in the following general equilibrium model calculations.

21.3 General Equilibrium Model and Simulations

CGE models are an efficient tool to analyze the impact of different policies and market shocks to an economy under study. By considering all financial flows in this economy, including inter-sectoral economic relations and public and private budgets, the influence of changes on the different economy agents can be estimated.

21.3.1 CGE Model Calibration

The standard CGE model was used in the scenario analyses (Lofgren et al. 2001). This model represents various aspects of an economy with considerable detail. For the calibration, additional data such as the elasticity coefficients for the Armington function and elasticities of substitution for constant elasticity of substitution (CES) functions were taken as previously developed for Uzbekistan (Müller 2006). After setting the different exogenous parameters, the model was calibrated in such a way

Table 21.4 Meso-SAM for Khorezm in 2005 (in billion UZS)

Macro	Name	Meso	№	1	2	3	4	5	6	7	8	9	10	11
	Agriculture	AAGR6	1							185				
	Industry	AIND6	2								217			
	Construction	ACON6	3									51		
	Trade	ATRD6	4										44	
ACT	Transport and communication	ATCM6	5											46
	Other services	AOTS6	6											
	Agriculture	CAGR6	7	13	90		0		1					
	Industry	CIND6	8	78	65	9	1	3	8					
	Construction	CCON6	9											
COM	Trade	CTRD6	10											
	Transport and communication	CTCM6	11	2	9	3	2	11	4					
	Other services	COTS6	12	7	8	2	6	2	16					
	Trade	TRCD6	13							8	13	4		4
TRD	Transport and communication	TRCM6	14								1			
	Other services	TRCE6	15								8			
LAB	Labor	LAB6	16	128	32	30	18	16	65					
CAP	Operating surplus	CAP6	17	112	16	7	17	14	81					
HHO	Households	HHO6	18											
	Shirkats	SHR6	19											
ENT	Farmer	FER6	20											
	Dehqon	DKH6	21											
	Other enterprises	ENT6	22											
GOV	Government	GOV6	23											
	Indirect taxes	ITD6	24							6	7	3	1	2
Ti	Import taxes	ITM6	25								1			
	Export taxes	ITE6	26								4			
Td	Direct taxes	DTX6	27											
Tf	Factor taxes	RES6	28											
Ts	Subsidies	SUB6	29						−2					
S-I	Capital account	S-I6	30											
ROW	"Rest of the world"	ROW6	31								62			
		Total	32	341	220	51	44	46	172	199	312	57	46	52

12	13	14	15	16	17	18	19	20	21	22	23	24	25	26	27	28	29	30	31	32
						156														341
						3														220
																				51
																				44
																				46
172																				172
						95														199
						45													103	312
						1												56		57
	37	1	8																	46
						21														52
						75					69								0	184
7																				37
																				1
0																				8
																				289
																				246
				289			9	28	71	115	13									525
					9															9
					28															28
					71															71
					130															130
										7		24	1	4	24	8	−2			65
5																				24
																				1
0																				4
						16				8										24
				8																8
																				−2
						114					−16								−42	56
																				62
184	37	1	8	289	246	525	9	28	71	130	65	24	1	4	24	8	−2	56	62	

that the model's equilibrium in the base scenario for 2005 reflected the observed equilibrium as represented by the micro-SAM.

21.3.2 Macroeconomic Closure Rules

The standard CGE model includes three macroeconomic balances: the current account of the government, the current account of the balance of payments, and the savings-investment balance, as previously suggested (Lofgren et al. 2001). A set of closures (macroeconomic constraints) determines the way how these accounts come into a balance (Müller 2006). Considering the specific properties of the Uzbek economy, flexible government savings with fixed direct tax rates were assumed as closure for the current account of the government (Müller 2006). The closure rule of flexible foreign savings with fixed real exchange rate was specified for current account of the balance of payments. Fixed investment and government consumption absorption shares (flexible quantities) with uniform marginal propensity to save (MPS) point change for selected institutions closure were admitted for the savings-investment balance.

21.3.3 Simulations

Single shocks after which the model economy reached a new equilibrium conditionally are called "items" here. The scenarios may consist of neither or several items according to the purpose of the experimental setting.

Item 1 – Liberalization of the Uzbek cotton market considered a 50% decline in subsidies for raw cotton producers, 50% decrease in indirect taxes for the cotton processing industries compared to the baseline scenario and omitting the state order that sets cotton production targets.

Considering the high contribution of cotton production to agricultural income and the presence of several agricultural policies which may hinder functioning of free markets in the agricultural sector – such as (i) an enforcement of the cotton production targets, (ii) provision of input subsidies to producers which in turn would influence the welfare of the cotton-processing industry due to lower prices for raw cotton, and (iii) existence of export taxes in the cotton processing sector, it would be interesting to analyze the impacts of the relaxation of these conditions on the incomes of the government and households.

Except for these policy changes, the effects of improved total factor productivity of animal production and main crops were simulated as well, taking into account the key role of the agricultural sector in the economy of Uzbekistan.

Item 2 – Improvement of total factor productivity of animal production builds on the observation of low animal productivity (Djanibekov 2008). The total factor

21 A Computable General Equilibrium Analysis of Agricultural Development Reforms... 361

Table 21.5 Overview of the scenarios (Sc.) related to the agricultural sector

Scenarios	Items		
	1. Liberalization of cotton market	2. Improvement of total factor productivity of animal production	3. Improvement of total factor productivity of main crops
Base	–	–	–
Sc. 1	+	–	–
Sc. 2	–	+	–
Sc. 3	–	–	+
Sc. 4	+	+	–
Sc. 5	+	–	+
Sc. 6	–	+	+
Sc. 7	+	+	+

Note: Items in scenarios: + included, – not included

productivity of the livestock sector is increased by 10% to gain insights into the relative importance of this sub-sector.

Item 3 – Improvement of total factor productivity of main crops considered the improvement of total factor productivity (increase by 10%) for the main crops cotton, cereals, and rice at the same rate.

The three items formed seven scenarios as given in Table 21.5.

To maintain the *ceteris-paribus* characteristic of the simulations, improvements of total factor productivity are here assumed to take place exogenously, thus without considering the sources of the increase.

21.3.4 Results

Table 21.6 shows the major macroeconomic results for the given policy scenarios. In this section, policy change impact results at the national scale are discussed first, followed by the analysis of the simulation results for regional scale.

When comparing the impact of the first three non-cumulative scenarios, the following observations were made: First, under partial cotton market liberalization for the entire national economy (Scenario 1), the domestic market prices for raw cotton and cotton fiber would increase by almost 10% and 6.5%, respectively. At the same time, this would trigger an increase in the production of raw cotton and cotton lint by 1.0% and 0.4%, respectively. Yet, cotton market liberalization would also cause a 1.2% decrease in average wages while providing an additional 2.7% employment in the cotton production sector. Consequently, total income of the private sector would decline by 0.8% owing to the 1.2% drop in wages, and an 8.1% decrease in government transfers. In contrast, the national revenues would decline by 0.2% only.

The increased livestock productivity in scenario 2 has a small negative impact on wages at national scale, but the largest positive impact on the operating surplus.

Table 21.6 Scenario (Sc.) results for cotton market liberalization and agricultural technology change simulations

Macroeconomic indicators	Uzbekistan Billion UZS	Sc. 1	Sc. 2	Sc. 3	Sc. 4	Sc. 5	Sc. 6	Sc. 7	Khorezm Billion UZS	Sc. 1	Sc. 2	Sc. 3	Sc. 4	Sc. 5	Sc. 6	Sc. 7
	Base				Change to base (in percent)				Base				Change to base (in percent)			
Generation of national income									*Generation of or regional Income*							
Wages	5,160	−1.2	−0.2	0.1	−1.5	−1.1	−0.2	−1.4	289	0.7	1.0	1.0	1.6	1.8	2.0	2.8
Operating surplus	9,070	0.7	3.0	0.6	3.7	1.3	3.7	4.4	246	−1.1	3.8	1.8	2.7	0.5	5.7	4.5
Indirect taxes less subsidies	1,055	−0.5	2.1	−0.3	1.6	−0.9	1.8	1.2	23	−1.1	3.4	0.5	2.4	−0.7	4.0	2.9
Imports	4,828	−0.5	2.2	−0.8	1.7	−1.4	1.5	0.8	62	0.8	2.4	3.0	3.2	4.0	5.5	6.4
Total generation	20,113	−0.2	2.0	0.1	1.8	−0.1	2.1	1.9	620	−0.1	2.3	1.5	2.3	1.4	3.9	3.8
Usage of national income									*Usage of regional income*							
Investment	3,499	−2.3	2.4	−3.0	0.1	−5.7	−0.7	−3.4	56	−18.9	6.8	−20.7	−11.3	−42.2	−13.1	−33.9
Exports	6,382	1.2	0.3	1.5	1.5	2.8	1.8	3.1	104	10.0	−2.4	17.3	7.3	28.7	14.4	25.7
Government consumption	2,225	−0.5	0.4	−0.5	−0.1	−1.2	0.0	−0.7	69	−1.6	−0.6	−2.8	−2.1	−4.6	−3.3	−5.0
Private consumption	8,006	−0.3	3.5	0.6	3.2	0.4	4.1	3.9	392	0.2	3.5	1.2	3.6	1.5	4.8	5.0
Total Usage	20,113	−0.2	2.0	0.1	1.8	−0.1	2.1	1.9	620	−0.1	2.3	1.5	2.3	1.4	3.9	3.8
Current revenues of the national government									*Current revenues of the regional government*							
Indirect taxes less subsidies	1,055	−0.5	2.1	−0.3	1.6	−0.9	1.8	1.2	23	−1	3.4	0.5	2.4	−0.7	4.0	2.9
Other tax and non-tax revenues	1,553	0.5	−0.3	0.9	0.2	1.0	0.5	0.7	36	−10	0.9	3.5	−9.2	−8.2	4.2	−7.1
Total revenues	2,608	0.1	0.6	0.4	0.8	0.3	1.0	0.9	58	−7	1.9	2.3	−4.7	−5.3	4.2	−3.3

Current expenditures of the national government								*Current expenditures of the regional government*							
Government consumption	2,225	−0.5	0.4	−0.5	−0.1	−1.2	0.0	−0.7	69	−2	−0.6	−2.8	−2.1	−4.6	−3.3	−5.0
Transfers to households	274	−8.1	−0.9	−2.9	−9.1	−10.6	−4.4	−12.3	13	1.8	11.6	6.4	13.2	8.5	17.9	19.6
Governmental savings	109	33.5	8.3	26.4	42.9	57.1	35.9	67.5	−23	13	0.2	−10.5	13.1	4.5	−10.1	4.4
Total expenditures	2,608	0.1	0.6	0.4	0.8	0.3	1.0	0.9	58	−7	1.9	2.3	−4.7	−5.3	4.2	−3.3
Current revenues of the private sector								*Current revenues of the private sector*							
Wages	5,160	−1.2	−0.2	0.1	−1.5	−1.1	−0.2	−1.4	289	0.7	1.0	1.0	1.6	1.8	2.0	2.8
Governmental transfers	274	−8.1	−0.9	−2.9	−9.1	−10.6	−4.4	−12.3	13	1.8	11.6	6.4	13.2	8.5	17.9	19.6
Other income	7,767	−0.3	−0.5	0.1	−0.8	−0.2	−0.4	−0.7	223	2.2	−0.3	1.4	2.0	3.8	1.1	3.5
Total revenues	13,201	−0.8	−0.4	0.1	−1.2	−0.8	−0.4	−1.2	525	1.4	0.7	1.3	2.1	2.8	2.0	3.5

This indicates that the improvement of the total factor productivity in the livestock sector may induce labor-savings on national scale. In contrast, technical progress in the production of main crops had a generally positive influence on the increase in various macroeconomic indicators.

An increased livestock productivity caused an income decrease of 0.4% in the private sector due to decreased wages (−0.2%) and governmental transfers (−0.9%). In addition, the increase of animal production efficiency stimulated the national income by 2.0% due to increased imports (+2.2%) and operating surplus (+3%). Total government revenues raised by 0.6% as a result of the 2.1% increase in indirect taxes and fewer subsidies.

Increase in the total productivity of the crop production sectors provided only a 0.1% increase of the national income and a 0.4% increase in government revenues. However, all stakeholders benefited from this policy change although at different magnitudes.

The findings of the same scenarios for the Khorezm region differed from those for the national economy (Table 21.6). Under the assumptions of partial market liberalization (Scenario 1), the general regional income of the study region Khorezm did not differ much from the baseline scenario. However, there was substantial change in the components of the regional income. For example, exports increased by 10% and investments reduced by 18% on regional level. At the same time, governmental revenues decreased by 7% and private income increased by 1.4% due to the 1.8% increase in governmental transfers and 0.7% shift in wages.

Although the income of the private sector rose in all scenarios examined, the level of increase differed among the seven scenarios: from +0.7% in scenario 2 to +3.5% in scenario 7. The highest overall growth of the regional income of 3.9% was found in scenario 6. An increased livestock productivity would provide much higher growth in total national income than an increased crop productivity (+2.3% and +1.5% growth in scenarios 2 and 3, respectively), while the latter option would contribute less than the second scenario simulations to the revenue of the government (+1.9% and +2.3% increase in scenarios 2 and 3, respectively).

In Tables 21.7 and 21.8, land and water allocations due to partial market liberalization, increased plant growth and livestock sector productivity are given. Market liberalization brought about little change in cropland allocation at the national scale. However, increased livestock productivity may cause an increase of fodder-cropped lands by 11%. In the case of increased plant productivity, lands cropped to grain and rice declined by 17% and 12%, respectively.

The same observations were revealed at regional scale with similar directional change, although at a different degree than in the case of the entire country. Cotton market liberalization led to an increase in cotton area by 6%, whereas the area for all other crops would decline (Table 21.7). The livestock productivity increase brought an increase in fodder area by 14%. Under scenario 3, grain and rice areas reduced by 13% an 11%, respectively.

The changes in water allocation under various scenarios showed the same directional trends as the change in land allocation in response to these scenario simulations (Table 21.8).

Table 21.7 Land allocation scenarios (Sc.) due to simulations related to the agricultural sector

Crops	Uzbekistan								Khorezm							
	Base	Sc. 1	Sc. 2	Sc. 3	Sc. 4	Sc. 5	Sc. 6	Sc. 7	Base	Sc. 1	Sc. 2	Sc. 3	Sc. 4	Sc. 5	Sc. 6	Sc. 7
	1,000 ha	Change to base year (in percent)							1,000 ha	Change to base year (in percent)						
Cotton	1,472	1	0	0	−2	11	0	7	110	6	0	0	2	16	0	12
Grain	1,473	−1	5	−17	1	−13	−12	−11	69	−6	5	−13	−2	−18	−8	−15
Rice	53	−1	5	−12	1	−7	−7	−5	19	−5	2	−11	−4	−16	−9	−15
Fruit and vegetable	329	−1	2	0	−2	6	1	4	9	−7	2	2	−6	−7	4	−5
Fodder	290	0	11	−3	8	2	8	11	6	−3	14	3	10	−1	17	12
Other crops	223	−1	1	0	−3	5	0	3	10	−7	2	1	−6	−7	3	−6

Table 21.8 Water allocation scenarios (Sc.) due to simulations related to the agricultural sector

| Crops | Uzbekistan | | | | | | | | Khorezm | | | | | | | |
| | Base | Sc. 1 | Sc. 2 | Sc. 3 | Sc. 4 | Sc. 5 | Sc. 6 | Sc. 7 | Base | Sc. 1 | Sc. 2 | Sc. 3 | Sc. 4 | Sc. 5 | Sc. 6 | Sc. 7 |
	km^3	Change to base year (in percent)							km^3	Change to base year [in percent]						
Cotton	8	1	−3	7	−2	9	4	6	0.61	7	−4	13	3	19	9	16
Grain	6	−1	2	−12	1	−14	−10	−12	0.27	−5	4	−12	−1	−16	−8	−12
Rice	1	−1	2	−7	1	−8	−5	−6	0.51	−4	1	−10	−3	−14	−9	−13
Fruit and vegetable	2	−1	−1	5	−2	4	4	3	0.06	−6	1	3	−5	−4	4	−3
Fodder	2	0	8	3	8	1	11	9	0.05	−2	13	4	11	2	18	16
Other crops	1	−1	−2	5	−3	4	3	1	0.06	−6	1	2	−5	−4	3	−3

21.4 Discussion

Salient findings of the analyses are the high impact of technical progress in the livestock sector and the substantial differences of policy options on regional and national level.

21.4.1 Livestock Productivity Change

Increasing livestock productivity provided substantial increase in overall income, operating surplus, and government earnings on both country and regional levels. High positive welfare impact of and an enormous potential for increasing livestock rearing productivity were also estimated by Djanibekov (2008) for the Khorezm region, owing to the present low productivity of the livestock sector in this region. Particularly remote rural districts that regularly face water scarcity would benefit from a development in livestock rearing. Since livestock at present is in particular kept by rural households, livestock productivity improvement may increase incomes of the majority of the rural population. Increased private incomes, at regional scale, may promote adoption of the advanced water saving technologies of plant production in water scarce areas, which at present are not widely accepted due to high initial investments (Bekchanov et al. 2010). Although livestock rearing has become key to livelihood security in rural Uzbekistan (Kan et al. 2008), it is hampered because the demand for high quality feed is presently surpassing its production (Djumaeva et al. 2009). Hence, a livestock productivity increase requires more and better fodder. Replacing the cotton monoculture with high feed value crops such as alfalfa would bear much room for promoting new varieties of fodder crops, and introducing intercropping systems and crop rotations (Djanibekov 2008). In addition, to avoid forage price increases which would in turn decrease the present comparative advantages of the livestock sector, the inadequate feeding practices could be improved, for example, by mixing tree foliage with the present low quality feed diets. With the support of a specially elaborated least-cost-ratio model, recent findings showed that mixing tree foliage with common feed bears the potential of reducing the dependence on the presently available feed or fodder while increasing farmers' profits by 53% at the season-onset, by 38% in mid-season and by 34% at the end of the season (Djumaeva et al. 2009).

21.4.2 Policy Change Impacts – National Versus Regional Scale

The regional results partly coincide with, and partly contradict the impact of policies on the national economy because of structural differences in the economies between the region and the nation as a whole. For instance, partial market liberalization has the potential to only slightly increase governmental revenues but even to decrease private income on national scale. In contrast, the same policy at the regional scale would clearly benefit the private sector due to increased wages and government transfers, but it would decrease government revenues due to reduced taxes and non-tax revenues. A substantial positive impact of market liberalization on income earnings and overall welfare was shown in studies by Bobojonov (2009) and Djanibekov (2008). However, both studies focus on Khorezm and the liberalization policies they analyzed differ from those discussed here.

A plant productivity increase would result in higher governmental (state budget) income change than a livestock productivity increase at regional level. At the national level, the plant productivity increase would bring fewer benefits compared to those in scenario 2. The differential impacts predicted at various scales can be explained by the difference in natural conditions and economic structure between the region and the whole country. For example, agriculture provides almost half of the regional income but only one third of the national income. Export revenues from raw and lint cotton are dominant in the regional economy, while energy commodities are substantial for national exports (CEEP 2006). Moreover, the share of the industrial sector is substantial at the national scale but of only modest importance at the regional scale. Thus, a partial cotton market liberalization may substantially decrease government earnings at regional level while only slightly influencing budget revenues at the national level. These and findings of Lovo et al. (2010) for Italy with the employment of national and multi-regional CGE models, indicate that in case of such structural differences between various scales, a diversified regional policy should be promoted rather than a single national agricultural reform policy.

21.5 Summary and Conclusions

The results of the simulations based on the general equilibrium model provided insight into the potentials of possible growth in the national and regional economies owing to policy changes in the agrarian sector. The impact of the policies of only a (partial) liberalization of the cotton market at national scale was estimated to be only small. However, due to the high dependence of regional economies on agriculture and cotton export revenues as well as the dominance of cotton in crop cultivation, such kind of policy would significantly decrease governmental earnings and provide increased private income at regional scale. In contrast, the increased livestock productivity scenario showed high welfare improvement potentials at both the national and regional scales. Livestock rearing productivity may have positive impacts on environmental indicators as well, since it promotes sustainable crop rotations and maintains rural household income, which can be used to finance advanced water saving technologies in the agricultural sector. The livestock sector has enormous potential to increase welfare level particularly in remote, poor rural areas.

Cotton market liberalization causes substantial decline in government (state budget) revenues at regional scale while its effects on the level of regional income are rather insignificant. Policy simulations, which related to the agricultural sector of the whole country, indicated that increasing total productivity of the livestock sector has the potential to provide more incomes than increasing total productivity of crop production; although it influences negatively on the income of private households, it does so to a small degree only. However, the situation differed at the regional level. Both simulations of increasing productivity in livestock and crop producing sectors predicted positive and significant income effects for all economic agents, but policies in crop production influenced the incomes of the private sector and revenues of the government slightly more than those in the livestock sector. The different outcomes of the same simulations for the economies of the region and country underline the importance of a regionally diversified policy that should account for the special features of the economies of each region and, thus, can provide more benefits for the stakeholders on both the national and regional levels.

Although the results presented are based upon actual economic statistics for the Republic of Uzbekistan and Khorezm, one of the 13 administrative regions of Uzbekistan, the findings have been interpreted with caution. This was needed since, in the absence of some data sets, these gaps had been filled while considering various assumptions. Because the construction and application of huge models such as CGE is a rather new experience in the context of Uzbekistan and the Khorezm region, this study is considered as a first step towards building both a comprehensive dataset and a model for macro-economic policy analysis in Uzbekistan. Future challenges include the improvement of the quality and credibility of data as well as multi-regional disaggregation of the Social Accounting Matrix for Uzbekistan.

References

Abdullaev I, De Fraiture C, Giordano M, Yakubov M, Rasulov A (2009) Agricultural water use and trade in Uzbekistan: situation and potential impacts of market liberalization. Int J Water Resour Dev 25(1):47–63

ADB (2006) Key indicators 2006: measuring policy effectiveness in health and education. pp 374–378. http://www.adb.org/Documents/Books/Key_Indicators/2006/pdf/UZB.pdf

Bekchanov M, Lamers JPA, Martius C (2010) Pros and cons of adopting water-wise approaches in the lower reaches of the Amu Darya: a socio-economic view. Water 2:200–216

Bobojonov I (2009) Modeling crop and water allocation under uncertainty in irrigated agriculture: a case study on the Khorezm Region, Uzbekistan. PhD dissertation, Bonn University, Bonn

CEEP (2006) Uzbekistan economy: statistical and analytical review. Centre for Effective Economic Policy, Ministry of Economy, Uzbekistan

Djanibekov N (2008) A micro-economic analysis of farm restructuring in Khorezm region, Uzbekistan. PhD dissertation, Bonn University, Bonn

Djumaeva DM, Djanibekov N, Vlek PLG, Martius C, Lamers JPA (2009) Options for optimizing dairy feed rations with foliage of trees grown in the irrigated drylands of Central Asia. Res J Agric Biol Sci 5:698–708

FAO (2003) Fertilizer use in Uzbekistan. Rome, Italy. http://www.fao.org/DOCREP/006/Y4711E/y4711e4704.htm

FAO AQUASTAT (1997) Internet database. http://www.fao.org/nr/water/aquastat/countries/uzbekistan/index.stm

International Crisis Group (2005) The curse of cotton: Central Asia's destructive monoculture. Asia report no. 93, 28 Feb 2005

Kan E, Lamers JPA, Eschanov R, Khamzina A (2008) Small-scale farmers' perceptions and knowledge of tree intercropping systems in the Khorezm region of Uzbekistan. For Trees Livelihoods 18:355–372

Lofgren H, Harris RL, Robinson S (2001) A standard computable general equilibrium (CGE) model in GAMS. International Food Policy Research Institute (IFPRI), Trade and Macroeconomics Division (TMD), discussion paper 75. Washington, DC

Lovo S, Magnani R, Perali F (2010) A multi-regional general equilibrium model to assess policy effects at regional level. Paper prepared for the 116th European association of agricultural economists seminar "Spatial dynamics in agrifood systems: implications for sustainability and consumer welfare", Parma, pp 127–130, Oct 2010

Müller M (2006) A general equilibrium approach to modeling water and land use reforms in Uzbekistan. PhD dissertation, Rheinische Friedrich-Wilhelms-Universität Bonn, Bonn

Müller M (2008) Cotton, agriculture, and the Uzbek Government. In: Wehrheim P, Shoeller-Schletter A, Martius C (eds) Continuity and change: land and water use reforms in rural Uzbekistan, vol 43, Studies on the agricultural and food sector in Central and Eastern Europe. Leibniz-Institut für Agrarentwicklung in Mittel- und Osteuropa IAMO, Halle, pp 183–203

Oblstat (2006) Regional department of the ministry of macro-economics and statistics of the republic of Uzbekistan in Khorezm Oblast; socio-economic indicators for Khorezm. 1991–2005, Urgench

Robinson S, Cattaneo A, El-Said M (2001) Updating and estimating a social accounting matrix using cross entropy methods. Econ Syst Res 13(1):47–64

Rudenko I, Grote U, Lamers JPA (2008) Using a value chain approach for economic and environmental impact assessment of cotton production in Uzbekistan. In: Qi J, Evered KT (eds) Environmental problems of Central Asia and their economic, social, and security impacts, NATO science for peace and security series – C: environmental stability. Springer, Dordrecht, pp 361–380

UNDP (2006) Modeling of the infuence of a change in tax rates on macroeconomic indicators, united nations development program in Uzbekistan, Tashkent Project report on Reforms on tax system and preparing new edition of Tax Codex (In Russian)

Wehrheim P (2003) Modeling Russia's economy in transition. Ashgate Publishers, Series Transition and Development, Aldershot

Chapter 22
State Order and Policy Strategies in the Cotton and Wheat Value Chains*

Inna Rudenko, Kudrat Nurmetov, and John P.A. Lamers

Abstract Despite the establishment of private farms in Uzbekistan after its independence from the Soviet Union in 1991, agricultural production and the respective decision making remain to a large extent centrally managed by the national administration. Maintaining the state order system on cotton and wheat reflects the gradual approach that Uzbekistan is taking to reforms, possibly with the aim of cushioning potential financial and social shocks and preventing the collapse of the agricultural and industrial production systems. However, due to a wide range of market constraints and administrative barriers, the efficiency of the present resource use remains low. One pathway to identify opportunities for increasing those efficiencies is to analyze entire value chains for agro-commodities – including the agricultural and agro-processing sectors and to identify the deficiencies in the chains. Here, we apply a value chain analysis approach to the two strategic crops in the Khorezm region of Uzbekistan. The results of the analyses show that the cotton chain plays a significant role in the regional economy. It earns 99% of the export revenues, contributes 16% to the regional GDP, creates considerable output value and value added, supports employment (30–40% of the total regional labor force) and social security, and creates positive fund flows to the state budget (taxes minus subsidies). Developing the agro-processing sector in the cotton chain would bear the potential for generating higher export revenues (double in the case of export of processed cotton products such as T-shirts) at the same level of resource use or for maintaining the present export revenues while reducing the use of resources (cropland, irrigation

---

*This chapter draws heavily on the Ph.D. dissertation of Rudenko (2008)

I. Rudenko (✉) • K. Nurmetov • J.P.A. Lamers
ZEF/UNESCO Khorezm Project, Khamid Olimjan Str., 14,
220100 Urgench, Uzbekistan
e-mail: inna@zef.uznet.net; kudrat@zef.uznet.net; j.lamers@uni-bonn.de

C. Martius et al. (eds.), *Cotton, Water, Salts and Soums: Economic and Ecological Restructuring in Khorezm, Uzbekistan*, DOI 10.1007/978-94-007-1963-7_22,
© Springer Science+Business Media B.V. 2012

water, labor and capital) by 30%. Whereas the state does not earn much from the wheat value chain, the declared Uzbek policy of food self-sufficiency after independence in 1991 helped to reduce food insecurity and alleviate poverty. This chapter gives some insights on the state order system, provides an overview of value chains of cotton and wheat and offers options for improving agricultural resource use, and the welfare and environment of the country in general and the Khorezm region in particular.

Keywords Uzbekistan • Value chain analysis • State order system • Cotton revenues

22.1 State Order Continues in Uzbekistan

In the Soviet Union, agricultural production was centrally managed, in particular by the Ministry of Agriculture and the Ministry of Water Resources. While the former determined the area to be allocated to each crop and set targets for expected yields, the latter consigned the amount of irrigation water for the vast agricultural fields and the production costs for the *kolkhozes* (collective farms) and *sovkhozes* (state farms), which were the main production units during that era (Bloch 2002). Central planners at that time thus influenced not only the structure of agricultural production (area and crops), but also the use of production factors (Abdullaev et al. 2007).

During the first 16 years after independence in 1991, the agricultural production system in Uzbekistan was gradually reformed (cf. Djanibekov et al. 2011). The state and collective farms were transferred into *shirkats* (Uzbek for agricultural cooperatives) and finally into private farms. The gradual reformation and privatization process took place between 1992 and 2006, with finally almost all *shirkats* being substituted by private farms. However, this privatization did not include land, which still belongs to the state and is only rented out to land users. Consequently, until today, farming decision making has been influenced by a centralized structure, and the state controls a considerable part of the agricultural production through state quotas on cotton and wheat output and area, prices, crediting and financing, marketing and control on farm inputs (Bloch 2002; Abdullaev et al. 2007; World Bank 2007). Also, the farm optimization process (consolidation of small-scale private farms into bigger ones), which started in November 2008 (cf. Djanibekov et al. 2011) but had not been completed by 2009–2010, has not changed these aspects.

Maintaining a state order system in Uzbekistan during the on-going transition from a command to a market economy helped the country to cushion potential shocks and prevent a collapse of the entire production capacities developed during the Soviet era, including the agricultural and processing sectors (Spoor and Visser 2001). Cotton was meant to generate and maintain stable export earnings for the national budget (UNESCAP 2005) and to keep the cotton processing industry functioning (including the ginning industry, some textiles, and agricultural machinery building plants).

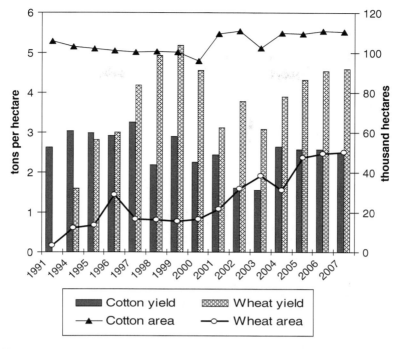

Fig. 22.1 Yield and area dynamics of cotton and wheat in the Khorezm region, 1991–2007

In addition, domestic wheat production became a declared objective of the national government[1] in the early years after independence and one component of the state order system to achieve self-sufficiency in food (Fig. 22.1). Consequently, the main objective was to rapidly increase grain production while reducing the dependency on food imports, mainly from Kazakhstan. Including wheat in the agricultural production system was also a strategy for preventing social unrest in a rapidly growing population.

Despite the retention of the state order system after independence, agricultural production in Uzbekistan has been subject to reformation/liberalization processes (Spechler et al. 2004). Although some authors argue differently (e.g. Veldwisch et al. 2011), the agrarian policies since 1991 have not only led to substantial change through a restructuring of the agro-producers' typology and privatization, but also:

- *Reduced the number of crops subject to the state order.* In 1991, the entire agricultural output had to be sold to the state, whereas several years later, state quotas

---

[1] According to the Decree of the Cabinet of Ministers of Uzbekistan No.450 from 10.09.1993 'On the measures for increasing cereals harvest from irrigated area'.

were removed except those for cotton, wheat and rice. The latter crop was removed from the state order system in the late 1990s.

- *Lowered quotas for the remaining state crops cotton and wheat.* Although the state quota for wheat was clearly lowered to 50% in the mid 1990s, the change in the state quota for cotton was less obvious. According to some sources, the state quota for cotton was reduced to 30% in the mid 1990s (FAO/WFP 2000; Spechler et al. 2004), whilst others reported this to be reduced to 50% (Abdullaev et al. 2007; Rudenko 2008). In any case, in 2002, the government adopted a decree that allowed farmers to sell up to 50% of their cotton output either domestically or abroad. However, there still is no functional mechanism that would allow this process to begin, and thus the government keeps a de-facto monopoly on cotton marketing (Swinnen 2005; UzReport.com 2005). However, a commodity exchange has been set up throughout the country, and it is foreseen that farmers may in the future gain the right to directly sell their cotton surplus, i.e., the cotton production exceeding the state quota, through this type of exchange (Rudenko 2008).
- *Raised and differentiated procurement prices.* State procurement prices for cotton and wheat are determined anew every agricultural season (Spechler et al. 2004), and allegedly (by the state) with a 50–55% increase annually (Ismailov 2003), an intention that is not, however, always met (Rudenko and Lamers 2006). For both cotton and wheat, state procurement prices are differentiated between a fixed price, which is paid for state quota volumes and which ranges according to cotton/wheat classes and grades and at a negotiated price (20% premium) for the cotton and wheat surplus that is above the state quota obligations. Additional opportunities exist for wheat marketing, and farmers may enjoy price differentials, whereas cotton producers have to cope with the monopsonic power of the state and have limited alternative marketing opportunities.
- *Reduced subsidization (and thus involvement of the state).* Following independence, the agricultural sector remained highly subsidized by the state both explicitly and implicitly (World Bank 2005). Explicit subsidies included maintenance costs for irrigation and drainage networks, free irrigation water, debt write-offs for *shirkats* (until 2005), and the provision of agricultural inputs such as fertilizers and fuel at low prices without VAT. Furthermore, implicit subsidies comprised of preferential credits at low interest rates and lower cotton oil prices for agricultural producers. The latest trends show, however, that along with privatization and a restructuring of the agricultural sector, various subsidies have gradually been omitted. For instance, since 2006 farmers have had to pay higher prices (with VAT) for the main agricultural inputs largely due to a partial liberalization of the input markets.
- *Crediting and financing.* Most farmers lack cash and other assets, which increases the demands for credits to produce agricultural commodities. The state finances the production of cotton and wheat through preferential bank credits at a low interest rate (3% annually since 2005; Abdullaev et al. 2007; Rudenko 2008).

It would be too undifferentiated to conclude that all farmers are obliged to produce cotton and wheat under the state procurement system. According to the 'Law on Private Farms', farm units can be established for different purposes and with specialization on, for example, cotton and/or wheat production, gardening, horticulture, animal husbandry, or aquaculture. But once farmers have chosen a specialization (e.g., cotton/wheat production) and concluded a contract with the state, they are obliged to follow the arrangements stipulated in the lease and land use agreement, which could mean, in the case of cotton and wheat farmers, that the strategic crops cotton and wheat have to be planted on up to 98% of the land (International Crisis Group 2005). However, if the farm was established for other purposes such as horticultural production, orchards or animal husbandry, farmers are free in their decision making. Even with cotton and wheat farmers, one part (although only a small part) of their cropland under the contract is frequently exempted from the production of state order crops (cf. Djanibekov et al. 2011; Oberkircher et al. 2011).

There is a paradox embedded in the state procurement system. On the one hand, this system dictates the production of cotton, which has to be sold at fixed and often low procurement prices. On the other hand, the cotton procurement system provides an extended social security net to the farmers (Abdullaev et al. 2007) and protects their interests (UNESCAP 2005). Consequently, the gradual modifications to the state order system, which is often seen a dysfunctional remnant from the past, can be considered a strategy for a transitional country to achieve gradual change. Various studies showed the potential to increase the efficiency of agricultural production and of the present land-use system and resource endowment (Bobojonov 2008; Djanibekov 2008; Rudenko 2008). But this would need additional reforms. These should not be separate for the agricultural and processing sectors, but should cover the entire value chains of agro-commodities including agro-processing, handling and transport (Rudenko 2008).

After independence, Uzbekistan aimed at becoming a partner in the international community, and thus entered bi- and multilateral cooperation with international organizations that tried to support the ongoing reforms in Uzbekistan through research and formulation of policy strategies for the government. However, it has lately become evident that the proposed policy strategies did not always coincide with the main course of reform activities in the country, especially in the cotton and wheat chains. Information shared on cotton from Uzbekistan is often incomplete, simplistic and debatable. When issued by international or multinational institutions, it may even reduce effective decision making. For example, propositions to support isolated measures and reforms (e.g. World Bank 2005) that are based on incomplete information and targeted mostly at only a few components of a complex production system will not bring about the anticipated changes.

To obtain a full understanding of the state procurement system and the potential of cotton and wheat production and processing for national and regional development, we suggest a more holistic assessment such as a value chain analysis, which can provide a comprehensive picture of the cotton and wheat chains in Uzbekistan, including the flows of products, the actors involved and their interrelationships, the costs of production, and the income distribution along the chains, and which may serve as a basis for-better informed strategy formulations (Rudenko 2008). We here develop this approach using the Khorezm region as a case study from which important lessons for the whole country can be drawn.

22.2 Cotton and Wheat Value Chains

A value chain analysis of cotton and wheat was conducted in 2005–2008 in the Khorezm region following the methodology of Kaplinsky and Morris (2002) and FAO (2006a, b, c). It used a stepwise approach to conduct a functional and institutional analysis of the involved stakeholders as well as financial and economic analyses of the activities along the chains. The functional analysis identified the flows, both in physical and monetary values along the chains, and the production stages, and provided the basis for developing virtual maps. The institutional analysis described the role of all agents in the chains with their functions. Furthermore, when complemented with a financial analysis, the value added and efficiency indicators of the individual actors as well as of the entire value chains could be assessed. An additional economic analysis pointed at the economic accounts corresponding to the activities of the actors and policy implications, and also provided the basis for policy simulations (for further details see Rudenko 2008).

'... *Value chains ... do not exist in a vacuum, but within a complex matrix of institutions and supporting industries. At the most basic level, it should be pointed out that value chains at every stage and in every location are sustained by a variety of critical inputs, including human resources, infrastructure, capital equipment, and services...*' (Sturgeon 2001: 10). Value chains thus require an operating environment that sets the required conditions and the framework for interaction of the chain actors. The operating environment includes the chain itself, the governing structure, and other service-providing institutions. Both the cotton and wheat chains in Khorezm involve several sectors of the economy, thus extending beyond the agricultural sector while including agro-processing and other industrial sectors (Figs. 22.2 and 22.3).

The governing structure of both the cotton and wheat chains consists of the Ministries of Agriculture and Water Resources, Finance, Economy, Foreign Economic Relations, Trade and Development, and departments thereof such as UzStandart, Cottonseed Corporation, Grain Management Office and State Bread Inspection. This structure is responsible for decisions on how many hectares of cotton or wheat to plant each year, how much and which varieties to produce each year and where. Furthermore, this structure defines prices for cotton and wheat products and elaborates agro-commodity product balances, regulates export and import operations, sets standards for agricultural production and processing, as well as quality standards.

The first and direct actor in both chains is the crop production sector. In the cotton chain, the main producers are private farms and *shirkats* (until 2005), and in the wheat chain, they are these two groups plus *dehqon* farms. The second actor is the agro-processing industry, in the cotton chain represented by the ginning industry (Fig. 22.2) and in the wheat chain by the flour milling industry (Fig. 22.3). The next actors are the light (textile) industry in the cotton and the food industry (bakeries and pasta producers) in the wheat chain. The cotton chain includes another group of actors, namely the oil-crushing and chemical industries mandated to process the major cotton by-product, i.e., cotton seeds.

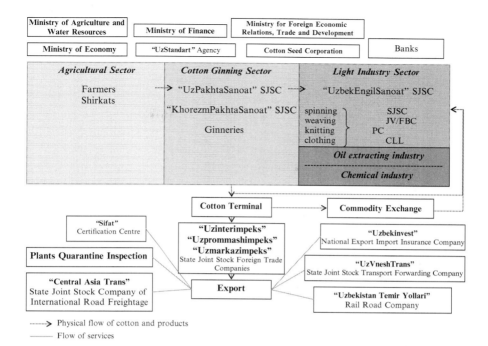

Fig. 22.2 Institutional map of the cotton value chain (Source: Rudenko 2008)

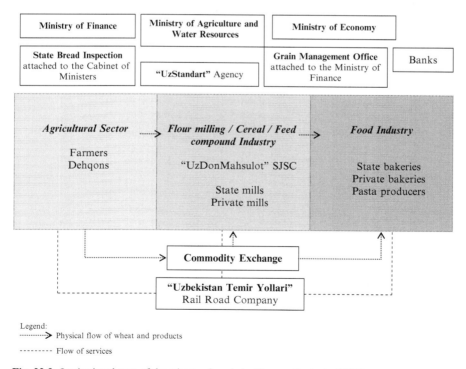

Fig. 22.3 Institutional map of the wheat value chain (Source: Rudenko 2008)

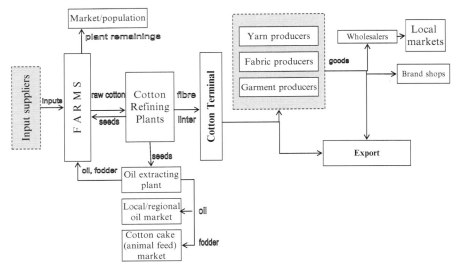

Fig. 22.4 Cotton value chain with the involved actors and flows of products (Source: Rudenko 2008)

Besides institutions governing and monitoring the chains, there are organizations and institutions aiming at facilitating the functioning of the entire chains and providing various services, including banking, marketing, exchange transactions, transportation, certification and quality control, insurance and other services (boxes below the grey bars in Figs. 22.2 and 22.3).

The cotton chain in Khorezm consists thus of cotton producers, ginneries, textile enterprises, and oil extracting plants (Fig. 22.4). The flows of cotton products along the chain begin with raw cotton coming from the farmers going to the ginneries. Cotton fiber from the ginneries then flows to the textile enterprises and is then, to a large extent, exported.

Cottonseed is partly returned to the producers as seeding material for the next agricultural season and partly to the oil extracting plant. Cotton oil and cottonseed meal and cake from the oil extracting plant are then purchased by the population or exported to neighboring countries. Finally, textile products from textile producers are consumed within the region or exported. The peculiarity of the cotton chain in Uzbekistan is the presence of the intermediate storage and distribution outlet, the Cotton Terminal.

Wheat also has to go through a sequence of transformations before it finally reaches the consumers. First, wheat produced by farmers and rural households (directly or through the grain preparing stations) flows to the wheat mills, which produce flour and fodder for cattle. The flour is forwarded to the food industry, which produces bread and pasta. Wheat and wheat products may also be sold via the

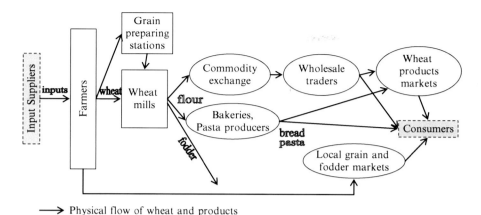

Fig. 22.5 Wheat value chain with the involved actors and flow of products (Source: Rudenko 2008)

commodity exchange (for grain and flour), via private traders (for flour, bread, pasta), or directly via local markets (Fig. 22.5).

The financial and economic analyses of the chains show that the cotton chain has played an important role in the regional economy of Khorezm, but also that it has concealed a considerable growth potential. In 2005, it contributed as much as 16% to the regional GDP, earned virtually 99% of the total export revenues of Khorezm, created value added slightly over 79,000,000 US$ and had market-oriented production objectives. The major contributing factors to the total value added were the wages and salaries paid along the chain (39%) as well as the taxes (35%). The major contributing actors were the agricultural component (44%), which provided the highest amount of wages, profits and depreciation costs and the ginning component (39%), which provided the highest amount of taxes, financial charges and remittances to the non-budgetary funds. All activities within the cotton chain resulted in surpluses, i.e., all agents received profits, with the crop production component of the cotton chain accruing the highest profit. However, on the other hand, this component was highly subsidized by the state. Should cotton producing farmers during this period face the real production costs including, for example, the operation and maintenance costs of the irrigation and drainage systems, water charges, or full costs of the main agricultural inputs with VAT inclusive, the received profit would be different and much likely much lower. According to the survey results, the total value of both explicit and implicit subsidies[2] provided to the cotton chain in Khorezm reached in 2005 about 26,000,000 US$.

The total amount of subsidies provided to the cotton chain was slightly less than all the taxes accrued by the state from the cotton chain in the same year. Results

---

[2] Methodology for calculating subsidies was adapted from the World Bank (2005).

from the analyses show that the cotton sector brought net economic gain to the government, albeit to a lesser extent than previously reported (World Bank 2005).

With regard to wheat, the value chain analysis showed that the wheat chain had the potential to alleviate poverty, to reduce food insecurity and was less controlled by the state. Furthermore, its contribution to economic performance of the Khorezm region was moderate compared to the cotton chain in terms of contribution to the regional GDP and creation of value added and total output value. Subsidies provided to the wheat chain (mainly to agricultural producers) by the state exceeded the taxes accrued and thus created negative flows to the state budget. The state has thus until now supported this chain, which worked inefficiently under the production conditions in Uzbekistan and given the world price levels of 2005 (Rudenko et al. 2008). It should be recognized, however, that irrigated wheat production in Uzbekistan has a short history, as it has become a major crop in the state order system only since Uzbekistan's independence. Nevertheless, considerable wheat gain yields have been achieved in this very short period, (cf. Ibragimov et al. 2011), and further improvements can be anticipated. Thus, wheat continues to be the second most important crop in the country as it provides food and social security. Also, the soaring food prices since 2007 show that the strategy of supporting domestic wheat production even under inefficient production conditions can in the end be profitable (Kienzler et al. 2011). Should world grain market prices continue to rise as in 2007 and 2008, additional opportunities would arise for supporting the present land-use pattern in Uzbekistan in favor of wheat production. Moreover, an increase in wheat production for subsequent exports can become an option for Uzbekistan if the baking quality can be improved (Kienzler et al. 2011).

22.3 Policy Simulation

The higher complexity of the cotton chain compared to the wheat chain is due to the various goods produced from cotton. This involves diverse industries that require the integration of the cotton chain in other sectors of the economy. However, the processing capacities along the cotton chain in the Khorezm region in 2005 were in general underutilized. This was especially true for the textile producers, who only processed about 9,000 tons of cotton fiber (or 11% of the total fiber output), despite the existing processing capacities of up to 20,000 tons per year. The by far largest share of cotton fiber output (89% of the total fiber output) was exported. Other cotton products, such as cotton yarn, fabrics, ready-made garments, and absorbent cotton were locally produced and exported from the region in small amounts.

Furthermore, in 2005 the ginneries operated with a low ginning efficiency as evidenced by the 30–33% raw-cotton-to-fiber output ratio. This ratio is lower than estimated in other cotton producing countries in West and Central Africa, where ginning efficiencies have reportedly been 40–43% (Rudenko 2008).

Three deterministic policy simulation models, i.e., two for the cotton chain and one for the wheat chain, were set up to support the analysis and predictions of

22 State Order and Policy Strategies in the Cotton and Wheat Value Chains

Table 22.1 Simulation results for Model 1: potential increase in cotton export revenues

	Export revenue, million US$	Potential increase in export revenue, million US$	Fiber export, % of total fiber output
Baseline	86.66	0	89
Scenario 1. Total fiber export	109.23	22.57	100
Scenario 2. Increased ginning efficiency	119.69	33.03	100
Scenario 3. Yarn export	121.34	34.68	80
Scenario 4. Fabrics export	133.68	47.02	80
Scenario 5. T-shirt export	164.55	77.89	80

Source: Rudenko (2008)

Table 22.2 Simulation results of Model 2: potential decrease in raw cotton production

	Export revenue, mln US$	Required raw cotton, tons	Required cotton area, ha	Reduction in cotton area, %	Irrigation water level, mln m^3	Explicit subsidies to agriculture, mln US$
Baseline	87.6	287,154	109,891	0	824	20
Scenario 1. Total fiber export	87.6	238,551	91,750	17	688	17
Scenario 2. Increased ginning efficiency	87.6	218,672	84,105	23	631	16
Scenario 3. Yarn export	87.6	206,684	79,494	28	596	14
Scenario 4. Fabrics export	87.6	173,058	66,561	39	499	12
Scenario 5. T-shirt export	87.6	88,991	34,227	69	257	6

Source: Rudenko (2008)

possible improvement strategies (Rudenko 2008). The models were designed to reveal (i) if increased processing of cotton fiber inside the Khorezm region could contribute to the generation of higher export revenues, (ii) if the same export revenue as obtained in 2005 could have been gained with a lower agricultural resource endowment, (iii) if there is scope for eliminating subsidies to agricultural producers and for increasing returns to the Uzbek cotton growing farmers, and (iv) if the import of wheat could increase economic feasibility and efficiency compared to local production of wheat.

Five scenarios were analyzed and compared to the baseline scenario of 2005 (Tables 22.1 and 22.2) to assess the potential impact and benefits from development of the cotton chain. Various variables were changed across the scenarios including the amount of cotton fiber forwarded for local processing as opposed to exports, ginning efficiency, stages of production along the chain, and the products to be exported. Scenario 1 assumed that the textile industry in the region is not functioning. In scenario 2, the ginning efficiency was increased by 3% (from 33% to 36%). Scenarios 3–5 allowed local textile enterprises to process more cotton fiber into cotton products for subsequent exporting. In scenario 3, along with export of cotton fiber, local textile enterprises produced and exported cotton yarn. In scenario 4, cotton fabrics were manufactured and exported along with the rest of the cotton fiber. And finally, scenario 5 assumed production and export of the ready-made garments (T-shirts).

The simulation results of Model 1 (Table 22.1) show the potential for generating higher export revenues with the same resource endowment by the cotton chain if more cotton fiber had been processed into products with higher value added, e.g., from cotton fiber to cotton yarn, fabrics and ready-made garments. Export revenues from textile products in the baseline scenario were as low as 2,300,000 US$. In the case of increased domestic fiber processing and subsequent export of textile products, additional export revenues of almost 34,680,000 US$ could be generated through yarn, 47,020,000 US$ through cotton fabrics, and almost 77,890,000 US$ through ready-made garments (Table 22.1). The increase in the ginning efficiency alone (scenario 2) could achieve up to 33,030,000 US$ compared to the baseline scenario.

Model 2 (Table 22.2) showed that development, upgrading and streamlining of the cotton chain in Khorezm would allow a reduction in the use of raw cotton production resources (land and water) and would require less subsidies from the state.

The more processing stages cotton fiber undergoes, the higher the prices the produced cotton products may fetch, and thus, the lower the amounts of these products (and in the end lower amounts of cotton fiber and raw cotton) required to obtain the same level of regional export revenue. Because of such price differentials, involvement of the local textile enterprises in processing cotton fiber into cotton yarn would have the potential to earn the same regional export revenue on 30,000 ha less land for cotton plantations, while saving 228,000,000 m^3 of irrigation water and also reducing the government's explicit subsidies by about 6,000,000 US$ (Table 22.2, scenario 3). The best case scenario (Table 22.2, scenario 5) with textile enterprises of Khorezm being engaged in more advanced processing of cotton fiber into ready-made garments could decrease the main critical inputs for raw cotton production by more than 66%, or by about 76,000 ha of land (69% of the baseline cotton area), 567,000,000 m^3 of irrigation water and about 14,000,000 US$ of subsidies.

Since a large share of cotton export revenues was absorbed by all the actors along the cotton chain, the farmers received the remaining part or the so-called farm-gate price (export price minus processing and other costs along the cotton chain). The economic and financial analyses of the cotton chain checked if the general assumption that the income of the Uzbek cotton growing farmers would be substantially increased if some of the governing structures were to be removed from the cotton chain. The analyses revealed furthermore that an elimination of the ginning governing structures, for example, the "UzPakhtaSanoat" state joint stock company for example would only marginally increase the prices paid to cotton producers, e.g., by about

16 US$ per ton of raw cotton (using 2005 prices and exchange rates) or about 5% of the farm-gate price. Findings reveal that a better approach to increase the returns to farmers would be to first improve the financial efficiency of the ginneries, for which immense potential exists through lowering processing expenses and reducing the taxes (Rudenko 2008). These savings could then be made available to producers. For instance, processing expenses in 2005 accounted for about 20% of the export price, while the ginning industry was heavily taxed: 73% of all taxes generated in the cotton chain and about 64% of all remittances to the non-budgetary funds were accrued by the ginning branch. Financial charges comprised 43%, indicating that ginneries depended heavily on credits. The low depreciation costs strongly indicate that the ginning equipment is worn out, which obstructs higher productivity, reduces fiber quality and increases expenses for maintenance and spare parts. Furthermore, privatizing the ginning sector and giving the ginneries the freedom to market their main produce (cotton fiber) could lead to a reduction in the revenue shares absorbed by the state foreign trade companies.

The wheat chain analysis revealed that, under a certain world price level for wheat, Uzbekistan could afford to import wheat. Under favorable world grain market prices, the import of wheat would be profitable, since the low wheat prices would require relatively low foreign exchange expenditures, while allowing the introduction of a land-use policy that gives room for crops more economically beneficial than wheat (Rudenko et al. 2008). Given the present demand for wheat, the funding of wheat imports could have easily been met by cotton fiber export revenues, the main hard currency earner in Uzbekistan. Or, as an alternative, the Uzbek government could use export earnings from locally manufactured textile products to cover the expenses of wheat imports. In both cases, a significant amount of land could be freed and concurrently a substantial amount of agricultural inputs could be saved to meet domestic wheat requirements. A change in land-use policy, however, demands an enabling legislation in particular with regard to customs regulations, import duties and other barriers for wheat import. However, the increased world market wheat prices observed since 2007/2008 dampen these expectations, and also a certain national independence in staple food production is favorable for any country (cf. discussion in Christmann et al. 2009). Should prices not return to pre-2007/2008 levels (as is predicted) or even increase, wheat would become an expensive import commodity, which would support the present land-use policy in Uzbekistan of promoting domestic wheat production.

Theoretically, Uzbekistan could also become a wheat exporter, as production could exceed the national consumption, and countries in the region such as Tajikistan or Afghanistan often face food crises. However, at present, exporting Uzbek wheat seems not feasible unless it's baking quality is improved, since the wheat is of low quality according to both national and international standards, which reduces the quality of wheat products such as flour and bread (Kienzler et al. 2011). Mixing imported high-quality wheat with Uzbek wheat is needed to raise the quality of the domestically produced wheat and thus, at least for now, imports would remain necessary. Kienzler et al. (2005) underscored that increasing the baking quality could be feasible by (1) introducing better wheat varieties, (2) creating incentives for the farmers to increase wheat quality through price differentials for better quality wheat, and (3) improving on-field agricultural practices, like better irrigation scheduling, and in particular nitrogen fertilization management. Flanking educational or awareness programs about the quality of wheat/wheat products would also be required.

22.4 Policy Strategies, Discussion and Conclusions

The fact that market constraints and administrative barriers contribute to the presently low efficiencies of resource use in Khorezm bears manifold opportunities for improvements. Although there has been frequent debate on the question whether the present state policy with its heavy involvement in agriculture takes decision-making power from agricultural producers and in turn hampers farmers' income opportunities (cf. Djanibekov et al. 2011; Veldwisch et al. 2011), a better question may be how to develop and support the entire chains of agro-commodities. Since the prosperity of Uzbekistan lies in the economic pursuit of a strategy of export-led growth (gas oil but also agro-commodities such as cotton fiber and value-added products), the analyses of the entire commodity chains for cotton and wheat show that opportunities for development exist within the agro-processing industries for improving marketing opportunities. As a consequence, the funds saved could be used to increase farmers' incomes. Opportunities lie in further development of the textile industry, in particular by increasing its efficiency while focusing on a complete cycle of commodities of competitive export-oriented outputs. The manufacturing of higher-value products at competitive prices conceals opportunities within both the cotton and wheat chains.

When striving for both economic and ecological improvements, options exist for intensifying the production of raw cotton on the most suitable lands and at the same time for increasing domestic fiber processing capacities. The subsequent export of products from cotton produced more efficiently on less land, and with a higher value added through post-harvest processing, could increase overall revenues while releasing the marginal land from cotton production. The land such freed could be allocated to alternative uses, either the production of more profitable (cash) crops (e.g. Bobojonov et al. 2011), or the establishment of less intensive and less expensive land-use systems such as tree plantations (cf. Khamzina et al. 2011). Furthermore, substantial amounts of irrigation water could be saved, which would contribute to overall water efficiency strategies (Bekchanov et al. 2010). Moreover, subsidies can be saved and funds thus freed re-allocated to rural development.

Such overall development strategies offer great potential to support ongoing attempts to prevent further aggravation of the current environmental crisis without adverse effects on the economic performance of the cotton chain. This represents a triple-win situation: for the farmers, the environment and the state. Yet it will take time to change not only the structure of the cotton chain, but also the present mind-set of the stakeholders in the commodity chains, a pre-condition for sustainable change.

In addition, the institutional ground would need to be prepared for such changes, as for example, setting favorable export environment, supporting process or industrial upgrading of local producers, or product upgrading, which could lead to higher competitiveness and world recognition of the Uzbek (cotton) products. Some lessons could be learnt from the textile and clothing sectors in the European Union, which has responded to a highly competitive and demanding world market through the quality of its products and their fashion content, through the capacity to develop highly demanded brands, the ability to promptly and reliably deliver the products, and finally through the sustainability and safety of industrial systems for the environment and the employers (Commission of the European Communities 2003).

Stakeholders along the cotton chain need to receive incentives to develop marketing and other capacities, and to learn how to do this, so that the value chains can function effectively in terms of time and monetary flows. Furthermore, privatization and upgrading of the main components in the cotton value chain may lower transaction costs by eliminating unnecessary intermediate agents. Under such conditions, farmers will get better prices for the produced raw cotton and thus increase their incomes and their capacity for building farm capital, which could in turn be invested in improving irrigation (cf. Bekchanov et al. 2011).

The declared Uzbek policy of food self-sufficiency after independence in 1991 has reached its goal in a relatively short time span, but only with large subsidies. As cotton fiber has been the main export earner in the Khorezm region, its revenue has been used to cover the import of goods, including wheat and flour. The analyses show that cotton revenues would have allowed the purchase of wheat volumes higher than those domestically produced, since the return from 1 ha of cotton is higher than that from wheat. However, assuming that the current high world market prices for wheat will be maintained in the future, the present land-use policy for domestic wheat production seems economically justifiable. In this case, however, a higher domestic wheat quality should be the target for both agricultural producers (with price incentives to produce higher quality wheat) and decision makers alike. In contrast, lower world market prices for wheat would justify wheat imports, and this in turn could offer a scope for crop diversification and alternative land uses while aiming at the production of other and more profitable crops such as fruits and vegetables, as postulated recently (Bobojonov 2008; Bobojonov et al. 2011).

References

Abdullaev I, Giordano M, Rasulov A (2007) Cotton in Uzbekistan: water & welfare. In: Proceedings of the international conference. Cotton sector in Central Asia: economic policy & development challenges. SOAS, University of London, London

Bekchanov M, Lamers JPA, Martius C (2010) Pros and cons of adopting water-wise approaches in the lower reaches of the Amu Darya: a socio-economic view. Water 2:200–216

Bekchanov M, Lamers JPA, Karimov A, Müller M (2011) Estimation of spatial and temporal variability of crop water productivity with incomplete data. In: Martius C, Rudenko I, Lamers JPA,

Vlek PLG (eds) Cotton, water, salts and *Soums*: economic and ecological restructuring in Khorezm, Uzbekistan. Springer, Berlin/Heidelberg/New York/Dordrecht

Bloch PC (2002) Agrarian reform in Uzbekistan and other Central Asian countries. Land tenure centre University of Wisconsin-Madison. Working paper no. 49

Bobojonov I (2008) Modeling crop and water allocation under uncertainty in irrigated agriculture: a case study on the Khorezm region, Uzbekistan. PhD dissertation, Bonn University, Bonn

Bobojonov I, Lamers JPA, Djanibekov N, Ibragimov N, Begdullaeva T, Ergashev A, Kienzler K, Eshchanov R, Rakhimov A, Ruzimov J, Martius C (2011) Crop diversification in support of sustainable agriculture in Khorezm. In: Martius C, Rudenko I, Lamers JPA, Vlek PLG (eds) Cotton, water, salts and *Soums*: economic and ecological restructuring in Khorezm, Uzbekistan. Springer, Berlin/Heidelberg/New York/Dordrecht

Christmann S, Martius C, Bedoshvili D, Bobojonov I, Carli C, Devkota K, Ibragimov Z, Khalikulov Z, Kienzler K, Manthrithilake H, Mavlyanova R, Mirzabaev A, Nishanov N, Sharma RC, Tashpulatova B, Toderich K, Turdieva M (2009) Food security and climate change in Central Asia and the Caucasus. Program for sustainable agricultural development in Central Asia and the Caucasus (CAC). Online publication available at: http://www.icarda.org/cac/files/food_security_en.pdf. p 80

Commission of the European Communities (2003) Economic and competitiveness analysis of the European textile and clothing sector in support of the communication "The future of the textiles and clothing sector in the enlarged Europe". Commission staff working paper, SEC (2003) 1345, Brussels

Djanibekov N (2008) A micro-economic analysis of farm restructuring in Khorezm region, Uzbekistan. PhD dissertation, Bonn University, Bonn

Djanibekov N, Bobojonov I, Lamers JPA (2011) Farm reform in Uzbekistan. In: Martius C, Rudenko I, Lamers JPA, Vlek PLG (eds) Cotton, water, salts and *Soums*: economic and ecological restructuring in Khorezm, Uzbekistan. Springer, Berlin/Heidelberg/New York/Dordrecht

FAO (2006a) EASYPol. On-line resource materials for policy making. Analytical tools. Module 043. Commodity chain analysis. Constructing the commodity chain, functional analysis and flow charts. www.fao.org/docs/up/easypol/330/cca_043EN.pdf. Accessed Nov 2006

FAO (2006b) EASYPol. On-line resource materials for policy making. Analytical tools. Module 044. Commodity chain analysis. Financial analysis. www.fao.org/docs/up/easypol/331/CCA_044EN.pdf. Accessed Nov 2006

FAO (2006c) EASYPol. On-line resource materials for policy making. Analytical tools. Module 045. Commodity chain analysis. Impact analysis using market prices. www.fao.org/docs/up/easypol/332/CCA_045EN.pdf. Accessed Nov 2006

FAO/WFP (2000) Special report. Crop and food supply assessment mission to the Karakalpakstan and Khorezm regions of Uzbekistan. http://www.fao.org/docrep/004/x9188e/x9188e00.htm

Ibragimov N, Djumaniyazova Y, Ruzimov J, Eshchanov R, Scheer C, Kienzler K, Lamers JPA, Bekchanov M (2011) Optimal irrigation and N-fertilizer management for sustainable winter wheat production in Khorezm, Uzbekistan. In: Martius C, Rudenko I, Lamers JPA, Vlek PLG (eds) Cotton, water, salts and *Soums*: economic and ecological restructuring in Khorezm, Uzbekistan. Springer, Berlin/Heidelberg/New York/Dordrecht

International Crisis Group (2005) The curse of cotton: Central Asia's destructive monoculture. Asia report 93. http://www.crisisgroup.org/home/index.cfm?id=3294&l=1. Accessed Sept 2007

Ismailov J (2003) Problems of price and income support of the agrarian sector of Uzbekistan. International agricultural magazine, vol 5

Kaplinsky R, Morris M (2002) A handbook for value chain research. http://www.ids.ac.uk/ids/global/pdfs/VchNov01.pdf. Accessed Mar 2005

Khamzina A, Lamers JPA, Vlek PLG (2011) Conversion of degraded cropland to tree plantations for ecosystem and livelihood benefits. In: Martius C, Rudenko I, Lamers JPA, Vlek PLG (eds) Cotton, water, salts and *Soums*: economic and ecological restructuring in Khorezm, Uzbekistan. Springer, Berlin/Heidelberg/New York/Dordrecht

Kienzler K, Lamers JPA, Ibragimov N, Vlek PLG (2005) Quantity, quality and economics: reviewing the use of mineral fertilizers on key crops in the Aral Sea Basin. In: Conference proceedings. XV international plant nutrition colloquium, plant nutrition for food security, human health and environmental protection, Beijing, China

Kienzler K, Rudenko I, Ruzimov J, Ibragimov N, Lamers JPA (2011) Winter wheat quantity or quality? Assessing food security in Uzbekistan. Food Sec 3:53–64. doi:10.1007/s12571-010-0109-9

Oberkircher L, Haubold A, Martius C, Buttschardt T (2011) Water patterns in the landscape of Khorezm, Uzbekistan. A GIS approach to socio-physical research. In: Martius C, Rudenko I, Lamers JPA, Vlek PLG (eds) Cotton, water, salts and *Soums*: economic and ecological restructuring in Khorezm, Uzbekistan. Springer, Berlin/Heidelberg/New York/Dordrecht

Rudenko I (2008) Value chains for rural and regional development: the case of cotton, wheat, fruit, and vegetable value chains in the lower reaches of the Amu Darya River, Uzbekistan. PhD dissertation, Hannover University, Hannover

Rudenko I, Lamers JPA (2006) The comparative advantages of the present and future payment structure for agricultural producers in Uzbekistan. Cent Asian J Manag Econ Soc Res 5(1–2):106–205

Rudenko I, Lamers JPA, Niyazmetov D (2008) Wheat value chain in Uzbekistan. When irrigated wheat becomes a feasible land use policy option? In: Proceedings from the international conference 'Institution building and economic development in Central Asia'. International School of Economics (ISE), Almaty, Kazakhstan

Spechler MC, Bektemirov K, Chepel S, Suvankulov F, (2004) The Uzbek paradox: Progress without neo-liberel reform. In: Ofer, G and Pomfret, R (eds), The Economic Prospects of the CIS, Sources of Long Term Growth. Edward Elgar: Cheltenham, UK and Northampton, MA. pp. 177–197

Spoor M, Visser O (2001) The state of agrarian reform in the former Soviet Union. Eur Asia Stud 53(6):885–901

Sturgeon T (2001) How do we define value chains and production networks. Industrial Performance Centre, Massachusetts Institute of Technology. IDS Bulletin 32:3

Swinnen JFM (2005) When the market comes to you or not. The dynamics of vertical coordination in agri-food chains in transition. ECSSD. The World Bank Report EW-P084034-ESW-BB, 084024 February 082005

UNESCAP (2005) Trade facilitation in selected landlocked countries in Asia. http://www.unescap.org/tid/publication/tipub2437_uzbek.pdf

UzReport.com (2005) New measures of state support for attraction of foreign investments into the textile industry. Uzbekistan news website, 14 Mar 2005

Veldwisch GJ, Mollinga P, Zavgorodnyaya D, Yalcin R (2011) Politics of agricultural water management in Khorezm, Uzbekistan. In: Martius C, Rudenko I, Lamers JPA, Vlek PLG (eds) Cotton, water, salts and *Soums*: economic and ecological restructuring in Khorezm, Uzbekistan. Springer, Berlin/Heidelberg/New York/Dordrecht

World Bank (2005) Cotton taxation in Uzbekistan. Opportunities for reform. ECSSD working paper no. 41. http://siteresources.worldbank.org/INTUZBEKISTAN/Resources/COTTON_TAX_NOTE.pdf. Accessed Dec 2006

World Bank (2007) Country brief 2007. http://www.worldbank.org.uz/WBSITE/EXTERNAL/COUNTRIES/ECAEXT/UZBEKISTANEXTN/0,,contentMDK:20152186~menuPK:202941 95~pagePK:20141137~piPK:20141127~theSitePK:20294188,20152100.html

Chapter 23
Prospects of Agricultural Water Service Fees in the Irrigated Drylands Downstream of Amudarya

Nodir Djanibekov, Ihtiyor Bobojonov, and Utkur Djanibekov

Abstract The limited availability of surface water, low water use efficiencies, a deteriorating irrigation network and land degradation aggravated by the impact of climate change are among the factors constraining agricultural production in the irrigated drylands of Central Asia. Recurrently, an introduction of water service fees has been suggested as one option to increase water use efficiency, which is analyzed here at the example of the downstream Amudarya Khorezm region, Uzbekistan. Underlying issues in introducing fees for water services in irrigated dryland agriculture given the state procurement policy are in Uzbekistan of crucial importance. Therefore, the impacts of different levels of water service fees were simulated with a mathematical programming model. The analysis and conclusions are based on changes in regional welfare, cropping pattern, export structure and economic attractiveness of crops to agricultural producers. Although the conclusions refer to the case study region, they help understanding the potential impact of water service fees on the national agricultural sector and add to the discussions on where opportunities for a (partial) cost recovery for the operation and maintenance of

---

N. Djanibekov (✉)
ZEF/UNESCO Khorezm Project, Khamid Olimjan Str., 14,
220100 Urgench, Uzbekistan
e-mail: nodir79@gmail.com

I. Bobojonov
Department of Agricultural Economics, Farm Management Group,
Humboldt-Universität zu Berlin, Philippstr. 13 – Building 12 A,
D-10115 Berlin, Germany
e-mail: ihtiyorb@yahoo.com

U. Djanibekov
Center for Development Research (ZEF), Walter-Flex-Str. 3,
53113 Bonn, Germany
e-mail: utkur@uni-bonn.de

C. Martius et al. (eds.), *Cotton, Water, Salts and Soums: Economic and Ecological Restructuring in Khorezm, Uzbekistan*, DOI 10.1007/978-94-007-1963-7_23,
© Springer Science+Business Media B.V. 2012

the irrigation system of Central Asian countries may exist. It is argued that the introduction of water service fees may indeed generate sufficient funds to recover costs for operation and maintenance of the irrigation network. However, as the current institutional setup constrains a significant reduction of agricultural water demand, the introduction of water service fees as an isolated measure is not likely to achieve the expected benefits unless flanked by additional measures such as reduction of the state production targets on crops.

Keywords Irrigation water use • Cost recovery • Partial equilibrium model • Policy analysis

23.1 Introduction

Global climate change has become a major environmental concern and research topic. Kazakhstan, Kyrgyzstan, Tajikistan, Turkmenistan and Uzbekistan, the five former Soviet republics comprising Central Asia, are located in arid and semiarid regions. Agriculture in the Central Asian countries, which relies on an irrigation network constructed during the Soviet Union era, is highly dependent on irrigation water diverted from the Amudarya and Syrdarya rivers. Due to a lack of investments after independence, this irrigation network has been poorly maintained and its operational period has been expiring, leading to significant water losses and low water productivity (Bucknall et al. 2003; Bekchanov et al. 2011; Tischbein et al. 2011). The agricultural sector uses 83.6% of the entire water resources in the region (Abdullaev et al. 2006) from which only 50–70% of water reaches crops (WARMAP 1996). Furthermore, the institutional set-up for water use and management in Central Asia does not encourage agricultural producers to use their water efficiently (Bucknall et al. 2003). As a result, Central Asia is prone to severe water scarcity (Alcamo et al. 2003). The irrigation water scarcity constrains the regional agricultural production particularly in downstream areas; since 1980 an increase has been observed in the probability of facing a water scarcity in Khorezm, a region situated downstream on the Amudarya (Müller 2006).

The current economic/institutional water scarcity can in the long run be exacerbated by the impacts of climate change (McCarthy et al. 2001). It is predicted that the availability of water in the two rivers may decrease by as much as 30% and 40% (Perelet 2007), and water needs of upstream users may increase, too (Martius et al. 2009). The estimated rise in temperatures will also increase crop water requirements (Fischer et al. 2002; Perelet 2007), although this is not yet substantiated by the findings of the last four decades (Jarsjö et al. 2007; Conrad et al. 2011). Water demand for non-agricultural uses is also likely to increase.

The agricultural sector employs a large share of a population and generates a significant share of GDP in most Central Asian countries. For instance, in Uzbekistan, agriculture accounts for about 25% of GDP and 32% of labor employment, and directly provides 70% of domestic trade (2006 data; IMF 2008). Furthermore, population growth requires increases in food production, which would be a challenge even with a functional water management system (McCarthy et al. 2001; Perelet 2007). The future of the regional economy and people's welfare under decreasing availability of water will depend on well-managed water distribution and allocation systems (cf. Oberkircher et al. 2011; Veldwisch et al. 2011) and a well-functioning irrigation and drainage system (cf. Tischbein et al. 2011), the latter requiring large investments (Bucknall et al. 2003).

It recurrently has been suggested that operation and maintenance costs can be recovered via the introduction of proper water service fees (WSFs) (Cornish et al. 2004). The introduction of WSFs in agriculture, subject of past and on-going discussions, is also part of the agricultural reforms in Uzbekistan (Bucknall et al. 2003; MacDonald 2003). However, studies by Bobojonov (2009) indicated that WSFs may not reduce actual water use in the agricultural sector unless very high fees are introduced. Given the controversial discussion on the impact of WSFs, the effectiveness of such a reform and its impact on water use, production patterns and regional welfare, are factors that need to be well understood. The development and introduction of adequate mechanisms of WSFs is a challenge to national policy makers under the increasing likelihoods of regional warming and water scarcity, ´the more so as agricultural producers in Central Asia typically operate at low income levels (Bucknall et al. 2003).

We discuss here the role of water service fees using the peculiar example of irrigated agriculture in the Khorezm region of Uzbekistan, as a representative case study for Central Asia as a whole. First, we describe the modeling approach chosen for testing the impact of different levels of WSFs on regional welfare, cropping pattern, trade and crop gross margins and present the results of the simulations. Based on those, we propose policy options to promote the adoption of WSFs.

23.2 The Case of Irrigation Water Use in the Drylands of the Khorezm Region

23.2.1 Irrigation Water Supply and Demand

In Khorezm, water from the Amudarya is the only source of irrigation and agricultural production in this region (Abdolniyozov 2000; cf. Conrad et al. 2011). The region consumes around 13% of the Amudarya water intake in Uzbekistan (2001 data; Djalalov 2003), around 4.5 km^3 per year (cf. Tischbein et al. 2011). The river water, before arriving in Khorezm, is collected in the upstream Tuyamuyun water reservoir, which had initially a total capacity of 7.8 km^3 (now reduced by sediments accumulated in the reservoir). The water is rationed by the water management organizations depending on the monthly water demand in the region (Müller 2006). Agriculture in Khorezm consumes up to 97% of the regional water intake, which is higher than the national average. In Khorezm, as in most parts of Uzbekistan, water is channeled through a hierarchically arranged irrigation network including main, inter-farm, and on-farm channels (FAO 2008). In the 1960s, the Soviet government heavily invested in an expansion of irrigated agriculture (cf. Tischbein et al. 2011) and consequently cropland in Khorezm has been expanded from 146,300 ha in 1960 to 276,000 ha in 1999 (Conrad 2006). Flood and surface furrow irrigation are the main irrigation techniques used, leading to a high gross water use (Bobojonov 2009). The average water application rate has declined from about 21,000 m^3 ha^{-1} (in 1940–1990) to about 16,000 m^3 ha^{-1} in (1991–2008). However, soil salinity and water logging are still widespread partly resulting from intensive irrigation caused also by a lack of incentives for producers to increase water use efficiency (Abdolniyozov 2000; Abdullaev et al. 2009).

Despite the large water discharges from the Tuyamuyun water reservoir (cf. Tischbein et al. 2011), long-term irrigation water supply in Khorezm has mostly but not always met crop water demand in the past, and some areas in the region face regular deficits even during years of sufficient water availability (Müller 2006). Water supply in Khorezm is unreliable in terms of volume, duration and periodicity; the likelihood for water-shortages in Khorezm has increased as evidenced in the probability of attaining sufficient amounts of water, which fell from 82% in 1992 to 74% in 1999 (Müller 2006).

Based on their location relative to the river, the districts can be aggregated in upstream and downstream districts. Upstream districts are those that have immediate access to irrigation water, thus usually receiving sufficient water even in drought years. Downstream districts are distant from the river and have historically been most affected during drought years (Müller 2006). For instance, during the 2000–2001 drought, the average irrigation rate per unit of area decreased by 57–65% in downstream districts, while the losses in upstream districts were 'only' 48–57% (Müller 2006). The probability of obtaining the expected amount of water was high in areas bordering the river, and low in the distant areas (Conrad 2006). In the large-scale irrigation system, which lacks any effective controlling mechanism, the limited surface water could not be equitably distributed to the numerous users which resulted from the split-up of the few large collective farms to many private farmers (cf. Djanibekov et al. 2011). As a result, inefficient water use in the upstream districts contributes also to downstream water scarcity (Djanibekov 2008).

23.2.2 Implication of Water Service Fees in Agriculture

After the water management and operation and maintenance (O&M) of canals had been transferred to the newly established water users associations (WUAs), state subsidies and budget expenditures directed to these activities were reduced. WUAs need to recover the incurred O&M and associated administrative costs fully or partially. They can do this by charging the farmers within their boundaries with membership fees and request direct payments for services provided for water delivery and distribution. In addition to the intention of cost recovery, WSFs may be accumulated for future activities of the WUA (Djanibekov 2008).

In the highly saline soils of Khorezm leaching is an important part of the field management and its water demands are significant, using up to 25% of the total annual water use (Forkutsa 2006). Therefore, a local peculiarity is that WSFs have to reflect also the pre-sowing season costs of leaching to fully recover the costs.

The introduction of WSFs has been advocated based on various arguments. The benefit expected from WSFs is to create sufficient funds for improving the existing irrigation and drainage systems, which in turn would affect both canal efficiency and crop yields. First, the collected WSFs could be invested into a series of agronomic, technical, managerial and institutional improvements such as cleaning the supply and drainage canals and developing more precise irrigation schedules (Kyle and Chabot 1997; Wallace and Batchelor 1997; Batchelor 1999). Furthermore, the improved managerial and institutional practices of WUAs and investments in O&M of canals would increase crop yields through a more timely supply and better quality of irrigation water, particular at stages of crop development when crop response to irrigation is most sensitive (Dinar and Latey 1991).

Worldwide, two mechanisms for water charging, volumetric and non-volumetric, are applied, each with their pros and cons depending on the situation (Tsur et al. 2004). WSFs based on the actual volumes of water supplied are the most obvious and widely studied economic instruments (Hellegers and Perry 2004). The volumetric approach requires regular information on the quantity of water used by each agricultural producer below a point of measurement (Dinar et al. 1997). However, during the installation of the irrigation infrastructure in 1960s in Khorezm there was no evidence of water scarcity. The production units then, *kolkhozes* and *sovkhozes*, were large. With the progress of farm restructuring that resulted in many smaller private farms (cf. Djanibekov et al. 2011), volumetric measurement and differentiation of water services and control at farm and field level become more important, but will only be possible if water use is measured at many points. The installation and maintenance of a large number of measurement devices would raise costs.

Because of the large number of water users and imperfect information on actual water supply (cf. Tischbein et al. 2011), the WUAs in Khorezm currently impose a WSF at a fixed rate for all farms within their boundaries, using non-volumetric area- or crop-based water use estimation methods. The non-volumetric fees are relatively simple to administer and can provide a stable income source for the WUAs to recover O&M costs.

23.3 Methodology

With the intention to understand and evaluate the effects of the introduction of WSF on regional welfare, production and consumption patterns, a mathematical model was developed according to the framework presented by Hazell and Norton (1986) via the integration of a well-behaved demand system.[1] The model is a static spatial programming model with endogenous commodity prices. In the model, we assume that production and consumption occur in five district aggregates spatially grouped according to their location relative to the Amudarya river. To trace the cross-effects between commodities, the model includes 11 production and consumption activities and one activity of consumption of the "rest of commodities not covered in the model". The products were grouped into wheat (which represents also other grains, except for rice and maize), rice, maize, potato, vegetables (all vegetables), fruits (fruits and melons), fodder (annual fodders, short-period maize for fodder and alfalfa), milk, eggs and meat.

The indices, parameters, variables and equations of the model are given in the Annex Tables 23.1–23.4. The objective function of the model maximizes regional welfare, which is a measure of summed consumer and producer surpluses. The regional welfare (Eq. 23.1; all equations are presented in Annex Table 23.4) is a sum of net-trade of commodities, producers' revenues from food consumption by consumers, money-metric utility of consumers less transportation costs, and agricultural production costs, which includes also the simulated values of WSFs.

The *supply part* of the model includes the production components such as production activities, commodity flows and constraints. The supply function (Eq. 23.2) is considered as linear and defined in the supply part of the model. *Total land availability constraint* (Eq. 23.3) consists of a single soil type and is specified at district aggregate according to the official statistics on sown area in 2007. The cropping calendar is introduced into the land constraint to cover double-cropping characteristics of regional agriculture. It allows two crops being grown and harvested from the same field during one agricultural year. The *monthly crop water demand constraint* (Eq. 23.4) is fixed at monthly volumes of surface water demanded for crop cultivation in each district aggregate according to cropping pattern observed in 2007. The *state procurement constraint* (Eq. 23.5) requires that activity levels for cotton production in district aggregates are not less than the assigned area in 2007. *Commodity market balance equation* (Eq. 23.6) imposes that demand equals supply in each district. This means that production, imports and purchases from other districts must be balanced with food consumption, animal feeding, export and sales to other districts. *Net trade earnings of agricultural producers* (Eq. 23.7) are defined as export revenues less expenditures for import. *Producer revenues* (Eq. 23.8) are equal to the product of commodity prices and quantities sold to local consumers. *Producers' commodity transportation costs* (Eq. 23.9) are defined as per-kilometer transportation fees multiplied by the transported quantity of commodities and distance. The commodity prices differentiate between districts according to the distances and the transportation costs. *Producers'*

---

[1] The model used in this study is a part of a forthcoming effort of developing a sector planning model which would incorporate a risk component for the agriculture of Khorezm (cf. Sommer et al. 2011).

23 Prospects of Agricultural Water Service Fees in the Irrigated Drylands... 395

production costs (Eq. 23.10) are defined as costs for labor, fertilizer, diesel, seeds and pesticides for crops and veterinary services for livestock. In the scenario simulations, production costs include also the product of WSF and total crop water demand.

As a starting point for the final model development and since observations on the shape of a non-linear Engel curve[2] (i.e. for Uzbekistan) were not available, we used the linear approach via deriving parameters for Normalized Quadratic Expenditure System (NQES), which is linear in income (Britz 2003). According to the criteria listed by Britz (2003), NQES is flexible in calibration, globally well-behaved, fulfill the microeconomic conditions and simple in computation. Consequently, in each modeled district aggregate consumption is given by NQES functions. The *demand part* of the model is based on the indirect utility function depending on consumer prices and per capita income (Ryan and Wales 1996; Britz 2003). Using Roy's identity,[3] the following *Marshallian demand*[4] was derived as presented in Britz (2003) (Eq. 23.11), where the functions D, Gi, G and F are presented in Eqs. 23.14–23.16. *Consumer expenditure (income) for consumption per capita* is defined as production of regional consumption and commodity prices divided by population (Eq. 23.12).

To attach a welfare measurement to the areas under the ordinary demand function, the demand function should be integrable. The general problem of incorporating demand functions of complex form, e.g. NQES, directly into the objective functions of the mathematical programming models, is that these functions are difficult to integrate. Consequently, a *money-metric utility function* (Eq. 23.13) was specified to measure the welfare effects of price changes and shaped according to the indirect utility function depending on consumer prices and per capita income (Britz 2003).

The supply and demand parts of the model were calibrated to one point, e.g. to the observed production and consumption information for district aggregates in 2007. The supply part of the model was calibrated via imposing quadratic costs function via the Positive Mathematical Programming approach (Heckelei 2002: 6–8) using own-price supply elasticities. The technical details of calibration of the demand system parameters to given elasticities and properties of NQES are as suggested by Britz (2003): derivation of demand system parameters to given elasticities in which (1) deviation between initial elasticities and the ones derived from the parameters are minimized and (2) the resulting parameters define a "well-behaved" function which satisfies the requirements on (a) homogeneity, (b) adding up, (c) symmetry and (d) negative own-price elasticities.

---

[2] An Engel curve describes how household expenditure on a particular good or service varies with household income (Deaton and Muellbauer 1980; Mas-Colell et al. 1995).

[3] Roy's Identity provides means of obtaining a demand function from an indirect utility function (Deaton and Muellbauer 1980; Mas-Colell et al. 1995).

[4] A Marshallian demand function specifies what the consumer would buy in each price and wealth situation, assuming it perfectly solves the utility maximization problem (Deaton and Muellbauer 1980; Mas-Colell et al. 1995).

23.3.1 Data Acquisition

For the supply part of the model the data was acquired from field observations as well as secondary data from official governmental agencies, e.g. Regional Department of Statistics (OblStat) and Regional Department of Agriculture (OblSelVodKhoz). Annual reports from 2007 by OblStat were the main source of data on crop and animal production, cropping area and animal stock, crop and animal product yields. The input prices in the model are exogenous and their values were obtained from field observation as average annual prices. The values of annual crop water demand in districts were obtained from OblSelVodKhoz as official recommendations (norms) for different hydromodule zones of soil developed for Khorezm (more details in Bobojonov 2009). It includes monthly water demand during leaching as well as irrigation water. The annual crop water demand is calculated for each modeled district according to the hydromodule structure of soils prevalent in each district aggregate.

Due to the lack of data on regional consumption patterns, the information on consumption in Uzbekistan for 2003 was taken from the Supply Utilization Accounts (SUA) and Food Balance Sheets (FBS) in the database of FAO's Statistics Division.[5] First, items presented in the FAO database were aggregated into nine main food categories according to the author's discretion. The total value of 'rest goods' was proportionally inferred using their share in total food expenditures (81.6%) for Khorezm as reported by OblStat (2008). The initial price vectors for food commodities were obtained from the official statistical reports of OblStat as average annual market prices observed in Khorezm in 2007. Initial prices for 'rest goods' are assumed to be equal to their unit values, i.e. 1,000 US$. The exchange rate of US$ in 2007 was 1,276 Uzbek *Soums* (UZS). The primary values for demand elasticities were obtained from Djanibekov (2008), which were generated from the WATSIM[6] model's base-run dataset on the rest of the world. Due to the absence of information on total export and import values of crops and animal production, their values were generated within the base run solution of the transportation model.

23.3.2 Characteristics of Regional Production in 2007

The average profit per 1 cubic meter of crop demanded water was 0.12 US$ in 2007. Almost 60% of regional profits are generated by livestock and poultry (Table 23.1). The largest gross margins, using output-input prices and yields of 2007, were observed

---

[5] No commodity price information for Uzbekistan was available in the database of FAO's Statistics Division by 01.04.2011.

[6] The WATSIM is a recursive-dynamic spatial world trade model for agricultural commodities. It is mainly applied for the medium-term analysis of trade policy changes (Kuhn 2003).

23 Prospects of Agricultural Water Service Fees in the Irrigated Drylands…

Table 23.1 Regional production in 2007

Production activities	Units	Observed activity level, 10^3 units	Average yield, t unit^{-1}	Annual water demand, m^3 unit^{-1}	Gross margins, USD unit^{-1}	Gross margins per water demand, USD 10^{-3} m^{-3}	Share in sown area, %	Share in water demand, %	Share in regional gross margins, %
Cotton	ha	111	2.49	9,860	38	3.9	51.4	52.8	1.5
Winter wheat	ha	51	4.56	3,650	974	266.7	23.7	9.0	12.6
Paddy rice	ha	15	4.43	26,200	1,740	66.4	7.1	19.3	6.4
Potato	ha	4	14.11	12,863	1,949	151.5	1.9	2.5	3.4
Vegetables	ha	12	19.82	12,913	1,745	135.1	5.4	7.2	4.9
Melons	ha	4	15.69	8,238	4,235	514.1	2.0	1.7	5.2
Maize	ha	1	3.49	9,545	544	57.0	0.4	0.4	0.1
Fodder crops	ha	17	16.86	8,340	781	93.6	8.1	7.0	6.8
Cows	head	245	2.07		265				15.6
Poultry	head	1,922	0.07 (10^3 eggs)		5				2.2
Bulls	head	379	0.24		448				41.3

Source: OblSelVodKhoz (2004), OblStat (2008), Own calculation

from fruits, vegetables, potato and rice. The state-prioritized crop, cotton, occupied almost half of the sown area and composed half of crop water demand in 2007, but generated only 1.5% of the total producer's profit.

The most water demanding crop cultivated in Khorezm, rice, had the second lowest gross margin per cubic meter of annual water demanded, after maize that is cultivated mostly as fodder. Rice comprised almost one-fifth of regional crop water demand and occupied 7% of the total sown area. According to the official rates of crop water demand, rice demands 26,200 m^3 ha^{-1}. These rates were developed to plan spatial and timely distribution of water. As such their values differ from the real volumes at which farmers apply water. For instance, Kyle and Chabot (1997) reported that the annual water application for rice in Khorezm could be greater than 50,000 m^3 ha^{-1}.

In Khorezm, the annual consumption expenditures in 2007 amounted to 281 US$ per capita, of which around 78% were spent for the consumption of agricultural food products. In the expenditure structure, meat and wheat products comprised the largest share, while the most water intensive commodity, rice, was the second lowest after eggs. Estimated regional foreign (outside of the regional borders) trade turnover was about 392,200,000 US$, of which export accounted for around 96%; creating a trade surplus of 363,200,000 US$. Locally produced cotton is almost entirely exported from the region and was the largest contributor to the regional budget. However, meat and rice exports generated about 50% of the export earnings of farmers. In 2007, half of the locally produced meat and 70% of the rice were exported from the region. Wheat is the main food commodity and contributes strongest to energy and protein intake per capita in Khorezm (Djanibekov 2008), but local production could cover only 88% of the regional demand in wheat, obligating an import. The region also depends on import of potato and fruits, which generated 14% and 52% of local demand, respectively.

23.4 Results

In the base run solution, the calibrated model replicates the agricultural production activities of Khorezm in 2007. The model solved for the optimal values of production, consumption, input use and commodity trade by consumer/producer groups and districts, while generating equilibrium commodity prices. To validate the developed model and to monitor the effects of introduction of WSFs into the regional crop production levels, WSFs are introduced stepwise. Since an introduction of volumetric WSFs under the present conditions is technically very difficult, we simulated the introduction of crop-based WSFs according to the officially fixed norms on crop water demand. We ran 50 simulations using a step of crop-based WSF of 0.8 US$ per 1000 m^3 for each loop, bringing the WSF up to 40 US$ 10^{-3} m^{-3} in the final 50th loop. In the model, the irrigation service fees were introduced into the equation of agricultural production costs of farmers. Cotton, which is subject to the state procurement system, had the lowest gross margin already, this could be worsened if

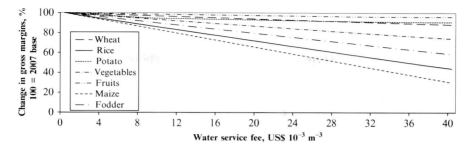

Fig. 23.1 Changes in crop gross margins under introduction of water service fees (WSF) in Khorezm, Uzbekistan

cotton producing farmers would be charged WSFs. Therefore in the simulation we set the WSF for cotton as zero. The rate of WSF is comparable to water charges and prices which were practiced in the agricultural sector of Morocco in 2003 (Chohin-Kuper et al. 2003), Namibia, Algeria, Tunisia, Brazil, Portugal, United States and Spain (Dinar and Subramanian 1997) in 1996. The model was programmed using the GAMS modelling language; it was then calibrated and solved as a non-linear optimization, using the numerical solver CONOPT3.

23.4.1 Change in Crop Gross Margins

The gross margins of all crops except cotton would decrease with the introduction of WSFs. Figure 23.1 shows this decrease at various rates of WSFs, up to 40 US$ 10^{-3} m^{-3}, in comparison to the base situation, i.e. without fees. When the WSFs are introduced at equal per-cubic-meter rates, the largest decrease in gross margins is observed for maize, rice and fodder. As rice is the most water demanding crop in Khorezm, the introduction of WSFs would significantly impact profits of farmers involved in rice cultivation: gross margin from rice would drop from 1,740 to 766 US$ ha^{-1} at 40 US$ 10^{-3} m^{-3} of water. Maize and fodder crops are mostly produced for livestock and poultry feeding and, thus, their profit loss would be offset via an increase in livestock production.

23.4.2 Change in Cropping Pattern

The introduction of WSFs would decrease the total sown area and change the structure of crop cultivation. Even though profits from other cropping activities would drop due to the WSFs, the mandatory cotton production, which is not levied by WSFs, would still be kept at the level determined by the state procurement system. As a result, WSFs neither decrease nor increase the cotton cultivated area.

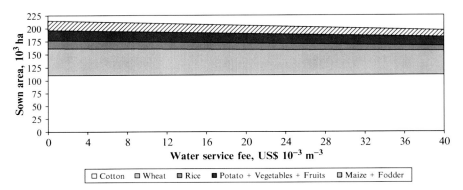

Fig. 23.2 Changes in regional cropping pattern under WSF introduction in Khorezm, Uzbekistan

Furthermore, the selected WSF rate (40 US$ 10^{-3} m^{-3}) guarantees that other crops remain attractive for farmers. Cotton remains less interesting even if WSFs are established (Fig. 23.2). The crop most affected by WSFs is rice. Rice cultivated area would drop by 40% compared to the no-WSF situation in 2007. Nevertheless, regionally, rice would be produced at levels that fully cover the local demand and allow exporting its surplus outside.

23.4.3 Change in Regional Welfare, Producer and Consumer Surplus

The regional welfare per capita in the base situation, i.e. in 2007 without WSFs, is 337 US$. According to the model, the producer surplus generates half of the regional welfare and is around 1,175 US$ ha^{-1} of sown area, or 0.12 US$ m^{-3} of water demanded annually for crops. An introduction of WSFs at a rate of 40 US$ 10^{-3} m^{-3} would decrease the regional welfare mostly at the expense of producer surplus (Fig. 23.3). This is because the prices of regional commodities do not increase drastically since in the model set-up the region is assumed to be sufficiently small to not affect the national commodity prices.

Under the base condition of the reference year 2007, the regional welfare is 244 US$ 10^{-3} m^{-3} of water demanded. As the simulation results demonstrate, when WSFs are introduced the crop water demand would decrease more rapidly than the regional welfare and, as a result, at a rate of 40 US$ 10^{-3} m^{-3}, welfare per cubic meter of water demanded would increase by 6%. This is largely caused by a relatively small decrease in regional consumer surplus. The price elasticity of annual crop water demand estimated from the simulation results is relatively weak (−0.04), and, thus, a given percentage variation in WSF induces a lower percentage variation in crop water demand. The state procurement policy determines that 51% of sown area should be occupied by cotton, which in turn would demand 53% of water. About

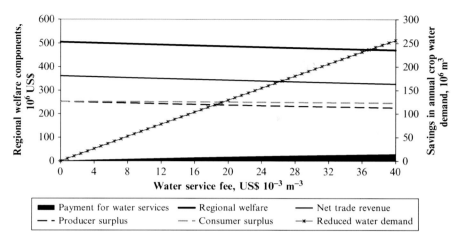

Fig. 23.3 Change in regional welfare, trade revenues and crop water demand

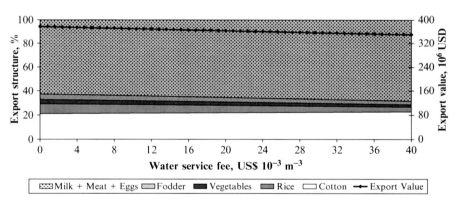

Fig. 23.4 Change in regional export structure

250,000,000 m^3 of water would be saved at WSF equal to 40 US$ 10^{-3} m^{-3}, a substantially high rate of WSF. However this amount of saved water accounts for only a small proportion (less than 10%) of what the region uses currently (see Fig. 23.3). Thus, there is not much room for WSFs to decrease the agricultural water consumption in Khorezm under the existing mechanism of cotton procurement (Fig. 23.3).

As expected, introducing WSFs would decrease the regional earnings from agricultural products exported outside the region (Fig. 23.4). However, the total export revenues would not drop significantly. The largest decline would be in rice exports.

Under the model assumptions, WSFs did not affect the cotton and livestock sectors, and hence their shares in the regional export earnings increased. The total production of food crops would however be decreased with the introduction of WSFs, which is unfavorable for regional food security given that already now Khorezm needs to import some food commodities (such as wheat and potato;

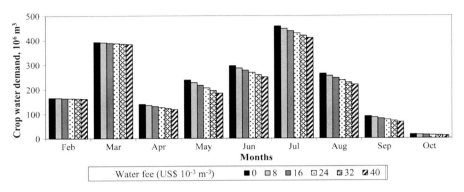

Fig. 23.5 Monthly crop water demand

Djanibekov 2008). Thus WSFs would result also in a drastic increase in the share of import commodities in total regional consumption, and thus would make the region more dependent on food imports. However, the net export revenue of the region is largely composed by the exports of raw cotton and livestock products, and thus would not suffer from drastic decline.

23.4.4 Change in Water Demand

According to the model results, the WSF at a rate of 40 US$ 10^{-3} m^{-3} of water demanded for crop cultivation when cotton producers are not charged would generate around 28,700,000 US$ of funds for WUAs, i.e. 1,800 US$ km^{-1} of irrigation canal or 147 US$ ha^{-1} of sown area. At the same time, the average annual O&M costs which enable full recovery of irrigation system would be about 450 US$ ha^{-1} for standard systems, and more than 680 US$ ha^{-1} for pump systems (FAO 2008), while in recent years the actual costs for O&M varied between 60 and 150 US$ ha^{-1} (Djanibekov 2008) declining to just 8.8 US$ ha^{-1} in 2008.

At rate of 40 US$ 10^{-3} m^{-3}, total annual crop water demand in the model would drop by 13% compared to crop water demand in 2007. The largest decrease in water demand is then observed in seasons when water is required for irrigation of rice fields. The amount of total decrease in crop water demands due to a service fee of 40 US$ 10^{-3} m^{-3} is around 256,000,000 m^3, which would be equal to a decrease in the rice-cropped area by about 9,800 ha. This amount of water could be allocated to promote other crops, improve crop yields or solve the problem of water scarcity in downstream districts, or simply it could leave the Khorezm region to the benefit of the Aral Sea (ecological use). In addition to the decrease in annual crop water demand, the structure of the crop water demand would change (Fig. 23.5). As in the base situation, 27% of water was demanded for leaching the fields, at a price of 40 US$ 10^{-3} m^{-3} this increased up to 30%. This may indicate that the introduction of WSF may require also a change in the services provided by WUAs, as farmers

23 Prospects of Agricultural Water Service Fees in the Irrigated Drylands... 403

may tend to grow crops, which require prior leaching. Also the model results show that WSF could reduce crop water demand in the water-intensive months (June, July), which in turn would increase the probability of receiving sufficient water.

23.5 Discussion and Conclusion

The results of the policy simulations in the study depend on model assumptions. These assumptions were mandatory to reduce the complexity of the situation and to be able to represent it in a relatively sophisticated model that is theoretically sound and also based on primary and secondary data. Therefore, the simulation results are illustrative but should not yet be taken as a policy action concept. Nevertheless, we believe these results to be reliable enough to contribute to the discussion on how policy changes would affect regional welfare, production patterns and factor use in agriculture, thus bringing new insights into the discussion on and formulation of water pricing policy.

Despite the fact that the introduction of WSFs as a single policy would decrease regional income, the model is comparatively static and the negative income effects may well turn out positive if long-term effects were taken into account, such as investments into overall reconstruction of canal network, operational improvement of the WUAs and positive environmental effects. Improving the efficiency of the irrigation system as well as improving the capacity of the WUAs will in turn reduce the losses associated with deteriorating canal infrastructure and management losses, which are shown to be very high in the region. WSFs may gain importance over time in water management as the constraining condition of the poor irrigation network will reduce the agricultural production on the limited arable land even further. Regional welfare, i.e. producer and consumer surpluses, might even increase under implementation of WSFs, as it might force upstream districts to use water in more efficient way and convey more water to downstream districts which suffer severe water shortage during 'drought' years. As model results show, the income loss may be offset by the expansion of the irrigated marginal areas and livestock sector in downstream regions.

As the regional timely and spatial demand exceeds its supply, the WSFs may have not only cost recovery purpose, but also allow managing the water scarcity between seasons and locations, and at least maintain the water productivity. As the model results show, the crop-based WSF can provoke a change in structure of water demand, which means that WUAs will have to increase their operations during the leaching seasons compared to the present situation. Additionally, the peak seasons when water demand is highest will be smoothened due to a decrease of rice fields. In general, the effect of WSFs on regional welfare is related largely to the area of rice cultivation. The selected rate of WSF could trigger a reduction in the area of rice cultivation to the level, which affects the regional exports, but still guarantees regional consumption. As the canal maintenance operations are not performed each year, WSF based on crop water demand should cover the allocated costs of WUA not based on 1-year estimations, but rather on a multi-year value. While this paper does not focus on impacts of WSFs on the environment, the reduced crop water demand due to a decline in rice cultivation and generated funds for O&M activities may make this approach part of a toolbox for adapting to and mitigating the negative effects of climate change.

The WSF rates should be high enough to allow WUAs function properly: fully cover the O&M costs and allow for further investments into improvement of irrigation system. The sole introduction of high WSFs will have the greatest negative impact on the incomes of agricultural producers. WSFs could therefore be promoted only gradually over a period of time to allow water users to adjust to higher production costs. The simulation results show that the livestock sector is important for livelihood security also when high WSFs are introduced for crops: it maintains export revenues of the region and keeps fodder crops in the agricultural system. Furthermore, as the largest share of the livestock is reared in backyards of rural households (Djanibekov 2008), it can be used by rural households as a shield against income decreases.

Due to several reasons, the present lack of payments for water services fees by farmers remains a central problem to generate funds for investments. The first problem is related to the limited capacity of the WUAs to serve all water users within its boundaries. Whilst the irrigated (served) area within a WUA remained the same size, the farm restructuring process increased the number of water users from tens of farms per *shirkat* (cf. Djanibekov et al. 2011) to several dozens of private farms per WUA. With limited human resources, the service delivery by the WUA declined (Abdullaev et al. 2009). A second problem is related to the insufficient capital of WUAs, which limits their capacity to provide adequate O&M services. In consequence, they can impose only very low fees for their services. Thus, WSFs are currently introduced only to recover the WUA expenses. As a result, the government still has to cover the larger part of the water management costs, e.g. in 2006, 89% of the total water management costs were provided via subsidies to organizations involved in the water distribution (Bobojonov 2009). As a result, there was no clear role for WSFs other than cost recovery and the spatial and temporal balancing of water demands. Therefore, a more complex mechanism than just increases in the WSF rate would be required to promote the adoption of WSFs by water users. For instance, Oberkircher et al. (2010) proposes to alter the WUAs into business units that also provide against payments management services, training, insurance as well as short-term credits to water users. Such additional offers can provide tangible benefits to water users and motivate them to pay WSFs.

The estimated price elasticity of annual crop water demand is relatively weak. This shows that given the current institutional setup of agricultural production in Khorezm, crop water demand is inelastic. Options to decrease crop water demand by the introduction of WSF are limited unless farmers are given more flexibility in their production decisions, i.e. through a modification in the state procurement of cotton. Currently, the top-down state procurement system defines the water management policy to guarantee the requested yields of cotton and wheat. Hence, currently WSFs are directly linked to the state procurement mechanism. A reduction in the production targets would increase the risk coping ability of farmers as long as the complete abolishment of the state target is not yet an option. For instance, a flexible mechanism of state policy is much desired when the probability of receiving enough water is low (Bobojonov 2009). The implementation of WSFs to address the dwindling water availability will only be effective if it is integrated with other policy interventions to enhance its uptake. For instance, if considered as a tool to control water demand, WSFs cannot significantly reduce agricultural water demand due to weak flexibility in the decision-making of farmers. Djanibekov (2008) shows that the abolishment of state procurement targets for cotton in Khorezm bears the danger that the regional agricultural producers would significantly increase their water consumption, as farmers are likely to switch to rice cultivation unless this is restricted (rice production, for example, is determined each year by the regional administration in Khorezm). In this context, once the cotton producers are partly or totally freed from state targets, a judicial introduction of WSF might be conducive to temper the regional water use.

23.6 Annexes

Annex Table 23.1 The indices of the model

Code	Description	Code	Description
x	Production activities	z	Production inputs
i,j	Commodities	o	Observed levels
r,rl	Districts aggregates		

Annex Table 23.2 Model parameters

Code	Indices	Description
ep	i	Border (outside) price of commodity
$dist$	r,rl	Distance between districts
tc	i	Transportation costs of commodity
hc	i	Handling costs of commodity
$yild$	r,i,x	Production yield per activity as observed in 2007
$sord$	r,x	State procurement constraint for cotton
inp	r,x,z	Input application rate per unit of activity
$inpp$	z	Input price as observed in 2007
$caln$	r,m,x	Cropping calendar
$area$	r,m	Land available for sowing as observed in 2007
$watr$	r,m,x	Monthly norms of crop water demand
$watr0$	r,m	Monthly total crop water demand as observed in 2007
$watp$	r,x	The studied rate of water service fee

(continued)

Annex Table 23.2 (continued)

Code	Indices	Description
D	r,i	Constant term of Marshallian demand ("minimum commitment levels" or consumption quantities independent of prices and income)
F	r	Part of generalized Leontief expenditure function
G	r	Part of generalized Leontief expenditure function
Gi	r,i	Derivative of G with respect to the product price
B	r,i,j	Price dependent terms of Marshallian demand function

Annex Table 23.3 Model variables

Code	Indices	Description
Welf		Regional welfare
MMU	r	Money-metric indirect utility function
Ntrad	r	Definition of net trade surplus of producers
Prev	r	Producer revenues from selling the local commodities to consumers
Tcst	r	Producers' commodity transportation costs
Pcst	r	Definition of agricultural production costs
Levl	r,x	Production activity levels
Supply	r,i	Supply (production) of commodity (not allowed for rest commodities)
Demand	r,i	Consumption level
Price	r,i	Endogenous commodity prices
Income	r	Total income (consumption expenditure) of consumers
Xflows	r,i	Export flows of commodities
Mflows	r,i	Import flows of commodities
WatrF	r,r1,m	Water flow to districts according to crop water demand and sown area
D	r,i	Constant terms of Marshallian demand function
G	r	Part of generalized Leontief expenditure function
Gi	r,i	Derivative of G of generalized Leontief expenditure vs prices
F	r	Part of generalized Leontief expenditure function
B	r,i,j	Price dependent terms of Marshallian demand function

Annex Table 23.4 Model equations

Eq.	Name of equation	Equation
23.1	Objective function	$Welf = \sum_r Ntrad_r + \sum_r MMU_r + \sum_r Prev_r - \sum_r Tcst_r - \sum_r Pcst_r$
23.2	Commodity supply	$Supply_{r,i} = \sum_x Levl_{r,x} yild_{r,i,x}$
23.3	Monthly sown land availability constraint	$\sum_x Levl_{r,x}\ caln_{r,m,x} \leq area_{r,m}$
23.4	Monthly crop water demand constraint	$\sum_x Levl_{r,x}\ watr_{r,m,x} + \sum_{r1} WatrF_{r,r1,m} \leq watr0_{r,m} + \sum_{r1} WatrF_{r1,r,m}$
23.5	State procurement constraint	$Levl_{r,x} \geq sord_{r,x}$ where x is "cotton"
23.6	Market balance for commodities	$Supply_{r,i} + MFlows_{r,i} + \sum_{r1} Flows_{r1,r,i} \geq \sum_f Demand_{r,i} + XFlows_{r,i} + \sum_{r1} Flows_{r,r1,i}$
23.7	Net trade surplus of producers	$NTrad_r = \sum_i \left(XFlows_{r,i} - MFlows_{r,i} \right) ep_i$, where i is not applicable for "rest of commodities"
23.8	Producer revenues	$Prev_r = \sum_i Price_{r,i} Demand_{r,i}$, where i is not applicable for "rest of commodities"
23.9	Producer commodity transportation costs	$Tcst_r = \sum_{r1} \sum_i Flows_{r,r1,i} \left(dist_{r,r1} tc_i + hc_i \right)$ where i is not applicable for "rest of commodities"
23.10	Agricultural production costs	$Pcst_r = \sum_x \sum_z Levl_{r,x} \left(inp_{r,x,z} inpp_z + watp_{r,x} \sum_m watr_{r,m,x} \right)$, where in the base-run solution $watp$ is set to zero

(continued)

Annex Table 23.4 (continued)

Eq.	Name of equation	Equation
23.11	Regional demand	$Demand_{r,i} = \left(D_{r,i} + \dfrac{Gi_{r,i}}{G_r}(Income_r - F_r) \right) pop_r$
23.12	Consumer income per capita	$Income_r = \left(\sum_i Price_{r,i} Demand_{r,i} \right) / pop_r$
23.13	Money-metric indirect utility function	$MMU_r = \left(\sum_i D_{r,i} Price^o_{r,i} + \dfrac{G^o_r}{G_r}(Income_r - F_r) \right) pop_r$
23.14	Definition of F	$F_r = \sum_i D_{r,i} Price_{r,i}$
23.15	Definition of G	$G_r = \sum_i \sum_j \left(B_{r,i,j} + B_{r,j,i} \right) \sqrt{Price_{r,i} Price_{r,j}}$
23.16	Derivative of G	$Gi_{r,i} = \sum_j \left(B_{r,i,j} + B_{r,j,i} \right) \sqrt{Price_{r,j} / Price_{r,i}}$

References

Abdolniyozov B (2000) Science-based agricultural production system in the Khorezm region. Khorazm publishing house, Urgench, p 310

Abdullaev I, Manthrithilake H, Kazbekov J (2006) Water security in Central Asia: troubled future or pragmatic partnership? In: International conference "The last drop? Water, security and sustainable development in Central Eurasia", Institute of Social Studies, The Hague, 2006

Abdullaev I, De Fraiture C, Giordano M, Yakubov M, Rasulov A (2009) Agricultural water use and trade in Uzbekistan: situation and potential impacts of market liberalization. Int J Water Resour Dev 25(1):47–63

Alcamo J, Döll P, Henrichs T, Kaspar F, Lehner B, Rösch T, Siebert S (2003) Development and testing of the waterGAP 2 global model of water use and availability. Hydrol Sci J 48:317–338

Batchelor C (1999) Improving water use efficiency as part of integrated catchment management. Agric Water Manag 40(2–3):249–263

Bekchanov M, Lamers JPA, Karimov A, Müller M (2011) Estimation of spatial and temporal variability of crop water productivity with incomplete data. In: Martius C, Rudenko I, Lamers JPA, Vlek PLG (eds) Cotton, water, salts and *Soums*: economic and ecological restructuring in Khorezm, Uzbekistan. Springer, Dordrecht

Bobojonov I (2009) Modeling crop and water allocation under uncertainty in irrigated agriculture: a case study on the Khorezm region, Uzbekistan. PhD dissertation, Bonn University, Bonn

Britz W (2003) Major enhancements of @2030 modelling system. http://www.agp.uni-bonn.de/agpo/rsrch/at2030/@2030_2003.doc

Bucknall J, Klytchnikova I, Lampietti J, Lundell M, Scatasta M, Thurman M (2003) Irrigation in Central Asia: social, economic and environmental considerations. World Bank, Washington, DC, p 52

Chohin-Kuper A, Rieu T, Montginoul M (2003) Water policy reforms: pricing water, cost recovery, water demand and impact on agriculture. Lessons from the mediterranean experience. Water pricing seminar (June 30–July 32, 2003) Agencia Catalana del Agua and World bank Institute Parallel session C: The impact of cost recovery on agriculture

Conrad C (2006) Remote sensing based modeling and hydrological measurements to assess the agricultural water use in the Khorezm region, Uzbekistan. PhD dissertation, University of Wuerzburg, Wuerzburg

Conrad C, Schorcht G, Tischbein B, Davletov S, Sultonov M, Lamers JPA (2011) Agro-meteorological trends of recent climate development in Khorezm and implications for crop production. In: Martius C, Rudenko I, Lamers JPA, Vlek PLG (eds) Cotton, water, salts and *Soums*: economic and ecological restructuring in Khorezm, Uzbekistan. Springer, Dordrecht

Cornish G, Bosworth B, Perry C, Burke J (2004) Water charging in irrigated agriculture: an analysis of international experience. FAO Water Reports 28, Rome, p 97

Deaton A, Muellbauer J (1980) Economics and consumer behaviour. Cambridge University Press, Cambridge, p 466

Dinar A, Latey J (1991) Agricultural water marketing: allocative efficiency and drainage reduction. J Environ Econ Manag 20(3):210–223

Dinar A, Subramanian A (1997) Water pricing experiences: an international perspective. Technical paper no. 386. World Bank, Washington, DC

Dinar A, Rosegrant MW, Meinzen-Dick R (1997) Water allocation mechanisms: principles and examples. World Bank: policy research working paper 1779. World Bank, Washington, DC

Djalalov A (2003) Rational water resources use in market economy conditions. The 3rd world water forum regional cooperation in shared water resources in Central Asia, Kyoto, p 2009, 18 Mar 2003

Djanibekov N (2008) A micro-economic analysis of farm restructuring in Khorezm region, Uzbekistan. PhD dissertation, Bonn University, Bonn

Djanibekov N, Bobojonov I, Lamers JPA (2011) Farm reform in Uzbekistan. In: Martius C, Rudenko I, Lamers JPA, Vlek PLG (eds) Cotton, water, salts and *Soums*: economic and ecological restructuring in Khorezm, Uzbekistan. Springer, Dordrecht/Berlin/Heidelberg/New York

FAO (2008) Water profile of Uzbekistan. In: Cutler JC (ed) Encyclopedia of Earth. http://www.eoearth.org/article/Water_profile_of_Uzbekistan

Fischer G, van Velthuizen H, Shah M, Nachtergaele F (2002) Global agroecological assessment for agriculture in the 21st century: methodology and results. RR-02-02, FAO and IIASA, Laxenburg

Forkutsa I (2006) Modeling water and salt dynamics under irrigated cotton with shallow groundwater in the Khorezm region of Uzbekistan. PhD dissertation, Rheinische Friedrich-Wilhelms-Universität Bonn, Bonn

Hazell P, Norton R (1986) Mathematical programming for economic analysis in agriculture. MacMillan, New York, p 432

Heckelei T (2002) Calibration and estimation of programming models for agricultural supply analysis. Habilitation dissertation, Rheinische Friedrich-Wilhelms-Universität Bonn, Bonn

Hellegers PJGJ, Perry CJ (2004) Water as an economic good in irrigated agriculture: theory and practice LEI. Rapport 3.04.12, Den Haag

IMF (2008) Republic of Uzbekistan: poverty strategy reduction paper, international monetary fund. Country report 08/34, Washington, DC, p 143

Jarsjö J, Asokan SM, Shibou Y, Destouni G (2007) Water scarcity in the Aral Sea drainage basin: contribution of agricultural irriation and a changing climate. In: Qi J, Evered T (eds) Environmental problems of Central Asia and their Economic, Social and Security impacts. Proceedings of the NATO advanced research workshop on environmental problems of Central Asia and their economic, sSocial and security impacts. Springer, Dordrecht/Tashkent

Kuhn A (2003) From world market to trade flow modelling – the re-designed WATSIM model. WATSIM AMPS final report. Institute for Agricultural Policy, University of Bonn. http://www.ilr1.uni-bonn.de/agpo/rsrch/watsim/watsim-amps-report.pdf

Kyle S, Chabot P (1997) Agriculture in the republic of Karakalpakstan and Khorezm oblast of Uzbekistan. Working paper no. 97–13. Cornell University Department of Agricultural Resource and Managerial Economics, Ithaca

MacDonald M (2003) Privatization/transfer of irrigation management in Central Asia. Final report, Department for International Development Knowledge and Research Services, p181

Martius C, Froebrich J, Nuppenau E-A (2009) Water resource management for improving environmental security and rural livelihoods in the irrigated Amu Darya lowlands. In: Brauch HG, Spring ÚO, Grin J, Mesjasz C, Kameri-Mbote P, Behera NC, Chourou B, Krummenacher H (eds) Facing global environmental change: environmental, human, energy, food, health and water security concepts. Hexagon series on human and environmental security and peace. Springer, Dordrecht/Berlin/Heidelberg/New York

Mas-Colell A, Whinston M, Green J (1995) Microeconomic theory. Oxford University Press, New York, p 1008

McCarthy JJ, Canziani OF, Leary NA, Dokken DJ, White KS (2001) Climate change 2001: impacts, adaptation, and vulnerability. Cambridge University Press, Cambridge, UK

Müller M (2006) A general equilibrium approach to modeling water and land use reforms in Uzbekistan. PhD dissertation, Rheinische Friedrich-Wilhelms-Universität Bonn, Bonn

Oberkircher L, Tischbein B, Hornidge A-K, Schorcht G, Bhaduri B, Awan UK, Manschadi A (2010) Rethinking water management in Khorezm, Uzbekistan - concepts and recommendations. ZEF working papers series no. 54, p 22

Oberkircher L, Haubold A, Martius C, Buttschardt T (2011) Water patterns in the landscape of Khorezm, Uzbekistan. A GIS approach to socio-physical research. In: Martius C, Rudenko I, Lamers JPA, Vlek PLG (eds) Cotton, water, salts and *Soums*: economic and ecological restructuring in Khorezm, Uzbekistan. Springer, Dordrecht/Berlin/Heidelberg/New York

OblSelVodKhoz (2004) Crop water use norms. Urgench

OblStat (2008) Socio-economic indicators for Khorezm. 1992–2003, Urgench

Perelet R (2007) Central Asia: background paper on climate change. UNDP, p 17

Ryan DL, Wales TJ (1996) Flexible and semiflexible consumer demands with quadratic engel curves. Discussion paper no. 96–30, Department of Economics, University of British Columbia, Vancouver

Sommer R, Djanibekov N, Müller M, Salaev O (2011) Economic-ecological optimization model of land and resource use at farm-aggregated level. In: Martius C, Rudenko I, Lamers JPA, Vlek PLG (eds) Cotton, water, salts and *Soums*: economic and ecological restructuring in Khorezm, Uzbekistan. Springer, Dordrecht/Berlin/Heidelberg/New York

Tischbein B, Awan UK, Abdullaev I, Bobojonov I, Conrad C, Forkutsa I, Ibrakhimov M, Poluasheva G (2011) Water management in Khorezm: current situation and options for improvement (hydrological perspective). In: Martius C, Rudenko I, Lamers JPA, Vlek PLG (eds) Cotton, water, salts and *Soums*: economic and ecological restructuring in Khorezm, Uzbekistan. Springer, Dordrecht/Berlin/Heidelberg/New York

Tsur Y, Roe T, Doukkali R, Dinar A (2004) Pricing irrigation water: principles and cases from developing countries. Resources for the future, Washington, DC

Veldwisch GJ, Mollinga P, Zavgorodnyaya D, Yalcin R (2011) Politics of agricultural water management in Khorezm, Uzbekistan. In: Martius C, Rudenko I, Lamers JPA, Vlek PLG (eds) Cotton, water, salts and *Soums*: economic and ecological restructuring in Khorezm, Uzbekistan. Springer, Dordrecht/Berlin/Heidelberg/New York

Wallace JS, Batchelor CH (1997) Managing water resources for crop production. Philos Trans R Soc Lond Ser B Biol Sci 352(1356):937–946

WARMAP (1996) Water resource management and agricultural production in the Central Asian Republics. Irrigated crop production systems, IV, TACIS, June 1996

Glossary of Uzbek and Russian words

Birja Commodity Exchange (Russian and Uzbek)

Bonitet Soil quality indicator; an aggregate of several parameters ranging from field characteristics (morphology, etc.) to results of laboratory analyses for various soil properties (fertility, chemistry, etc.) (Russian)

BUIS The Basin Department of Irrigation Systems (Russian abbreviation)

Dehqon Rural (peasant) households (Uzbek)

Dehqonbozor A local open market (Uzbek)

Fermer A private farmer (Uzbek)

Hokim Local governor, indicating the chairman of the local council and head of local administration at provincial, district and city levels (Uzbek)

Hokimiat The executive branch of local governments at provincial and district levels (Uzbek)

Kolkhoz Large-sized collective farm (Russian abbreviation for 'kollektivnoye khozyaistvo')

Pudrat A special form of organizing agricultural production with the direct participation of family members and labor on lands given by shirkats or other agricultural organizations based on *family* contract (Uzbek)

Shirkat Agricultural cooperative (Uzbek)

Sovkhoz Large-sized state farm (Russian abbreviation of 'sovietskoye khozyaistvo')

Soums Same as UZS (Uzbek currency)

TEZIM Informal for UIS (Uzbek)

Tugai Floodplain forests (Uzbek)

Tuman Administrative division, indicating a district (Uzbek)

UIS The Irrigation System Authority at sub level (under BUIS) (Russian abbreviation)

Viloyat Administrative division, usually indicating a province (Uzbek)

Zemlemer Local land surveyor (Russian)

Zveno Smaller labor units for carrying out field tasks under brigades in *kolkhoz*es and *sovkhoz*es (Russian)

C. Martius et al. (eds.), *Cotton, Water, Salts and Soums: Economic and Ecological Restructuring in Khorezm, Uzbekistan*, DOI 10.1007/978-94-007-1963-7,
© Springer Science+Business Media B.V. 2012

Index

A

Abdullaev, I., 69–90
Abrol, I.P., 47, 50
Afforestation, 9, 15, 38, 48, 49, 56, 236, 237, 239, 241–245
Agrarian change, 9
Agricultural Service Providing Organizations, 115, 116
Ahrens, C.D., 25–34
Akramkhanov, A., 37–56, 73
Alternative crops, 221–223, 230–232
Amudarya, 5–7, 11, 20, 26–28, 34, 38–40, 45, 46, 63, 64, 66, 72, 73, 80, 83, 98, 131, 137, 182, 227, 249–262, 268, 286, 287, 290, 311, 330, 348, 389–408
Aral Sea Basin (ASB), 4–7, 26, 34, 72, 198, 221, 252, 311, 312, 330, 340, 341
Atakhanova, Z., 158, 160, 161
Awan, U.K., 16, 69–90, 189, 309–324, 334, 340
Ayars, J.E., 189, 311

B

Begdullaeva, T., 219–232
Bekchanov, M., 88, 171–179, 329–342, 347–369
Biodrainage, 237, 239–240
Bio-economic modeling, 269–271
Bobojonov, I., 15, 69–90, 95–110, 219–232, 368, 389–408
Bos, M.G., 83, 84, 317
Botman, E., 249–262
Britz, W., 395
Buttschardt, T.K., 285–305

C

Carbon sequestration, 15, 237, 241–242, 262
Central Asia, 4–7, 17–19, 21, 26–29, 38, 72, 96, 97, 132, 147, 156, 163, 164, 172, 190, 197, 220, 242, 249–251, 312, 313, 322, 323, 348, 377, 390, 391
Choudhari, A., 198
Chub, E.V., 28
Climate change, 4, 5, 11, 18, 19, 26–28, 230, 236, 288, 289, 330, 390, 404
Climate change impact, 255–259
Conant, R.T., 52
Conrad, C., 11, 25–34, 59–90, 189, 334
Conservation agriculture, 5, 9, 11, 14, 15, 18, 48, 53–56, 195–215
Cost recovery, 393, 403, 404
Cotton, 3, 26, 41, 60, 72, 96, 121, 128, 159, 172, 182, 196, 221, 243, 251, 268, 287, 310, 334, 348, 372, 394
Cotton market liberalization, 348, 361, 362, 364, 368
Cotton revenues, 385
Crop residue (CR), 14, 51, 196, 198, 202–205, 207, 213–215
Crop specific water allocation, 331–333, 341
CropSyst, 183–185, 187, 269, 282
CROPWAT, 310, 312–316, 319, 320, 323

D

Davletov, S., 25–34
Deforestation, 15, 250, 252, 255, 256, 262
Diamond, J., 21
Diarrhoea, 142–144, 147, 148
Distribution outlets, 121–124

C. Martius et al. (eds.), *Cotton, Water, Salts and Soums: Economic and Ecological Restructuring in Khorezm, Uzbekistan*, DOI 10.1007/978-94-007-1963-7,
© Springer Science+Business Media B.V. 2012

415

Djanibekov, N., 13, 17, 95–110, 135, 219–232, 267–282, 367, 368, 389–408
Djanibekov, U., 389–408
Djumaniyazova, Y., 171–179
Dommergues, Y.R., 241
Drinking water quality, 145, 149
Dryland, 12, 17, 19, 55, 196, 198, 213, 339, 389–408

E

Economic and ecological benefits, 15, 220, 242, 262
Egamberdiev, O., 195–215
Electricity demand, 156–161, 164–166
Environmental monitoring, 60, 61
Ergashev, A.K., 219–232
Eshchanov, B., 155–166
Eshchanov, R., 3–21, 171–179, 219–232
Evapotranspiration (ET), 5, 10, 12, 26, 28, 30, 33, 60–62, 72, 83, 84, 87, 90, 182, 185, 187–190, 287, 310–313, 334

F

Farm-aggregate model, 269–274
Farmers' preference and motivation, 236–242, 244–245
Farm optimization, 6, 100, 102, 104, 109, 372
Farm restructuring, 13, 97–99, 101–104, 106, 108–110, 393, 404
Fayzieva, D., 141–151
Felitciant, I.N., 39
Fisher, F.M., 158
Forest degradation, 255, 256
Forkutsa, I., 69–90, 184, 187, 189, 323
Forkutsa, O., 37–56

G

Gatzweiler, F.W., 21
GDD. *See* Growing degree days (GDD)
General equilibrium model, 16, 357–366, 368
Geographic information system (GIS), 10, 15, 16, 18, 61, 67, 68, 72, 73, 81, 83, 196, 253, 269, 271, 282, 285–305, 313, 315
George, B.A., 322
Giese, E., 29
GIS. *See* Geographic information system (GIS)
Global warming, 26, 172, 177, 261
Governmental earnings, 368
Greene, W.H., 333

Grinwis, M., 155–166
Groundwater contribution, 16, 74, 81, 181–190, 311, 319, 324
Growing degree days (GDD), 11, 26, 28–32, 34
Guo, L.B., 242

H

Hafeez, M., 309–324
Halicioglu, F., 158, 160
Halvorsen, R., 158
Haubold, A., 285–305
Hazell, P., 393
Hbirkou, C., 49, 50
Herbst, S., 141–151
Hirsch, D., 127–139
Hoffman, M., 289
Holtedahl, P., 158
Hornidge, A., 303
Howell, T.A., 30
Howie, P., 158, 160, 161
Huby, M., 289
Huxman, T.E., 52
HYDRUS, 74, 75, 184, 310, 313–315
Hygiene, 13, 142–144, 146–151

I

Ibragimov, N., 32, 59–68, 171–179, 181–190, 219–232
Ibrakhimov, I., 69–90
Ibrakhimov, M., 59–68, 240
Income security, 232
Indigo, 222, 227–228, 232, 341
Informal rules, 120, 123, 124
Insam, H., 51
Interdisciplinary research, 305
Irrigated agriculture, 6, 11, 17, 38, 39, 41, 55, 172, 196, 220, 250, 252, 259, 269, 311, 330, 331, 391, 392
Irrigated crop production, 399
Irrigation performance, 60, 71, 78, 81–84
Irrigation scheduling, 12, 30, 71, 74, 78, 84, 86, 89, 90, 175, 309–324, 384
Irrigation system, 6, 16, 28, 30, 32–34, 63, 71, 75, 77, 84, 86, 88, 105, 108, 131, 230, 239, 280, 286, 287, 290, 297, 298, 300, 310, 313, 317, 322, 340, 342, 392, 403, 404
Irrigation water use, 268, 276, 280, 289, 331, 334, 341, 391–393

Index

J
Jabborov, H., 69–90
Jarsjö, J., 33
Joutz, F.L., 158
Jurriens, M., 316
Just, R.E., 331, 332

K
Kadirhodjayev, F., 189
Kamalov, P., 309–324
Kan, E., 244
Kaplinsky, R., 376
Karimov, A., 195–215, 329–342
Kaysen, C., 158
Khaitbayev, K.K., 189
Khamzina, A., 15, 235–245
Khepar, S.D., 322
Khorezm region, 4, 26, 38, 60, 71, 72, 78, 98, 114, 117, 128, 142, 148, 157, 158, 172, 173, 182, 183, 198, 221, 236, 251, 252, 268, 286, 311, 330, 348, 373, 390, 391
Kienzler, K.M., 14, 59–68, 171–179, 181–190, 219–232, 383
Kienzler, S., 195–215
Kistemann, T., 141–151
Kladivko, E.J., 54
Kohlschmitt, S., 224
Kuziev, R., 37–56
Kyle, S., 398

L
Lamers, J.P.A., 3–21, 25–34, 59–68, 95–110, 113–126, 171–179, 181–190, 195–215, 219–232, 235–245, 249–262, 329–342, 347–369, 371–385
Landscape, 14, 26, 34, 39, 48, 125, 236, 285–305, 322
Le Blanc, D., 289
Lerman, Z., 98
Livestock productivity, 361, 364, 367, 368
Lovo, S., 368
Löwy, I., 288
Lund, H.G., 255

M
Maize, 11, 27–28, 30, 31, 34, 101, 135, 185, 186, 222, 226, 228–230, 232, 243, 268, 269, 271, 272, 275–278, 280, 287, 298, 341, 394, 397–400
Manschadi, A., 3–21

Martius, C., 3–21, 37–56, 59–69, 127, 171, 195–215, 219–232, 249–262, 285–305, 309–324
Masharipova, H., 323
Massucati, L., 37–56
Messina, J.P., 289
Methane, 250
Micro-and macro-fauna, 52–55
Miller, P., 31
Mini-banks, 115, 116, 124–125
Mixed estimation method (MEM), 331, 333–336
Moderate resolution imaging spectroradiometer (MODIS), 11, 61, 64, 67
MODIS. *See* Moderate resolution imaging spectroradiometer (MODIS)
Mollinga, P.P., 127–139, 288
Moßg, I., 29
Müller, M., 34, 136, 231, 267–282, 329–342, 347–369
Multipurpose tree species, 237
Mung bean, 222, 228–229, 232

N
National and regional income, 362, 364, 368, 403
Natural resource use efficiency, 274–280
Net present value (NPV), 244
Nitrate, 14, 45, 47, 142, 143, 146–147, 149, 151, 178, 183–187, 190, 272, 273, 282
Nitrogen fixation, 237
Nitrogen rate, 175
Nitrous oxide (N_2O), 14, 15, 172, 177–179, 250, 255, 259–262
Niyazmetov, D., 107, 113–126
N_2O emissions, 14, 172, 177, 179, 259–261
NPV. *See* Net present value (NPV)
Nurmetov, K., 371–385

O
Oberkircher, L., 16, 285–305, 404
Öhlinger, R., 51

P
Panel data, 159, 160
Partial equilibrium model, 396–403
Paulo, A.M., 322
Pereira, L.A., 322
Policy analysis, 369
Poluasheva, G., 69–90

Potato, 100, 103, 135, 136, 220–222, 226–227, 230–232, 341, 394, 397–401
Pratharpar, S.A., 311
Private sector, 18, 258, 361, 363, 364, 368, 369
Productivity, 5, 9, 12, 14, 16–18, 20, 56, 101, 103, 109, 110, 197, 240, 278, 317, 329–342, 348, 349, 360, 361, 364, 367–369, 383, 390, 403
Property rights, 13, 98, 113, 120, 126
Pulatov, A., 56, 195–215

R
Rakhimov, A., 219–232
Remote sensing, 9–11, 18, 60–61, 64, 67, 72, 81, 83, 85, 90, 289, 291, 334, 340
Rhoades, J.D., 83, 84
Risk factor analysis, 143–145
Robinson, S., 352
Rudenko, I., 3–21, 113–126, 136, 322, 371–385
Ruecker, G., 59–68
Ruzimov, J., 171–179, 219–232

S
Salaev, O., 267–282
Salaev, S., 155–166
Salinity, 5, 39, 71, 143, 184, 196, 222, 237, 251, 270, 305, 312, 341, 392
SAM. *See* Social accounting matrix (SAM)
Sanchez, J.E., 208
Sanitation, 13, 141–151
Sayre, K., 195–215
Scheer, C., 171–179, 249–262
Schorcht, G., 25–34
Semi-arid climate, 64–67
Shi, Z., 340
Short-run elasticity, 161, 164
Social accounting matrix (SAM), 349–353, 350, 356–258, 360, 369
Socio-physical analysis, 295–296
Soil respiration, 38, 49, 51, 52
Soil tillage, 53, 198–205, 207, 208, 213
Sommer, R., 37–56, 181–190, 267–282
Sorghum, 27–31, 34, 220, 222–226, 230, 232, 243, 298, 341
Spoor, M., 97, 136
State-centric politics, 13, 128
State order system, 14, 372–375, 380
State procurement, 13, 20, 100–102, 105, 107–110, 268, 272–274, 276, 279–281, 374, 375, 394, 398–400, 404, 405, 407

Stolbovoi, V., 41
Sultonov, M., 25–34
Supplemental N-fertilizer, 190

T
Temperature, 11, 26, 28–30, 32, 34, 55, 61, 158, 172, 178, 222–224, 260, 312, 390
Texture, 11, 12, 38, 41–45, 50, 51, 55, 59–62, 64, 67, 73, 74, 198, 213, 255, 287, 303, 305, 313–315, 317
Theil, H., 331, 333
Tischbein, B., 10, 12, 25–34, 69–90, 309–324
Transition economies, 156, 160, 331, 349
Tree fodder, 243, 245
Tupitsa, A., 249–262
Tursunov, L., 39, 195–215
Tursunov, M., 195–215

U
Uzbekistan, 4, 28, 40, 73, 96, 114, 128, 142, 157, 172, 182, 196, 220, 236, 251, 268, 286, 311, 330, 348, 372, 390

V
Value chain analysis, 9, 17, 375, 376, 380
Vaughn, D., 29
Veldwisch, G.J., 101, 127–139, 335
Visser, O., 97
Vlek, P.L.G., 3–21, 181–190, 235–245, 249–262

W
Wahid, S.M., 289
Wallender, W.W., 311
Wassmann, R., 249–262
Water access, 16, 290, 292, 297, 298, 300, 301, 305
Water and salt balancing, 310, 322
Water reform, 13, 108, 130–133, 391
Water saving, 16, 20, 70, 72, 77, 78, 87–88, 239, 262, 279, 286, 287, 290, 298, 304, 305, 313, 317, 321–323, 330, 340, 342, 348, 367, 368
Water saving potential, 72, 77, 87–88
Water Users Associations (WUAs), 6, 10, 13, 16, 34, 64, 66–68, 83, 102, 115, 116, 128, 132–134, 137, 138, 269, 290, 291, 293–298, 303–305, 331, 342, 392, 393, 402–404

Index

Wegerich, K., 130, 131
Wehrheim, P., 10
Wood for fuel, 242–243
Worbes, M., 249–262
Wright, D.J., 289
WUAs. *See* Water Users Associations
(WUAs)

Y
Yalcin, R., 127–139

Z
Zakharov, P.I., 189
Zavgorodnyaya, D., 132

Printed by Publishers' Graphics LLC
SO20120726